ARMS POLITICS

Arms Politics

Becoming and Being a Weapon in the
Borderlands of Myanmar

Francesco Buscemi

SOUTHEAST ASIA PROGRAM PUBLICATIONS

AN IMPRINT OF CORNELL UNIVERSITY PRESS   ITHACA AND LONDON

Copyright © 2025 by Francesco Buscemi

All rights reserved. Except for brief quotations in a review, this book, or parts thereof, must not be reproduced in any form without permission in writing from the publisher. For information, address Cornell University Press, Sage House, 512 East State Street, Ithaca, New York 14850. Visit our website at cornellpress.cornell.edu.

First published 2025 by Cornell University Press

Librarians: A CIP catalog record for this book is available from the Library of Congress.

ISBN 9781501781735 (hardcover)
ISBN 9781501781742 (paperback)
ISBN 9781501781759 (pdf)
ISBN 9781501781766 (epub)

GPSR EU contact: Sam Thornton, Mare Nostrum Group B.V., Mauritskade 21D, 1091 GC, Amsterdam, NL, gpsr@mare-nostrum.co.uk.

To Angelo&Laura&
Ta'angStudentYouthUnion&
Ta'angWomenOrganization&
MoeSet&
YarNoon&
Lucia&Giovanni&Renata&Angelo&
Ciccio&Marianna&Matteo&Ornella&Niki&Antonio
Rossana&Marcello&Fabrizio&
Lorenzo&
NicMarsh&
Katia&Cristian&Sasha&
PascalSimone&
AlessandraRusso&
FrancescoStrazzari&
MatteoProto&
Parker&
Demo&
JohnPaul&Chris&M&KB&
LawiWeng&
AnthonyDavis&
Amelia&
DavidBrenner&
JasneaSarma&
ElliottPrasseFreeman&
Sarah.E.M.Grossman&
MatthewEvangelista&
TimothyRaeymaekers&
BenediktKorf&
*Stefan Collignon who, one year into the PhD, told me to desist
'cause what I was doing was wrong and unfeasible.*

Contents

Abbreviations

BGF	Border Guard Forces
CCP	Chinese Communist Party
CPB	Communist Party of Burma
CSO	civil society organization
ERO	Ethnic Revolutionary Organization
IED	improvised explosive device
INGO	international nongovernmental organization
Ka-Pa-Sa	Karkweye Pyitsu Setyoun (Defense of Defense Industries)
KIO/A	Kachin Independence Organization/Army
KNPP/KA	Karenni National Progressive Party/Karenni Army
KNU/KNLA	Karen National Union/Karen National Liberation Army
KMT	Kuomintang
LGF	Local Guerrilla Force
MANPADS	man portable air defense system
MOC	Military Operation Command—Sit Sin Ye Koot Ke Htar Na Chote (Sa-Ka-Kha)
MRE	mine risk education
MTA	Mong Tai Army
NCA	Nationwide Ceasefire Agreement
NDF	National Democratic Front
INGO	international nongovernmental organization
NMSP	New Mon State Party
PMF	People's Militia Forces
PNF	Palaung National Force
P-SAZ	Palaung Self-Administered Zone
PSLF/TNLA	Palaung State Liberation Front/Ta'ang National Liberation Army
PSLO/A	Palaung State Liberation Organization/Army
PWO	Palaung Women Organization
RCSS/SSA-S	Revolutionary Council of Shan State/Shan State Army South
RPG	rocket-propelled grenade
SALW	small arms and light weapons
SLORC	State Law and Order Restoration Council
SPDC	State Peace and Development Party
SSA	Shan State Army
SSPP/SSA-N	Shan State Progressive Party/Shan State Army North
T-GAD	TNLA General Administrative Department
TNP	Ta'ang National Party
TSYO	Ta'ang Students and Youth Organization
TSYU	Ta'ang Students and Youth Union

TWO	Ta'ang Women Organization
ULA/AA	United League of Arakan/Arakan Army
UNODC	United Nations Office for Drugs and Crime
UWSP/UWSA	United Wa State Party/United Wa State Army
WNO/A	Wa National Organization/Army

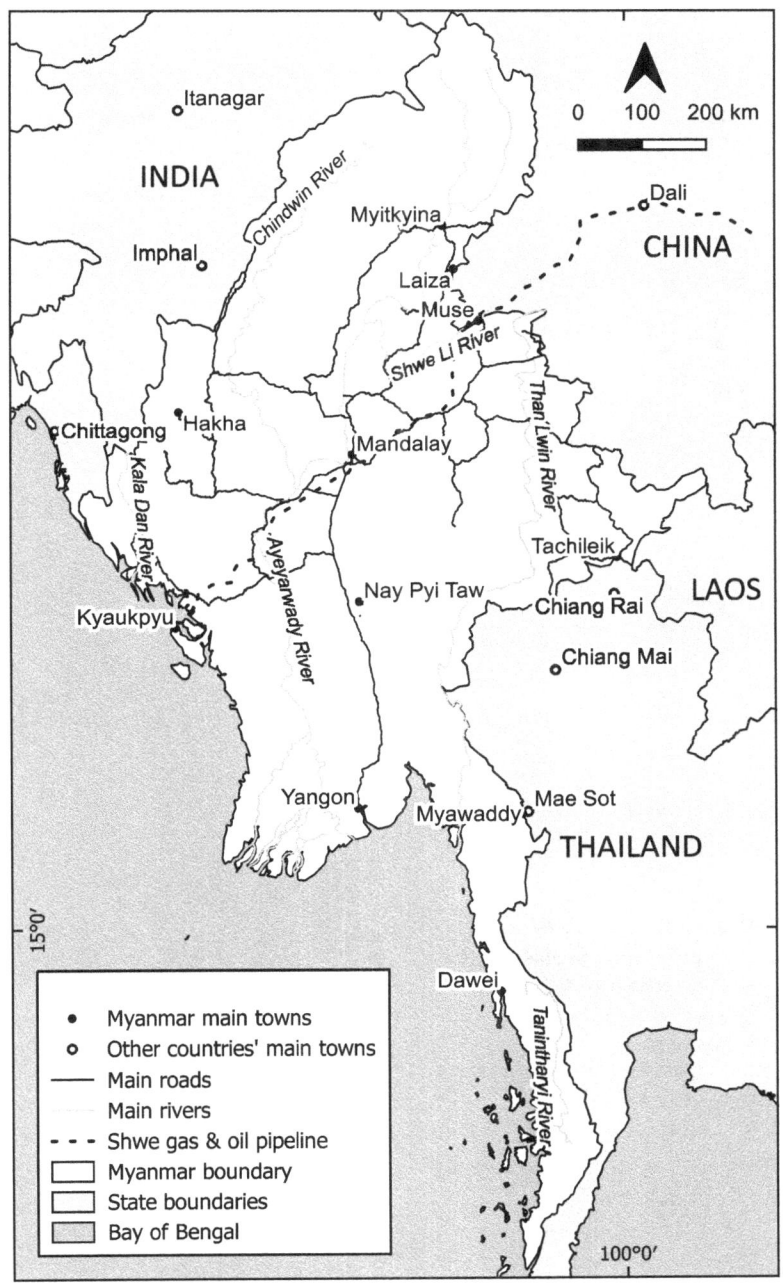

MAP 1. Shwe Gas and Oil Pipeline. Map by Dr. Michaela De Giglio, Department of History and Cultures—Geography Unit, University of Bologna.

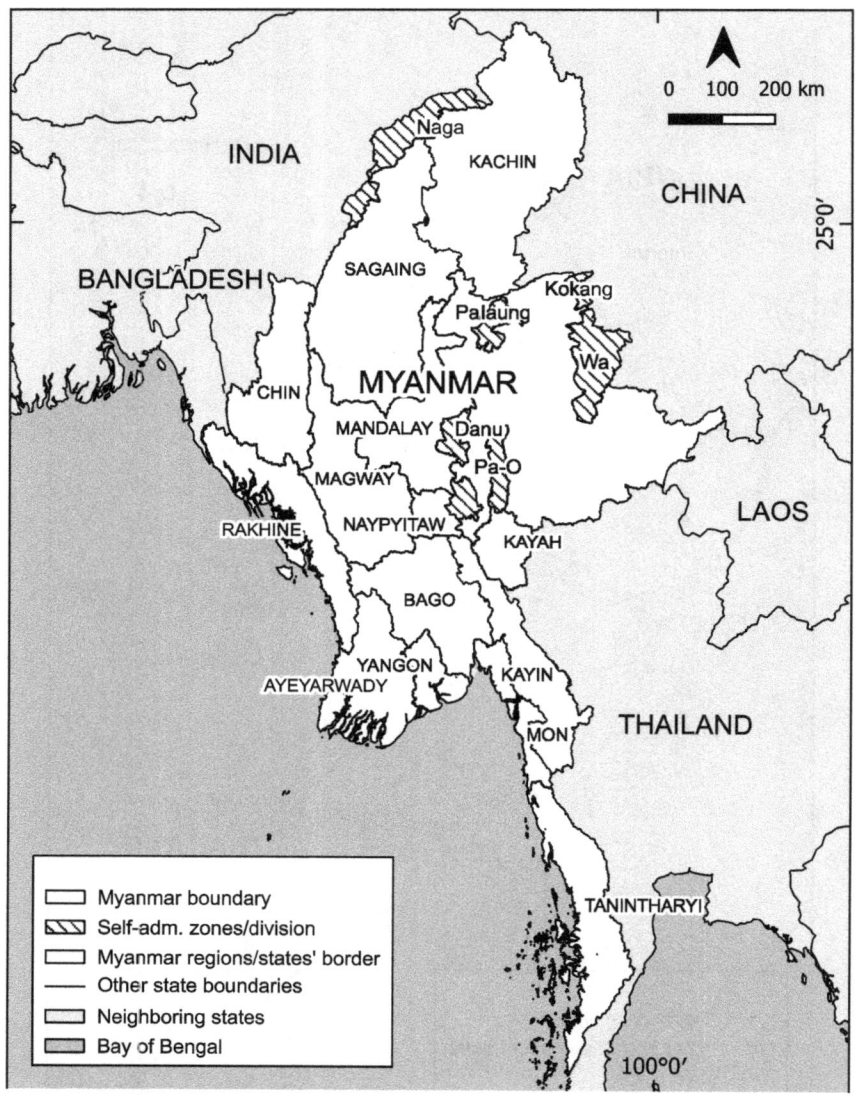

MAP 2. Political map of Myanmar according to the 2008 constitution. Map by Dr. Michaela De Giglio, Department of History and Cultures—Geography Unit, University of Bologna.

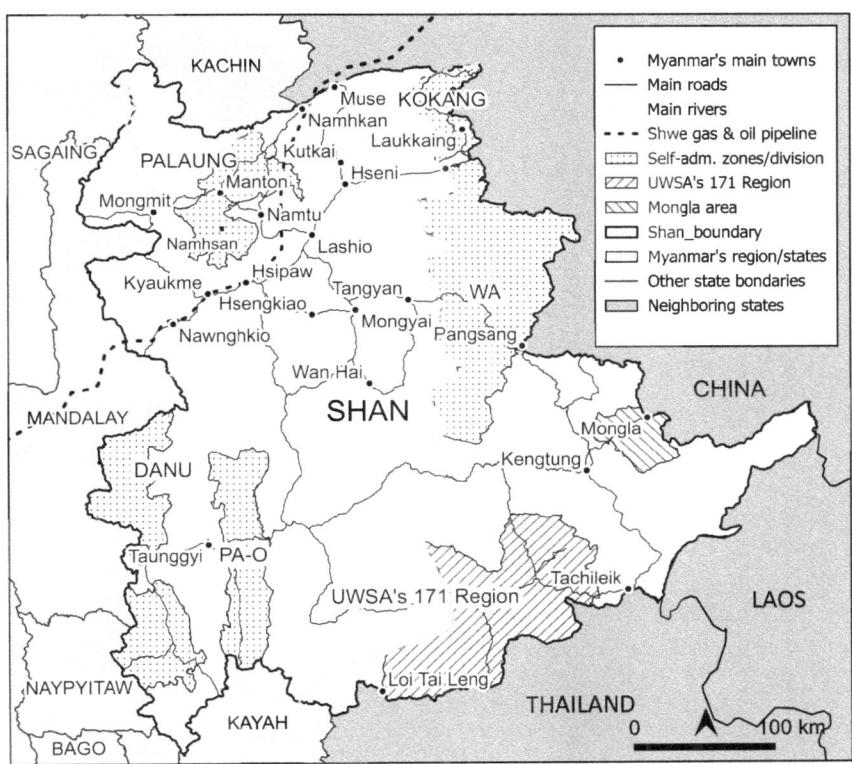

MAP 3. Political map of Shan State, with de jure/de facto self-administered areas according to the 2008 constitution of the Republic of the Union of Myanmar. Map by Dr. Michaela De Giglio, Department of History and Cultures—Geography Unit, University of Bologna.

ARMS POLITICS

FIGURE 1. Weapon seizure in Homein village, Namhsan township, Myanmar, One News, November 23, 2019.

"THOSE WEAPONS YOU USE TO FIGHT AGAINST THE AIR FORCE"

Like the cawing of the black crows ripping through Yangon's electric blue sky outside the window, fragments of a conversation continued to punctuate the lines I was reading: "those weapons you use to fight against the air force." After coming back to Myanmar's major city from a two-week stay in northern Shan State, I was checking the news when official press releases started circulating. The Sit-Tat (Myanmar armed forces) had just carried out what was defined as one of the largest weapon seizures in more than seven decades of armed conflict.[1]

A day earlier, on November 22, 2019, an army column operating some three hundred kilometers south of the border with China had run into troops of the Palaung State Liberation Front/Ta'ang National Liberation Army (PSLF/TNLA), one of Myanmar's ethnic revolutionary organizations (EROs).[2] Searching for rebel camps, the arms were found stored in three hidden caches spread throughout the hilly surroundings of Homein village—the heartland of the Ta'ang movement located at the northernmost edges of Namhsan township. A picture accompanying the press releases portrayed security personnel immersed in a typical northern Shan State mountainscape busy recording the variety of material discovered (figure 1): rocket-propelled grenade launchers, light machine guns, assault rifles, ammunition boxes, grenades, and other equipment. Among these, one object in particular catalyzed the attention of observers: an FN-6 surface-to-air missile. Normally manufactured in China by Poly Technologies, the FN-6 is a so-called man-portable air-defense system (MANPADS)—that is, "those weapons you use to fight against the air force."

The recovery of a MANPADS at these latitudes was a surprise to many. Up to then it was believed the only nonstate armed force holding such a weapon in the country was the United Wa State Army (UWSA) (Small Arms Survey 2013). Often compared to Lebanon's Hezbollah apropos of its military strength, the UWSA maintains a de-facto confederated state in Myanmar and is often understood as an alleged proxy of Chinese authorities (Ong 2023). For its part, the PSLF/TNLA was generally deemed a second-class rebel movement navigating these borderlands. Ten years after recreating its armed force in 2009, following a ceasefire and disarmament process, PSLF/TNLA's capacity to actually build a political entity and attend to the populations and political geographies it claimed to represent remained very much disputed. In northern Kyaukme township, for example, only a hundred kilometers south to its Namhsan-based headquarters, I remember how during a conversation with a village head about the PSLF/TNLA 2017 promotional calendar hanging behind him on the wall of his house, he had completely dismissed the administrative reach of the Ta'ang ERO.[3] Its influence and shifting territorial control undeniable, as the presence of more recent editions of the calendar in the house seemed to suggest, PSLF/TNLA's actual governing capabilities at a macro scale were all but evident. And yet, although negligible as a governing authority, the rebel movement had grown into a quite impressive war apparatus. Its latest show of strength dated to just a few months earlier when, in August 2019, it had orchestrated a spectacular attack on the Sit-Tat's military defense academy in Pyin Oo Lwin.

Thus, in the days following the seizure, analysts were quick in reacting to the event. International observers emphasized the technological properties of the FN-6. While the acquisition of surface-to-air missiles represented a potential "game changer" in the wars between the Sit-Tat and the rebel movements, should it fall in the wrong hands, such technology posed inherent threats to regional civil aviation.[4] Myanmar media instead shifted the attention toward the agencies and meanings behind its acquisition by the PSLF/TNLA. For some, the presence of an FN-6 in the pictures spoke of Chinese authorities' political consent and UWSA's logistic mediation to access gray and black arms markets. For others, given its exorbitant price, it constituted tangible proof of the involvement of the PSLF/TNLA in illicit activities vaguely subsumed under the triad of "drugs, extortion, and their own businesses."[5]

At the same time, the weapon triggered certain geopolitical imaginaries. In a few years, with Sit-Tat's carpet bombings becoming, once and again, a daily reality in the aftermath of the February 1, 2021, coup d'état (Myanmar Witness 2023), surface-to-air missiles would come to assume an almost mythical aura across public debates about the needs of the armed resistance (Aung Zaw 2022). But for now, the recovery of a MANPADS fueled an underlying image of the PSLF/

TNLA as a public enemy, and of the borderlands as a stateless, anarchic environment where any monopoly on the means of violence is lacking. The rebels' alleged involvement as a puppet of a foreign invading power, alluded to by some media outlets, depicted Ta'ang areas as another potential enclave under China's geopolitical sphere of influence. Similarly, the representation of the Ta'ang armed force as a depoliticized warlord formation enmeshed in narco trafficking portrayed these spaces as frontiers plagued by resources and greed. Different interpretations of the seizure suggested partial interpretations of the political geographies of these portions of the Myanmar borderlands. But outside of the horizon of the views offered by the news, two main questions remained: How are weapons and military formations controlled at the edge of the state? And how does control over weapons and military formations in turn shape the political geographies of rule in borderlands?

This book investigates these two questions. It tells a story about weapons, military forces, and space in frontiers at the borderlands. Narrating the trajectory of ceasefire, disarmament, and rearmament experienced by the Ta'ang rebel movement(s) in Shan State, it focuses on two aspects. First, the book analyzes how the process of governing weapons and forming armed forces molds rebel political communities and their territories. Second, it illuminates the ways in which the politics of governing military apparatuses produce other forms of space, in terms of networks, territorial scales, and places. The processes and practices of governing weapons and armed collectives contribute to shape social and spatial relations of rule at the margins of state authority. Echoing transdisciplinary contributions to the literature on Myanmar borderlands that have looked into the histories and politics of constructing different forms of collective identities and spaces at the margins (Leach 1959; Lehman 1967; Scott 2009; Chit Hlaing 2008; Sadan 2013a; Than 2016; Ferguson 2021; Than and May 2022), I take an analytical approach foregrounding the processes and practices that, by governing weapons and armed humans, reproduce borderland political entities.

From the "margins" (Cons and Sanyal 2013), the FN-6 and its acquisition stood in for much more than the geopolitical imaginaries suggested by analysts and journalists. The weapon was the epitome of complex political trajectories, in which recreating and governing the TNLA could not be disentangled from the production of a Ta'ang political community with its population and attendant political space, as the words of K.—a young Ta'ang officer—were now reminding me. Two weeks before the seizure, I traveled with him to Pansali, a village in Namhsan township not too far from the places where the Sit-Tat would discover the arms caches. After talking about the origins of the Ta'ang rebel movement in a number of conversations with the PSLF/TNLA leadership, K. had been ordered by his superiors to give me a guided tour of the small war memorial in

Pansali housing the grave of one of the movement's founding figures. For a good while, I had accompanied the bumpy motorbike journey from the headquarters to the village with questions, trying to get his view on the "territorial control" of his army in Ta'ang areas of Shan State. But eventually we had fallen silent—K. diligently preferring to focus on the rough road terrain, me silently repeating to myself his answers on "territorial control" so as not to forget them. As Pansali was slowly appearing in the distance, I asked if he worried about the recent escalation of hostilities across northern Shan State and the approaching dry season that historically in Myanmar coincides with an intensification of fighting. Surrounded by landscapes similar to those now silhouetted against the backdrop of the images of armaments and soldiers gathered at Homein (figure 1), K. pointed to the village and noted: "You know, they [Sit-Tat] do not dare to come here over land—it is difficult, and these are *our Ta'ang lands*. By the sky it is different, though; if they come with the plane or the helicopter, they can do what they want. But now things will change, though, because we have obtained *those weapons you use to fight against the air force*."[6] (Emphasis mine)

The remark K. had broken his silence with resurfaced vividly in my mind two weeks later. His words gave voice to a perspective totally glossed over by mainstream accounts of the seizure. Perhaps just a hint of ethnonationalist rhetoric for anyone unfamiliar with these borderlands, his reference to Ta'ang lands spoke of the PSLF/TNLA's primary objective upon rearmament in 2009: the liberation of all Ta'ang people from all forms of oppression, via the creation of a state of self-rule and autonomy inside a federal democratic union of Myanmar—that is, Ta'ang Land. K.'s view connected the acquisition of weapons and military means to the production of political space.

Far from being a fully articulated political technology, Ta'ang land has never ceased to be a fluid self-identified territory, existing at the borderlands between Myanmar, China, and Shan State. Since the late 1950s, the Ta'ang resistance developed in the midst of two main intertwined threads of conflict that, brutally simplifying, have characterized Myanmar's contemporary histories of war (Aung-Thwin 1985; Callahan 2002, 2003; Fink 2009; Than 2015; Thawnghmung 2019; MacLean 2022). One is the fight for the democratization of state institutions. A struggle that has seen pro-democracy movements from different walks of life opposing a series of military regimes cyclically seizing power. The other is the fight for the very shape of Myanmar as a political community and the postcolonial state itself (Htet Min Lwin and Thiha Wint Aung 2024). This line of conflict has materialized claims about federalization and the localization of political organization and authority (Brenner 2024).[7] In fact, ethnonationalist movements have not been confronting just the ethnocentrism of the Bamar-dominated Sit-Tat. They have also been battling against the ethnocentrism of pro-democracy elites envisaging Myanmar as the nation-state of the Bamar majority populations. But the PSLF/TNLA's struggle entailed yet another, additional layer. Its claim to a state of self-rule had to be articulated also vis-à-vis other competing ethnon-ationalist projects, the Palaung having long been considered—and by some even nowadays—a minority within the minorities. A minority somehow related to the Shan and living within Shan State (Ferguson 2021).

In turn, not only the Myanmar state, but also other rebel movements (Shan and Kachin in particular) have worked to include Ta'ang Land into their respective geographies of rule since the 1960s (Yawnghwe 2010; Sadan 2013a; Sai Latt 2016; Ferguson 2021). As subsequent Ta'ang politico-armed formations have opposed such state as well as rebel geographies, seeking to produce their own politico-territorial entity, Ta'ang Land has remained a highly contested geopolitical idea in the midst of protracted armed conflicts. Therefore, as the empirical chapters of the book will show, Ta'ang Land can best be understood as a frontier space at the borderlands that has been experiencing multiple frontierization and ter-ritorialization projects and moments (Tsing 2005; Faxon 2021), especially in rela-tion to narcotics trade and forms of extractive ceasefire capitalism from the late 1980s onward (Woods 2011). Frontier projects and moments that unfolded as the original Ta'ang rebel movement, the Palaung State Liberation Organization/Army (PSLO/A), was undergoing a fourteen-year ceasefire agreed to in 1991. Frontier projects and moments that were compounded by the official disarmament and disbandment of its armed force in 2005.

Thus, in retrospect, K.'s comments about "those weapons you use to fight against the air force" allowed me to read the munitions seized in Homein as part of a divergent political-geographic vision of Ta'ang Land. The weaponry was part of a broader process to produce the territory of a Ta'ang political community. A process that had unfolded through rearmament. By foreground-ing, with his air-mindedness, how the technical properties of the MANPADS extend different forms of agency to the sky, K. pointed to the ways in which the TNLA war apparatus produced the territory in which the "here" (of Pansali village and its memorial, for example) is embedded, and how managing mili-tary means could recalibrate rebel territories vis-à-vis military-state territorial scales (Ferguson 2018). His take on the acquisition of FN-6s at these latitudes suggested how a specific way of killing by downing the air machine of another political order (e.g., the Sit-Tat) would make a certain vertical articulation of territory possible. In other words, the management of a specific way of killing—made possible, legitimate, manifest, being ontologically inscribed in the MAN-PADS by design—carried with it also a discursive and material calibration of political space: in this case, the territory of the PSLF/TNLA polity, "our Ta'ang lands," as K. put it.

In the Homein seizure, regardless of the different views one would adopt—from those of international analysts to those of national outlets, from those of military-state apparatuses to those of rebel groups, from those of civilians to those of partisans—it appeared clear how the acquisition of weapons and the related formation of an armed force constitute fully political arenas of struggle. The violent histories of Ta'ang land as a frontier space could be largely read in terms of how various politico-armed actors that do not hold any bordered territory nonetheless still reproduce their political entities. That is, how rebel movements that do not necessarily maintain the apparatuses of government needed to attend to the populations they claim to stand for at a large scale yet do produce a collective identity and political geographies of territory by governing the means of violence. The Ta'ang case seemed an illustrative one in this sense. An ethnonational rebel movement identifying with two different names—the exonym Palaung, in the acronym of its political branch (PSLF); the endonym Ta'ang, in that of its military one (TNLA)—had been struggling to clearly define the population and space that it aimed to govern and foster. But at the same time, in a relatively short frame, by rearming the resistance movement and forming an armed force, it had been able to reproduce a Ta'ang polity in the borderlands. In the politics of becoming and being a Ta'ang weapon and army, one could see also the becoming and being of a Ta'ang polity with its political geographies.

Military Means in Rebel Governance and Spaces at the Margins

The acquisition of weapons and the formation of armed collectives are highly political processes. These are linked to, and shaped by, political rationalities and techniques of governing the entanglements between humans and weapons understood as technical objects. Such rationalities and techniques circulate throughout society and codify a certain way of doing violence. Through this argument the book contributes to the literature on rebel governance (Kasfir, Frerks, and Terpstra 2017; Duyvensteyn 2017; Cunningham and Loyle 2021; Malthaner and Malešević 2022). It shows the fully political nature of the governing of weapons and armed collectives at the edge of the state, and how this governing activity is shaped by and shapes historically and spatially complex relationships among rebel movements' leaderships, rank-and-file, civilians, civil society actors, and larger frontier assemblages (Cons and Eilenberg 2019).

Furthermore, rebel movements harness the acquisition of weapons, and the formation and governing of their armed collectives, to reproduce and shape the collective identity of their polity and the political geographies of vital space of their polity. With this second argument, this research contributes to the literature on spaces at the margins in political geography by providing an empirical analysis of how the political space of the borderlands and frontiers (in Myanmar) is produced by governing violence. It feeds as well into Foucauldian/poststructuralist accounts of rebel governance by qualifying forms of rebel rule in biopolitical terms (Hoffman and Verweijen 2018; Brenner and Tazzioli 2022). As explained in more detail later in the book, although certain forms of rebel rule do not maintain the governmentality apparatuses necessary to define and enhance biological life at aggregate scales, they still massify and divide the populations and political spaces of their polities by governing weapons and military means.[8]

Studying how governing military means shapes social and spatial relations of rule at the margins, the book necessarily intersects those fields of inquiry that, in the past decades, have looked into the production of social and political (dis)orders by armed actors and rebel governance in contexts of war (Schlichte 2009; Mampilly 2011; Arjona et al. 2015; Jentzsch et al. 2015; Arjona 2016; Kasfir, Frerks and Terpstra 2017; Hoffmann and Verweijen 2019; Brenner 2019; Harrison and Kyed 2019; Harrison 2020). The literature on rebel governance has not focused specifically on the acquisition of weapons and the governing of armed forces. Moreover, it has expressed an overall tendency to uphold a cartographic conception of territory, which it has replicated in its approach to the relations between weapons, humans, and the political geographies of conflict and rule.[9] Besides the rebel governance camp, literature in the ambit of civil wars studies has produced unprecedented contributions concerning the role of firearms and armed violence in conflict. Although not explicitly adopting the rebel governance paradigm, these studies have been concerned with how weapons and military means are used and controlled (Sislin and Pearson 2001, 2006; Marsh 2007; Bourne 2007; Beavan 2008; Duquet 2009; Krause 2009; Strazzari and Tholens 2010; Greene and Marsh 2012; Bartolucci and Kanneworff 2012; Ashkenazi 2012; Strazzari and Tholens 2014; Enomoto 2018; Hazama 2018; Buscemi 2019).

In both these ambits, scholars have usually (directly or indirectly) adopted an ontology of the weapon, and an understanding of the relations between weapons, humans, and sociopolitical phenomena, that can be described through two visions: the substantive/materialist vision, and the instrumentalist one.[10] The substantive/materialist vision understands weapons as technologies determining social phenomena, while the instrumentalist vision understands human actors as determining social phenomena (Bourne 2012). Substantivist positions characterize weapons as autonomous factors in determining social and political relations, while instrumentalist ones suggest the irrelevance of weapons and their subordination to material interests and rational agencies, the distribution of power, or the trajectories of intersubjective relations and norms (for a review see Marsh 2018; Bousquet et al. 2017, 2020).

Often overlapping or alternatively adopted, substantivist and instrumentalist approaches have been criticized for being equally affected by a "Cartesian" dualism and by a deterministic conception. They are dualist in that they conceptualize a separation between materiality and sociopolitical relations. And they are determinist because they tend to look at weapons as material causes determining certain configurations of sociopolitical relations, or remaining purely subservient to actors' interests and intersubjective relations (Bourne 2012). Focusing either on weapon technologies or actors, substantivist and instrumentalist approaches tend to isolate the linkages between weapons, military means, and sociopolitical relations from other material and social dimensions of life, such as the production of space and geography.[11]

The issue of control on weapons and military means aptly shows such tendency. The question of arms and armed forces control has been mostly seen as a prerogative of wartime institutions, armed and nonarmed authorities, or networks of authority, and the negotiations among them (Krause 2009; Greene and Marsh 2012; Kasfir 2015; Arjona 2016). Prevalent approaches have understood control over the means of violence as a quality of authority deterministically related to spatial forms of rule, especially territory (Kaldor 2012; Tilly 1990, 2003; Krause 2009; Greene and Marsh 2012; Kasfir 2015; Arjona 2016). Territory has become the primary spatial cornerstone through which scholars have read weapons and military means control: either focusing on how control on them is articulated inside a given territory; or focusing on the acquisition of weapons and control of military means as a precondition to establish territory (Marsh 2012, 15; Strazzari and Tholens 2014; Buscemi 2019). Through the lenses of the so-called territorial trap, the notion of territory has been conceptualized as a fixed and given entity preceding societies and separating armed authorities, —an entity that is instrumentally related to weapons, armed forces, and military capabilities (Kasfir 2015; for an exception and critique see Hoffman and Verweijen 2019). This has been connected to the adoption of blended substantive/instrumentalist approaches that have limited the analysis of processes and practices of control on weapons and military means, and the ways in which they unfold through and shape spatial relations of rule and political (dis)order.

But spatial relations, such as territory, are actually produced *through* the governing of weapons and violence. This book foregrounds an analysis of how sociotechnical networks of weapons and armed humans work simultaneously as an instrument of control and as a field of control infused with specific logics and techniques of government. The act of governing weapons acquisition, for example, contributes to shape state as well as rebel polities. Through the empirical chapters that follow, it becomes clear that weapons and armed humans control remain mutually related to space; and that social and spatial dynamics of rule other than territory—such as networks, scales, and places—are also reproduced.

In turn, this study brings the literature on rebel governance and civil wars into dialogue with that on spaces at the margins in political geography. The latter in fact has provided alternative geographical perspectives, capable of moving past dominant views of the relations between means of violence and sociopolitical orders at the margins, but it has only indirectly and partially concerned itself with weapons and military means. Developed at the intersections of political geography, anthropology, and area studies (Baud and Van Schendel 1998; Van Schendel 2002; Das and Poole 2004; Abraham and Van Schendel 2005; Korf and Raeymaekers 2013; Rasmussen and Lund 2018; Cons and Eilenberg 2019), in the last decades this scholarship has explored questions of authority and order in spaces at the limits of state sovereignty (Ballvè 2020). It has shown how forms of authority alternative to state ones display their own multiple histories and geographies (Van Schendel 2002; Scott 2009; Sadan 2013a), and how not all forms of authority are always related to one or all of the elements formulated by Weberian conceptions of sovereignty: territorial control; enforcement of norms and rule; and control over the means of violence (Lottholz and Lemay-Hèbert 2016; Goodhand 2013). These tenets have been coupled with the understanding that social and political transformations are not top-down and centrally driven events, but rather emerge from constant and diffused processes of contestation and negotiation of power that reshape the geometries of rule and authority at the putative margins of the state (Korf et al. 2018). This literature has adopted a conception of power as a diffused effect of practices, processes, and relations of authority and has combined it with the idea that space is at one time socially produced and an arena productive of social relations (Lefebvre 1991).

Yet, while it offers great potential for understanding weapons and armed forces as both produced by and productive of social and spatial relations, this scholarship has not focused on the governing of weapons and military means. The Myanmar borderlands exemplify this broader trend quite neatly. In line with recent increasing attention to how heterogeneous nonstate actors and organizations practice and reproduce spatialities of rule (in particular territory) that has characterized this field (Watts 2004; Peluso 2008; Korf, Hagmann, and Engeler 2010; Lund 2011; Rasmussen and Lund 2018), scholars have emphasized how in Myanmar a multiplicity of armed and nonarmed actors produce and navigate different forms of territorialization (Suhardiman et al. 2019; Mark 2016). They have looked at how such territories can be temporally and spatially multiscalar (Hong 2017), noting their fragmented and layered character (Woods 2019a, 2019b), as well as their relational, mimetic, and encompassed nature vis-à-vis long-standing state and nonstate institutions and practices (Harrison 2020). Here, though, control of the means of violence has been explored mostly through the lens of politicoeconomic competition to govern access to resources and rent systems via retribution tactics (MacLean 2008), patron-client relationships (Woods 2011), and the institution of "limited access orders" (Meehan 2011; McCarthy and Farrelly 2020). This scholarship has looked at the means of violence mainly in instrumental terms focusing on social, economic, and political relations among armed and nonarmed actors. Instead, I concentrate specifically on the role of technologies of violence—weapons above all, but not only—and the processes and practices of controlling both the technical objects and the human-nonhuman collectives of armed violence. This way, the research ultimately adds also to enquiries on territorialization processes in Myanmar's borderlands by focusing on the much messier and underresearched scenario of the Shan State and Ta'ang areas. Here, territorializations are not only multiple, overlapping, and nested but also constantly becoming in mobile, fluid, and shifting ways.

Becoming and Being a Weapon: Methodology

Besides contributing to the literature on rebel governance and spaces at the margins in political geography, this book addresses a gap in the study of rebel movements in Myanmar. Ta'ang areas and armed movements have remained underresearched in a landscape of relatively underresearched borderlands in peace and conflict, civil war, and rebel governance studies (for a critique: Brenner and Han 2021).[12] There are reasons for this. Ta'ang rebel movements have been located far from actual borderlines, even though they have maintained mobile presence and networks throughout the regions at the intersections of Myanmar, China, and Thailand. At the same time, as research communities have grown in relation to other nonstate epicenters—for instance, surrounding the Karen and Kachin rebel movements and struggles, or Rakhine State's conflictual politics—they have tended to replicate access and knowledge production dynamics (Chu May Paing and Than Toe Aung 2021; for some scholars working on different areas, see, for example, Kramer 2007, Than 2016, Xiaobo 2018, Sarma 2020, Prasse-Freeman 2020, Faxon 2021, Ong 2023). Moreover, the development of international humanitarian and social support networks linked to religious-based organizations at the Thai-Myanmar and the Kachin-Chinese border has further contributed to relegate other borderland areas, and in particular Ta'ang ones, to the background (Smith 2016).

Yet, Ta'ang areas and armed movements present important peculiarities. Recursive processes of mobilization, demobilization, and remobilization have become a characteristic trait of the struggles of many armed actors in the country (Smith 2007), so much so that it is often difficult to define what counts as disarmament and demobilization and what does not. For example, the Karen National Union/Karen National Liberation Army (KNU/KNLA) underwent a process of fragmentation along brigade lines during the second half of the 1990s and the 2000s. Around 2012, while the central commands were agreeing on ceasefire arrangements, some KNU/KNLA brigades—in particular the Mutraw-based Brigade 5—experienced a process of remilitarization (Brenner 2017b). Similarly, in 2011, the breakdown of the Kachin ceasefire was in part related to a process of remobilization and militarization of youth components of both leadership and rank-and-file (Brenner 2017a; 2019). On the western frontiers of Rakhine State, at the tri-border areas spanning from the Bay of Bengal up to the Tibetan plateau, Martin Smith has noted how an astounding variety of Rakhine and Rohingya politico-armed rebel movements has been mobilized, demobilized, and at times disarmed, only to then rearm in different shapes (Smith 2019, 45, 107–8). In southern Shan State, to add a further example, the 1995–96 Sit-Tat-orchestrated offensive that led to the surrender and disarmament of Khun Sa's Mong Tai Army (MTA) was followed by the constitution of different armed formations, among which was the Revolutionary Council of Shan State/Shan State Army-South (RCSS/SSA-S) led by Yawd Serk, a prominent commander of a Shan United Revolutionary Army (SURA) faction that had been part of MTA (Ferguson 2021). Nonetheless, throughout more than seven decades of ethnopolitical and ideological armed conflicts in Myanmar's borderworlds (Sadan 2013a; Han 2019), Ta'ang land has been the only area where a well-established ethnonational rebel movement was officially disarmed by Sit-Tat military regimes and later undertook a full-fledged rearmament process.[13]

As the first part of the book explains, the PSLO/A agreed to a ceasefire in 1991. Subsequently, part of the movement, together with new political figures, rejected the ceasefire and formed a different political front in continuity with the PSLO/A: the Palaung State Liberation Front, based at the Thai-Myanmar border. In 2005 the PSLO/A was forced to agree to a disarmament and disbandment process. The disarmament was officially sanctioned in a ceremony organized by the State Peace and Development Council military regime held in Manton, northern Shan State, at the presence of the then first secretary Thein Sein. The disarmament of PSLO/A was welcomed as a fulgid example of how EROs should have undertaken the path of "exchanging arms for peace" (New Light of Myanmar 2005). After the disarmament and disbandment, the PSLF undertook a lengthy rearmament process at the Thai-Myanmar border that later, in 2009, led to the creation of its armed branch, the Ta'ang National Liberation Army, and to its reconsolidation in Ta'ang areas of northern Shan. The PSLF/TNLA has continued to claim that a direct lineage can be traced from the Palaung National Force (PNF), the original predecessor of all Ta'ang politico-armed organizations formed in 1963, all the way to the PSLF/TNLA passing through the PSLO/A. This peculiar process offers a good case to study the research questions formulated earlier.

The book addresses such gaps on Ta'ang resistance movements by drawing on intensive multisited fieldwork in Myanmar, and for a small part in Thailand, for a total of fifteen months between late 2018 and August 2023. The core of the analysis is based on qualitative in-depth (unstructured and semistructured) interviews, informal conversations, and nonparticipant observation (Atkinson and Hammersley 2007). While informed by rich fieldwork research, vignettes, and interlocutors' life stories, the book is not an ethnography and may thus disappoint readers of the ethnographic genre.

Fieldwork was conducted in four different phases. In September-October 2018, I first carried out a preliminary fieldwork period at the Myanmar-Thai border, in the area of Mae Sot, as well as in collaboration with mountain guides in Kyaukme, northern Shan State. Exchanging English-language conversation hours with the possibility to hang out with a group of local mountain guides during their leisure time dedicated to explorations for future tours turned out to be a particularly fruitful practice to generate conversations on different aspects of life in the mountainous areas of northern Shan. From February to July 2019, I embedded myself with a humanitarian mine action organization active in the ambit of mine risk education, victims assistance, and nontechnical surveying of contaminated areas. Lastly, between September 2019 and January 2020, with an additional period in March 2020, I liaised with a local civil society organization (CSO) in Kyaukme carrying out voluntary work ranging from writing documents in English to performing errands in Yangon. Throughout all these phases (except for March 2020, for reasons of access and security), I alternated periods of approximately two weeks between northern Shan State and Yangon. Periods in northern Shan entailed living for days in mountainous villages and towns in the townships of Kyaukme, Hsipaw, and Lashio in particular. During two of these alternated periods I embedded myself with two EROs: PSLF/TNLA, in the areas of Namtu and Namhsan; and the Shan State Progressive Party/Shan State Army-North, in its headquarters area of Wan Hai. Eventually, I carried out two additional follow-up fieldwork periods of one month each, in Chiang Mai in 2022 and in Karenni Nationalities Defense Forces bases between Mese and Demoso townships throughout August 2023.

Interviews, conversations, and field notes were gathered by engaging with interlocutors as varied as EROs' leaderships and rank-and-file, village heads and dwellers, Buddhist monks and abbots, community militiamen, CSOs operating in mountainous areas, local humanitarian workers, mountain guides, mine action researchers and workers, field analysts, and journalists. The appendix provides an interview and fieldwork methods table in which I have profiled this material in more detail. The material presented in the appendix does not convey the importance of a conspicuous work of note-taking and nonparticipant observation that accompanied fieldwork experiences. For example, access to PSLF/TNLA's temporary mobile headquarters and bases in Namtu and Namhsan was granted upon an understanding that I would engage in certain activities while present. These activities consisted of holding sessions of talks on the politics of Italy after World War II and the socioeconomic North-South divides, to be delivered to TNLA cadets and officers; English-language conversation sessions with students in a new, PSLF/TNLA-sponsored women's leadership training program; and playing sports during off-duty evening hours. Some of these occasions turned into settings akin to focus groups insofar as exchanges and conversations would at times flow into extremely useful collective discussions. Similarly, during the time spent with the humanitarian mine action organization, I could take part in field activities such as mine awareness and education workshops, map-drawing exercises, victim-assistance meetings, survey discussions, or institutional meetings.

In gathering such material, I was seldom alone. Interpreters were essential, at times also facilitating access through their own social networks, but acting as gatekeepers as well. They did not just provide or limit access but also represented a form of constant accompaniment dictated by my very position as a researcher and observer. Their presence is an element that should be considered as an aspect of what I could see, listen to, and experience. My understanding of the topics covered by the book is thus necessarily partial—for roughly half of the material was gathered in English, while the rest was filtered through interpreters because my language skills did not allow me to conduct full interviews in Burmese. (And at times because interlocutors would feel more comfortable speaking in Shan or Ta'ang.)

Besides shaping access and information, these fieldwork methods have also reproduced power effects, silenced certain voices, amplified others, and normalized or objectified processes and dynamics remaining out of the spotlight (Burawoy 1998; Mitchell 2013). Although I made sure to clarify my presence and position, I was there for reasons that where hidden to many and that remained distant from the people I engaged with. My presence would quickly be integrated into, and reproduce, relations and practices of authority. Eating breakfast, lunch, and dinner with PSLF/TNLA leaders under the bamboo veranda overlooking the drill ground at the center of their headquarters, for example, I simultaneously confirmed and upheld norms of authority and hierarchy linked to culinary practices in the rebel base. My being there was inextricably enmeshed in relations of authority and power that were highly gendered. In other words, my presence upheld the power positions of some (e.g., male commanders with both a gender- and age-affected positionality) and was in part based on the subaltern position of others (e.g., women workers cooking meals in the base). The same goes for the potentially "extractive" relations I was part of (Mitchel 2013), such as those with translator-fixer figures, many of whom I became friends with. I attempted to avoid the creation of benefit gaps by remunerating work, creating long-term connections, and providing support to people. For example, I offered to help with the process of searching job or study opportunities abroad, linking up with contacts, commenting on English-written documents, or more simply nurturing friendships and discussing these life-trajectory topics. At the same time, however, staying in touch with the people whose stories populate these pages benefited me too, perhaps more than them. It constituted an invaluable opportunity for me to constantly reengage on a conceptual level too, providing a chance to keep questioning and refining the contributions formulated by this study as the manuscript was taking shape.

Structure of the Book

A retired TNLA soldier now living on his farmland north of Kyaukme town once made me reflect on something he understood as a key feature of firearms. Together with B., a friend who worked as a mountain guide and as an interpreter for me during fieldwork, we were walking back toward the former soldier's house after a brief visit to his maize field located just a couple of kilometers away. Adjusting the sling of the craft-manufactured musket hanging on his shoulder, he drew the weapon tighter to his body for a second and turned to B. to explain why he had brought it along on the very short hike. Weapons always move around; they never stand still. If you leave them somewhere, he added, people take them and move them to another place, even if they do not really want to use them. The idea that weapons have certain technical properties that cannot be disentangled from the politics of managing them, as well as from the ensuing spatial relations, is what makes the stories of disarmament and rearmament in Ta'ang land particularly interesting. This idea of motion as a technical feature inscribed in weapons—a technical feature that generates the need (and the politics) to manage them in space—is the key entry point from which the stories of this book begin.

The first part of the book—"Blunt Rebel Rule in Frontier Assemblages"— focuses on the comings and goings of weapons throughout the Ta'ang cease- fire, disarmament, and rearmament, as a perfect background to explore how, in "becoming and being" a weapon and an armed force, the becoming and being of a polity unfolds too. Chapter 2 delves into the role of weapons and armed collectives in the (un)making of the borderlands of Shan State, placing Ta'ang land into the broader historical context of northeastern Myanmar, southwestern China, and Thai and Lao border areas. Drawing mainly from secondary sources, this chapter frames the histories of weapon flows, their politicization, and implications for the reproduction of borderland territories and multiple frontier spaces. Chapter 3 analyzes the trajectories of the Ta'ang rebel movements throughout the processes of ceasefire, disarmament, and rearmament. Chapter 4 shifts toward the ways in which a Ta'ang polity, with its population as a biological body and its political geography of territory as a vital space, has been shaped via techniques of governing military means. The chapter illuminates also how different rebel polities overlapping with the Ta'ang one produce their own political geographies of territory.

Part Two—"Weaponscapes"—comprises three chapters that inquire into the ways in which governing military means produces different political geographies of rule beyond territory in frontier spaces. Chapter 5 foregrounds the specific contributions that weapons as technical objects of violence make in shaping the spaces of the frontier. Providing an analysis of how the spatialities and modalities of violence that are materially codified in weapons require to govern the encounters of humans with weapons in manners that are highly political, it shows the networked spaces that are produced by weapons and armed forces in the borderlands. Chapter 6 delves into the ways in which different territorial scales of rule are produced and contested by governing military means. And last, chapter 7 closes with an exploration of how processes and practices of governing weapons and armed collectives reproduce places understood as locations, locales, and sense of place.

This two-part structure mirrors the conceptual framework presented in chapter 1. The book combines assemblage thinking with biopolitical governmentality to conceptualize weapons and armed forces as sociotechnical ensembles that are key in the making of frontiers. These "armed assemblages" are governed via rationalities and techniques of managing violence. The governmentality of weapons and military means, together with the (un)making of frontiers, is harnessed by rebel rule to shape the biopolitics of their polities. That is to say, by governing their armed forces, and by reproducing their frontiers, rebel movements shape the collective identity of their polities' population and their political geographies of vital space in the borderlands.

FRONTIERS OF VIOLENCE

Conceptual Bearings

Two research questions stand at the core of this book: How are weapons and military formations controlled at the edge of the state? And how does control over weapons and military formations, in turn, shape the political geographies of rule in borderlands? These two questions point to the nexus between the governing of weaponry, the formation of military forces, and the production of rebel polities in political-geographical terms. In Myanmar, this nexus has become particularly relevant in the aftermath of the latest coup d'état. A nexus epitomized by a current debate concerning the question of what is a "just war" in the post-coup era.[1]

After roughly a decade of semidemocratic experimentations, the rearticulation of the (semi-)authoritarian system that occurred on February 1 2021 has been met with outright resistance throughout the country. This resistance quickly shifted toward armed struggle and irregular warfare in the ensuing months, as the Sit-Tat's interim government—the State Administration Council (SAC)—violently repressed peaceful demonstrations. With the SAC's counterinsurgency operations mounting, Bamar-majority populations based in the so-called Myanmar heartland experienced extraordinary waves of violence.[2] Such violence—very much ordinary in the borderlands even before the coup—was extraordinary here. These areas had had little to no exposure to the mass violence that the military state had been deploying since the 1950s in processes of frontierization and territorialization of border areas (Ong and Prasse-Freeman 2021). In this context, a debate concerning the idea of a "just war" has emerged across the complex landscape of post-coup resistance (Buscemi 2023, 2024).

While an unprecedented cross-sectional solidarity along class and ethnonational lines has animated these political milieus, progressive voices and ethnic minority activists have felt it important to note that, in the current conjuncture, the Bamar-majority populations of the heartland are essentially having firsthand experience of what minorities have been living for roughly seven decades. Today the Bamar-majority people think of Sit-Tat's wars as unjust, even though yesterday they, or at least some, may have considered violence at the nation-state's frontiers as part of a "just war." Today they think of the wars of resistance as "just wars," even though yesterday they may have struggled to understand the motives of rebel movements. Thus, it has been argued, Bamar people can now understand more easily why ethnonational rebel movements have long been fighting in the borderlands. Pan,[3] a leading Ta'ang activist, explained this point to me when, reflecting on the expression *ta-ya-thaw-siq'* (just war) during a conversation, she said:

> Before [the coup], . . . when . . . human rights violations or genocide or war crimes were happening in the ethnic areas . . . the media did not cover a lot about that case. And when the local media would raise the issue . . ., the people in central Burma did not have interest on this. In their mind, they thought the ethnic people do not really want peace, and that they do not have any suffering. . . . When we say discrimination, they [Bamar people from central areas] feel nothing: it means that they for example thought that we were making up stories. . . . [I]f we sometimes speak to our friends about our area, they would listen and comment "oh, really, is it really like that?" They would not recognize it. . . . And now, they do understand well what the military did to the ethnic people and what they have done. Now they see it, and they cannot accept, they cannot accept that kind of violence. They cannot think that the Sit-Tat does like that. They thought that the Sit-Tat was their institution that would protect all the people, but now they really really see the difference. So they cannot accept that [violence]. That is why they changed their mind, and that is why they embraced the gun to fight back.[4]

Referring to subaltern histories of living with violence and Ta'ang experiences of war—as Sarma, Faxon, and Roberts (2023) have put it—Pan's explanation suggests how, throughout past decades, the Ta'ang borderlands were more than borderlands. They were also frontiers of violence: spaces where different conceptions, practical modalities, and standards of what is deemed to be an acceptable form of violence encountered each other and became visible in the attempted formation of a military-state Myanmar polity, as well as in the attempted formation of ethnonational rebel polities. In what Pan sees as an awakening of the Bamar-majority "heartland," there is also the observation that managing the relations between weapons and people shapes political communities and spaces by shaping the frontiers and boundaries of them.

Frontiers are projects that mold specific spaces, in political and geographical terms, as zones where new forms of material or immaterial goods—be it natural resources, labor, populations, or knowledge—can be extracted (Tsing 2003). To make such extraction possible the frontier has to first be made legible and secure through violence. Besides being related to some form of extractive process, the frontier is first and foremost a politically defined space where different conceptions, practical modalities, and standards of violence constitute and embody a sociopolitical capital in the attempted formation of a polity. The processes and practices of governing weapons and military means, in Myanmar as elsewhere, are defining features of the making of frontiers. This is the reason why, in order to understand rebellions and political orders at the margins, it is important to study how military means are managed and how frontiers are reproduced by both state and nonstate polities.

This chapter presents the conceptual approach the book adopts to look at how weapons and military formations are governed at the edge of the state, and how this government activity shapes political geographies of rule in borderlands. I use the concept of assemblage to think about the borderlands of Myanmar as frontier spaces reproduced by a range of elements, weapons and military means in particular. Widely deployed throughout the social sciences in recent years (Tsing 2019), the core function of the concept of assemblage is that of decentralizing the role of human actors, behaviors, and agency in order to gear the analysis toward the complexity of social and natural compositions and apparatuses (Deleuze and Guattari 1987; Ozguc and Burridge 2023). In her work on the materiality of global capitalist processes and race, Anna Tsing has pointed to the several meanings through which the term "assemblage" has been used as an analytical grid in the literature (2019). To decentralize the role of human agency, in fact, one can focus, for example, on assemblages understood as complex discursive formations (Ong and Collier 2005); on assemblages as the ways in which different heterogeneous elements gather in place at different scales (Tsing 2015); or on assemblages as sets of relations that concatenate and jointly structure other relations (Latour 2005). The following sections bring together assemblage and biopolitical thinking by drawing on three possible understandings of assemblage: frontier assemblage (Cons and Eilenberg 2019); governmental assemblage (Dean 2010); and armed assemblage.

Thinking about the borderlands through the lens of *frontier assemblages*, the aim is to highlight how weapons and armed collectives play a key role in the making and unmaking of the political space of the frontier. Subsequently, I use the concept of *governmental assemblages* of violence to show how military means are at one time a referent object of government and also a key dimension of the governmental apparatuses that shape rebel polities' populations and political space in the borderlands. Lastly, via the concept of *armed assemblages*, I explain that armed forces can be understood as sociotechnical ensembles. These ensembles are managed via specific rationalities and techniques of governing the relations between humans and weapons. Looking at military means as assemblages in and of themselves allows us to shift the focus toward the politics of governing weapons and how it shapes political space—that is, territories, networks, scales, and places—at the margins of state authority.

Frontier Assemblages

Frontiers, as many have observed (Rasmussen and Lund 2018), are politically defined spaces that emerge from a double move of frontierization and (re)territorialization. Although analytically distinct, frontierization and territorialization are closely related processes that often overlap. Frontierization entails the violent dissolution of preexisting forms of rule, rights and norms, land management systems, and sociopolitical forms of administering, while territorialization entails the (usually violent) embedding of forms of authority, systems of economic and political administration, property rights, and related security apparatuses (Meehan and Dan Seng Lawn 2022).

Violence is thus central. It is central to the ways in which processes of politicoeconomic expansion operate to turn "illegible" and "insecure" frontier spaces into legible and secure borderland spaces (Woods 2019a; Cons and Eilenberg 2019; Ballvè 2020; Dean, Sarma, and Rippa 2022; Dan Seng Lawn 2022). Through the double move of frontierization and territorialization, scholars have noted how frontiers often remain under the rule of military authorities and administration (Peluso and Vandergeest 2011), how they constitute areas of legalized lawlessness and regulated exceptional violence (Hagmann and Korf 2012), how the militarization of frontiers represents a tool to generate public budget or becomes a service provided to (and by) mixed state-private entrepreneurs and business networks (Cons and Eilenberg 2019). Yet—although research on frontiers has illuminated the key nexus between violence and processes of frontierization and territorialization— weapons, armed forces, military, and security means have been approached as "pre-given" assets at the disposal of agency to transform frontiers into secure and legible areas, albeit in contexts of high political and armed contestation.

Building on such remarks, I use frontier assemblage thinking to focus the analysis on the role of weapons and armed collectives, and the politics of governing them, in processes of frontierization and territorialization. Elaborated by Cons and Eilenberg (2019), the concept of frontier assemblage understands frontiers as spaces that are constantly reproduced by a range of human and nonhuman agencies, forces, and processes. Through discursive as well as material technologies, these agencies, forces, and processes operate concertedly to turn space and place into calculable territory and populations, soil into owned lands, elements into extractable resources. Similar to what other technologies do (such as cartography; photography; telecommunication, road, and railway infrastructures; property regimes), weapons and armed formations play an important role in the making and unmaking of frontier spaces. They play an important role in the constant attempts to territorialize political rule through and over populations and space at the frontier. The introduction of weapons and armed collectives in the inherently contested space of the frontier requires governmental authorities to calculate populations and territory in relation to the presence, deployment, and control of armed violence.

Frontier assemblage helps us unpack the nexuses of weapons, military means, and frontierization-cum-territorialization processes in two primary ways. First, it decenters human agency (Sunberg 2014; Tsing 2019). In doing so, it provides a look at the relations among weapons, humans, and space in a way that neither attributes to the munitions any independent magical power, nor privileges the preeminence and control of human agency over weapons as mere instruments. In relation to armed collectives, a frontier assemblage approach sheds light on the coordination among weapons' materiality—as technical objects that codify violence—and other contributors to the production of frontiers in borderlands. Second, it problematizes the construction of the "human" (Chandler and Pugh 2022). By widening the analytical focus toward more-than-human dimensions of social processes, assemblage thinking allows for an interrogation of how the "human" and alterity identifications, such as ethnonational identity categories, are (un)made in relation to the nonhuman (Ozguc and Little 2023). In this sense, the concept of frontier assemblage foregrounds the use and management of weapons and violence as political arenas for the reproduction of a certain understanding of the human and of political communities in frontierization and territorialization processes.

Going back 136 years ago and looking at the historical linkages between the Maxim gun and colonial imperial expeditions, we find an example of how the symbolic and material realm of the weapon and military forces constitute an integral part of the production of frontiers. The Maxim gun was the first fully automatic recoil-operated machine gun in the world. Invented by Hiram Stevens Maxim in 1884, the Maxim gun found one of its very first deployments in the British invasion of Burma in 1885, before being deployed to other "colonizable" places such as today's Benin, Ghana, and further areas of the African continent (Thant Myint U 2006; Hicks 2020). Here the civilizational projects of colonial expeditions unfolded together with a certain modulation of violence via the Maxim gun's deployment and control. The configuration of these spaces as frontiers, as blank slates on the fringes of civilization populated by "barbarians," by less-than-human forms of life, was inseparable from how the Maxim gun (and the violence materially codified in the automatic fire of the machine gun) would be used and controlled both discursively and materially.

Similarly, as chapters 2 and 3 will illustrate, the processes of disarmament and rearmament of the Ta'ang rebel movement briefly sketched in the introduction have to be interpreted against the background of the attempted partial incorporation of borderlands' frontier spaces into the Myanmar polity by successive military-state regimes since at least the late 1980s. Along processes of neoliberal capital investment, extraction, and accumulation by dispossession (Woods 2011; Dan Seng Lawn 2022; Campbell 2022), the attempted frontierization and territorialization of Ta'ang areas of Shan State entailed the rearticulation of the security landscape in terms of weapons and armed forces control. As we will see in part one of the book, a range of forces, actors, technologies of violence, and processes pertaining to the relinquishment of weapons and the disbandment of a Ta'ang military force—as well as to the acquisition of the former, and the reconstitution of the latter later on—have been key political arenas to de-territorialize and reterritorialize the Ta'ang polity.

Governmental Assemblages

The vocabulary of frontier assemblage also draws attention to the question of how weapons and armed forces are governed. By foregrounding the role of weapons and armed forces in the making of frontiers, it suggests that sovereignty is not the sole form of power operating at the frontier. In turn, using the lens of *governmental assemblage* together with that of *frontier assemblage* shows a twofold dimension of weapons and armed collectives. Weapons and armed collectives are both technologies that are governed; and technologies of rule through which people and spaces are produced and governed.

The governmentality of military means in Myanmar has to be interpreted as a dimension of a historically and spatially contingent version of biopolitical governmentality (Robinne and Sadan 2007; Cheesman, 2017; Chu May Paing 2020; McCarthy 2023; Hsu 2019; Yuzana Khine Zaw 2022; Steinmüller 2022). As argued by Elliott Prasse-Freeman, Myanmar's state power regimes configure a "blunt" form of biopolitical power (2023a, 2023b). Blunt biopolitics displays a lack of interest in the promotion, care, and fostering of life at both individual and aggregate scales (Prasse-Freeman 2023a), which instead is a characteristic trait of biopolitics as a form of power (Foucault 1977, 2003, 2007, 2008; Esposito 2008; Lemke 2011). Long-term political and economic processes of savage extractive capitalism and coercive rule have geared Myanmar state apparatuses more toward the exploitation of "nature" than toward the care of populations to be turned into labor force, thus resulting in an overall lack of interest in the actual promotion of life. Yet, as state and nonstate authorities in the country still operate to protect the polity, the delineation of the polity's (ethnic) population bodies and vital space has remained an important object of government. The delineation of the populations whose lives are the primary object of government in Myanmar occurs through rather blunt categories, rationalities, and techniques that do not clearly know and define life. While biopolitical governmentality necessitates sophisticated, fine-grained apparatuses to know and define the forms of life that are to be fostered (Lemke 2011), Myanmar's blunt biopolitical governmentality has relied instead on confused and fluid classificatory schemes. First and foremost, those of the so-called *taingyintha* apparatus of rule (Cheesman 2017; Dunford 2019; Campbell and Prasse-Freeman 2021). *Taingyintha*—a term that can be translated as "sons of the soil" or "country friendly people"—constitutes a loose taxonomy. It designates a number of "national races" (usually 135) that are said to belong to the Myanmar polity. Such taxonomy also classifies the "national races" into major and minor ones, while construing the Bamar ethnic majority as the most advanced form of civilization securing and maintaining the integrity of the polity (IRN 2022).

In order to obviate to its bluntness and lack of well-oiled biopolitical infrastructures—such as census systems, citizenship apparatuses, or public welfare—blunt biopolitics heavily relies on different forms of violence. In this sense, blunt biopolitics brings a general trait of biopolitics to its extreme (Lemke 2011; Mbembe 2019). If biopolitics is a mode of power that cares for life by placing violence and death front and center as a modality to govern life (Dean 2002; Chatterjee 2004; Agamben 2005; da Silva 2007; Gregory 2008; Legg and Heath 2018; Mbembe 2019; Bergner 2019[5]), blunt biopolitics functions through violence. It harnesses violence as a key necessary mechanism to massify and divide the populations and spaces that it takes as its referent object of government.[6]

In the context of a state biopolitical power that does not operate governance systems to enhance life—although it still promotes the polity's life through violence—alternative (bio)political projects emerge at the margins of state authority (Prasse-Freeman 2023a, 62). Rebel rule, for example, articulates life-promoting projects in the borderlands. And—I argue—although some rebel movements do not appear to deploy the full array of governmental apparatuses needed to know, manage, and enhance life at large aggregate scales, "blunt" rebel rule still carries out its own massifications and divisions of populations and political space. And it does so by reproducing and managing armed collectives.

To draw such a biopolitical governmentality of weapons and military means, the conceptual frame proposed here foregrounds two aspects of the notion of biopolitical governmental assemblages and blunt rebel rule. First, biopolitical governmentality is characterized by an inherently spatial and geographical dimension. In fact, the forms of life and of population that constitute the main referent object of government for biopolitical power are also a demographic-spatial and political entity. Population is part of a living organism, with its living spaces to be secured and defended via various security mechanisms (Klinke and Bassin 2018; Esposito 2010, 2011a, 2011b).[7] As a form of politics and power, biopolitics can in fact be traced back to the thought of the Swedish political scientist Rudolf Kjellèn. Along the lines of the German geographer Friedrich Ratzel, Kjellèn delineated an organicist approach to the nation-state as a biological-political body with its biological-geographic "vital" spaces that it needs to preserve through the conduct of life and death in order to advance the societies and cultures of the polity (Klinke and Bassin 2018; Klinke 2019; Proto 2023).[8]

Second, biological populations and their conduct are not the sole referent object of biopolitical governmentality. Governmental assemblages operate through, and aim to govern, "the entanglements of men and things, the natural and the artificial, the physical and the moral" (Glouftsios 2020). In relation to weapons and armed groups, biopolitical governmentality refers to how the relations between humans and weapons are governed at multiple scales in two ways: one, the governing of the relations between humans and weapons in the process of constituting an armed force and regulating its use and control of violence; and two, the governing of the relations between humans and weapons in the process of delineating and reproducing the polity as a living entity—a living entity with an armed force and a living space. Thus, rebel rule can be seen as a form of governmentality that is not merely concerned with the conduct of biological life (Hoffmann and Verwijen 2018; Brenner and Tazzioli 2022), but also with the management of the entanglements of humans and objects (such as weapons). By governing weapons and military means, blunt forms of rebel rule reach out and cleave their polity apart, delineating their own political geographies (Prasse-Freeman 2023a, 25), with their territories and their frontiers.

The governmental apparatuses that manage the relations between humans and weapons shape frontier spaces, and in doing so they delineate the polity's political geography of territory.

Armed Assemblages

But how does blunt rebel rule shape the polity's population and political geography through the governmentality of weapons and military means? And what spatial relations emerge from the governing of them? While the perspective offered by the notions of frontier and governmental assemblage triggers these questions, the third understanding of *assemblage*—as a network of collective relations that shapes other relations (Law and Mol 2001)—allows us to unpack the politics of governing weapons and military means.

Armed collectives in Myanmar's borderlands are heterogeneous ensembles made of the relations among weapons, trained and disciplined human bodies, uniforms and ways to wear them, military and support infrastructures, and so on. The events of the Homein seizure, described in the introduction, exemplify this point. The munitions in Homein are intertwined with the politics of weapons production and supply chains: that is, weapons acquisition cannot be disentangled from the politicoeconomic orders of the borderlands. Some days after the seizure in Homein, a friend working as a mountain guide in Kyaukme relayed to me how rumors had spread concerning the possibility that the MANPADS did not belong to PSLF/TNLA: "Some say it was kept by them [PSLF/TNLA] on behalf of the Arakan Army before going to Rakhine."[9] The rumors resonated with broader dynamics of arms circulation in the borderlands. A 2021 Global Witness report, for example, documented how weapons have been transferred by the United Wa State Army (UWSA) to the Kachin Independence Organization/Army (KIO/A) as in-kind payments for taxes due for jade extraction operations in the Hpakant mines, Kachin State (Global Witness 2021, 45–46). Weapons take long detours before arriving in northern Shan and Kachin, and transfers may have to be approved at different territorial scales. Manufactured in factories in the Wa self-administered division, such as in the case of assault rifles, the arms need to receive the assent of Chinese authorities and move through highly contested and overlapping territories (Global Witness 2021). At the same time, the three hidden caches in the Homein area where the weapons were found required the mobilization of local people to act as liaisons. The physical infrastructure of the stash storing the weapons—sometimes a cave or waterproofed hole in the ground, other times an actual stock of sorts—operates together with the liaisons managing them and the record-keeping forms and procedures of the rebel army to control the munitions. All these elements have to link up to make the violence of the armed collective possible, to move and deploy it.

Governing weapon-human entanglements requires maintaining the interactions between all the various elements and processes that make an armed collective possible as a social and technical setting (Glouftsios 2020). The acquisition of weapons and the formation of an armed collective require the emergence of discursive fields (i.e., rationalities) that govern the entanglements between humans and weapons that are part of the armed collective, as well as the ways in which the armed collective relates to the body politic. Such rationalities condition practical ways to manage and govern weapons (techniques), while both rationalities and techniques are actively mediated and embodied by technical objects and material contexts.

This becomes evident, for example, in relation to the rationalities and techniques of control through which the technical properties of the landmines that often surround military positions are governed. In northern Shan State, armed actors make extensive (but most often nonsystematic) use of landmines, which may be triggered by farmers or their cattle. The Myanmar army deploys industrially manufactured landmines, while rebel movements usually assemble devices out of commercially available items. Local authorities impose verbal warnings on communities not to go to certain areas. When an explosive device explodes, armed actors often request monetary compensation by the victim's family or cattle owners for having destroyed their weapons. Yet, if the explosion were to occur near a base, and/or the affected families were of the same ethnicity as the armed actors, at times the armed actors may instead *provide* monetary compensation. The problem of landmine contamination oscillates among different overlapping rationalities for the governing of the relations among arms, violence, and humans, such as military technology preservation and humanitarian civilian safety. Moreover, it emerges via mutually constitutive relations between different technologies and techniques of control—from the manufacture of weapons to deployment and control methods such as warnings, spatial limitations, or compensations. Compensation *requests* by armed actors are informed by institutional rationalities of weapons wear and depletion management, while the *provision* of compensation responds to rationalities of ethnonationalism and/or humanitarian arms control. This is not to say that the provision of compensation is performed only vis-à-vis ethnic kins, as armed actors at times provide compensation to "other" subjects they include in their polities. Rather, providing compensation is a technique to manage the consequences of the technical properties of the landmine and its explosion. A technique that can be inflected by rationalities of ethnonationalism and humanitarianism. One that contributes to shape civilian populations and spaces to be differentiated and preserved by rebel movements.

Governing armed collectives targets sets of relations between weapons and armed humans on the one hand, and populations and space on the other. The tasks and acts of acquiring arms and governing an armed force are constantly related to the ideal and material qualities of space and populations, to calculations and materializations of space and subjects. As we saw in the Homein seizure, the weapon stashes require the organization of liaisons to manage them. At times the local liaisons are structured into so-called local guerrilla forces, that is, militias of the PSLF/TNLA that sustain the logistics of the ERO (for example, the movement and stashing of weapons). The Ta'ang rebel movement has constituted these militias in what are understood and reproduced as its borders and frontier spaces at the edges of "Ta'ang Land." Governing militias and arms caches contribute to reproduce different spatial relations. First, they reproduce rebel territory both materially and symbolically. Second, the collective ensemble of the arms stash can alter locations and senses of place in the village, for instance, depending on the interpretations that locals give about the stash as part of Ta'ang Land or not. Third, the arms caches are a catalyst for scale reproductions and struggles over scalar arrangements of territorial authority: they contribute to and are involved in the production of an autonomous ethnonational territorial scale by the rebel movement while also being the object of counterinsurgency at the scale of the nation-state reproduced by the Sit-Tat's projects of nation and state building (Prasse-Freeman and Ong 2021).

Rationalities: Narcotics Eradication, Ethnonationality, Humanitarian Security

Myanmar's frontier assemblages have been characterized by a number of rationalities through which governmental assemblages negotiate and shape the acquisition of weapons and the governing of armed assemblages. These rationalities are practical and discursive fields of rule and contestation for the governing of violence via weapons and military formations (Rose and Miller 1992; Rose 1999; Law 2007). This section anticipates and situates three main rationalities informing the governmentality of weapons and military means. They are the rationality of narcotics eradication, of ethnonationality, and of humanitarian security.

One of the most relevant of these political rationalities concerns the fight against narcotics and for their eradication. Although a full genealogy of a so-called war on drugs or antinarcotics rationality traces back to the Opium Wars (Yawnghwe 1993; Lintner, 1999; TNI 2012; Khun Moe Htun 2018), one could identify a first key conjuncture for its development as part of rebel regimes of power and rule in the early 1970s (Meehan 2011, 2016; Ko Ko and Braithwaite

2020; Dan Seng Lawn et al. 2021). In 1973 the Sit-Tat terminated the so-called Ka-Kwe-Ye (literally "home guards") militia program that entailed the creation of counterinsurgency local forces to quell the communist and ethnonationalist rebellions that emerged after independence (Buchanan 2016). The Ka-Kwe-Ye had grown into influential political and economic actors in the borderlands. They had assumed a key brokering role in the narcotics trade as well as in facilitating rebel access to weapons flows (Meehan 2011; Buscemi 2019).

Albeit aimed at reestablishing military-state authorities' control over narcotics and informal markets, the disbandment of the militia program was justified via the condemnation of the production and smuggling of drugs. In the decades to follow, for rebel movements the involvement in the narcotics trade continued to represent a necessary evil: a tool to locate much-needed resources, albeit one detrimental to their political legitimacy. Yet, as the waves of ceasefire agreements characterizing the 1990s were accompanied by an increase in Sit-Tat-sponsored narcotics smuggling that was starting to take a heavy toll on rebel polities' populations (Min Zaw Oo and Win Min 2007; Kramer 2007), narcotics eradication progressively emerged as a discursive and practical political domain. Since the late 1990s, different rebel polities, militias, civil society organizations, and social movements in Shan and Kachin States started to adopt measures to curb drug production and consumption (Kramer 2007; Dan Seng Lawn et al. 2021).

Concerning the Ta'ang rebel movements in particular, the rationality of narcotics eradication has been a paramount political field that PSLF and other actors navigated and reproduced to acquire weapons, enforce recruitment and militarized "drug rehabilitation" programs, or forge politicomilitary alliances. As chapters 2 and 3 detail, through techniques of governing military means related to the logic of narcotics eradication, the Ta'ang rebel polity also attempted a reformulation of the boundaries of its population and related vital space in ethnonational terms. Here, the rationality of narcotics eradication could hardly be disentangled from the logics of another paramount rationality of governing military means in Myanmar's borderlands: the rationality of ethnonationality.

Ethnonationality is a central political idea today expressed by the concept of national races, or *taingyintha* (Cheesman 2017; Dunford 2019; IRN 2022).[10] Such rationality has to be situated against the background of colonialism, although it cannot be identified as a governmental technique fully sharpened and reified by the latter. The governmental projects of British colonialism in Burma were characterized by attempts to massify and divide populations that nonetheless failed to produce an intense governmental system capable of knowing and conducting individuals and populations on the basis of a rationality and clear-cut categories of ethnicity (Prasse-Freeman 2023b and 2023c; Ferguson 2015; Cheesman 2017). The postcolonial state, particularly in its incarnations under socialist (1962–1988) and military/semimilitary rule (1988–present), made of *taingyintha*, a regime of truth to structure the polity of the Burma/Myanmar nation-state around a number of national races (Cheesman 2017, 5–7). These national races belong to and are united in a single political community while also being hierarchically and temporally organized according to a trajectory of political and social development (Cheesman 2017). Yet, the apparatuses of the postcolonial state regimes similarly fell short of accurately defining categories of ethnic people (Prasse-Freeman 2023a and 2023b). Ethnicity in Myanmar has thus remained mutable and evolving, a "reticulate" of relations, as Lehman has said (1967: cited in Prasse-Freeman 2023c), one in which subjects and groups take meaning as a certain ethnicity only in relation to other ethnicities and the political system and moments/circumstances of reference (Leach 1959; Sadan 2013a; Boutry 2016; Campbell and Prasse-Freeman 2021; Prasse-Freeman 2023c).

Amidst the inefficiencies of ethnicity as a governmental technique and its high political relevance, violence (in both military and legal declinations) has come to constitute an important governing mechanism to massify and divide polities into different malleable ethnic components with their inner fragmentations (Prasse-Freeman and Ong 2021). In turn, given its political relevance for different governmental actors, ethnonationalism has inflected techniques of control that conduct the conduct of weapons and armed humans (hence the violence they reproduce) on the basis of ideas and materializations of ethnonationalized spaces and collective subjects to be expelled, incorporated, molded, or secured (Prasse-Freeman and Ong 2021; Ferguson 2021; Ong 2023). Examples of such techniques can be found in the recruitment and military deployment of human bodies as military machines (such as in porterage, human minesweeping, and human shields (MacLean 2022); the imitation of territorial counterinsurgency techniques targeting specific ethnic communities on the basis of assumed links with armed actors (such as the infamous four cuts of the Sit-Tat); or the planting of landmines in specific areas on the basis of ethnonational community zoning, and the overall idea that a mapped geo-body includes different ethnic communities that have to be taken care of by a majoritarian ethnonational political entity and its army (Thongchai 1994; Ferguson 2021).

The aspiration to defend and foster the life of collective population groups that ultimately informs both antinarcotics and ethnonational rationalities of governing military means runs also through the last main logic the book accounts for, the rationality of humanitarian security. With *humanitarian security* I refer to the logics and techniques concerning the protection of a social body of "civilian" populations from armed hostilities, weapons proliferation, and contamination by "inhumane" and "uncivilized" weapons incapable of discrimination (Bourne 2012). Here, the politics of using and controlling landmines—a constant thread throughout the chapters to come—offers a good vantage point to understand how humanitarian security is harnessed to govern armed assemblages.

The problem of landmine contamination oscillates between different political instances and technical domains (Rose and Miller 1992, 192). After a landmine explodes, often armed actors operating in the area of the incident provide monetary compensation, as noted earlier. The provision of monetary compensation to victims, or their families, shows how rationalities informed by humanitarian ideas of weapons control are at stake. Although it may seem paradoxical and meaningless (and surely it is in the face of the tragic event)—the provision of compensation frames landmine contamination not as a tool to defend the polity but as a plague affecting civilian populations. This is especially the case when armed actors claim that the exploded landmines were "enemy" landmines: here the provision of compensation is associated with the condemnation of landmines use by rival armed actors.

Depending on the circumstances, though, at times rebel movements may instead request monetary compensation from either the victim and their family or the head of the village. Rebels request monetary compensation to cope with the destruction of the weapon. Here as well, albeit from a completely different angle, a logic of humanitarian security is at stake. While rebel movements' spokesmen reject extortion charges by arguing that their troops are not mandated to collect money from the victims, compensation requests are informed by techniques of internal weapons control, according to which arms have to be preserved and military personnel are responsible for their wear and depletion. These techniques are inflected by a rationality of protection. A civilian target population has to be protected from threats and insecurity. It has to be protected as a referent object of rebel government—one that is further delineated via rationalities of race and ethnonationality. Here the landmine—whose destruction should be compensated—is thus an effective weapon geared toward that end. Consequently, landmines have to be preserved to defend and preserve the polity's civilian population (De Larringa 2016).

Blunt Rebel Rule and Weaponscapes in Frontier Assemblages

The conceptual frame illustrated here considers borderlands not just as spaces crossed by borderlines—spaces that often have much more in common with what lays beyond the border than with the political centers that claim authority over them—but also as frontiers. Besides being a space and moment of extraction, the frontier is above all an ideological project. A project that defines a geographical space in political terms as an area characterized by violence and disorder that has to be territorialized and incorporated (with a view to bring that violence and disorder under control). While the idea of frontiers as assemblages makes us aware of the various human/nonhuman processes and entanglements that reproduce frontiers in borderlands, combining it with the lens of armed and governmental assemblages allows us to zoom in on the politics of managing weapons and military means.

The rationalities, techniques, and materialities emerging through the processes of governing weapons and their entanglements with the human, social, and natural world in order to create an armed force (i.e., armed assemblages) both require and reproduce a political calculation of the rebel polity's population and space (i.e., a governmental assemblage). Governing military means delineates, for example, why weapons should be acquired; how they should be acquired and managed; who can be part of the armed collective, how, and why; and what functions the armed collective has. At the same time, governing military means delineates whose life the armed collective should preserve, regulate, and foster. In the making and maintaining of an armed assemblage, implications are reproduced for what concerns the shaping of (biological) populations and related political space, for their massification and division by forms of biopolitical power.

Parts one and two of this book show how various rebel movements in the northeastern borderlands shape their polities at the margins of state authority, and how people living here navigate the politics of military means. In part one, the process of ceasefire, disarmament, and rearmament of the Ta'ang resistance movements provides the background for analyzing how blunt rebel rule produces the territory of the political community amidst frontier assemblages. That is to say, how rebel movements harness the governing of armed assemblages—through governmental rationalities and techniques—to shape the collective identity and political space of their polities. In part two, the focus shifts toward the networks, scales, and places—that is, the different "weaponscapes"—that armed assemblages and the politics of governing them produce in the borderlands.

Part I

BLUNT REBEL RULE IN FRONTIER ASSEMBLAGES

FIGURE 2. Myo Thit village, author's fieldwork, November 2019

FRONTIER HISTORIES OF WEAPON FLOWS

It was a cool morning in early November 2019, and I was in the family house of K. in Myo Thit. Myo Thit is a village located at the junction of Kyaukme and Namhsan townships (figure 2). Like many others in these areas, it is a place with more than one name besides its Burmese one: called Kung Ka Mai in Ta'ang, it is also known as Vain Moe in Shan.[1] Woken up by the 5:00 a.m. monks' chanting pouring out from the scratchy loudspeakers of the monastery just in front of the house, I was taking some notes when K., the TNLA officer we met in the introduction, entered the room where I had slept. His room as a child. We had arrived in the village the day before because he had been tasked to accompany me from the TNLA's headquarters in Namhsan to Myo Thit, to facilitate my trip back to Kyaukme town. He had changed clothes, left his camouflage military uniform in the camp, and placed his service weapons—a Colt-type Magnum revolver and an old AK-pattern M22 Chinese rifle—in the bamboo hut waterproofed with light green South Korean canopies that housed his squad. TNLA's field manual, he had explained, tying his shoes and grabbing the bike keys, allows members to be armed only during operations, never while wearing civilian clothes.

"I have never been used to spend[ing] so much time in the village actually, as my parents first sent me to study in Mandalay for secondary school and then to central Burma when I was 15," he said, as if to justify his early morning walk.[2] K.'s father had proposed to send him to Pathein, in the Ayeyarwady region of central lower Burma, to look for trading connections to expand the family tea business. The idea was to avoid brokers and their structurally imposed prices on tea leaf production and trade. As a matter of fact, since the second half of the 2000s, northern Shan State experienced a collapse of its tea industry, whose fulcrum has historically lain in the Ta'ang regions (Dunford 2024). After the PSLO/A concluded its ceasefire with the State Law and Order Restoration Council (SLORC) military regime in 1991, Ta'ang areas witnessed an unprecedented militarization process. This was accompanied by the establishment of military-private economic partnerships forged through business concessions in the ambit of mega construction projects and large-scale agribusiness investments, which aimed to consolidate state authority taking advantage of ceasefire arrangements (Woods 2011; Meehan 2016a; Woods 2019b). In more remote or militarily and economically less strategic areas, such partnerships worked to control the nodes between rural agricultural production and market distribution (Meehan 2016a, 381). In the case of the tea industry, this aggravated larger entrenched problems—like the lack of industrial manufacturing infrastructures and transnational economic competition across the border in Yunnan—into a situation in which a few brokers and companies could strangle local producers by imposing ever lower prices. For tea plantation owners, among whom K.'s father also figured, it had become logistically and financially difficult to directly access distribution markets in profitable areas like Mandalay and Yangon. Hence, sending one's children to lowland Myanmar had become a viable coping strategy to try to build connections and alternative trading arrangements.

In the country's heartland, though, the situation was not necessarily easier, as K. would soon discover. For "if one wanted to open up a company, one had to be part of the family of a crony, or leverage some connections with one of those families. And those cronies were actually related to the Sit-Tat."[3] After some years working in Pathein, Yangon, and Mandalay as an assistant manager or receptionist in tea-trading companies, K. moved back to Myo Thit. He had made up his mind about his professional future: "I took interest in political economy and information technology. So, when I went back home, my father spoke to me and told me, 'Maybe our army [TNLA] may be good for developing your interests; maybe they could offer you good opportunities.'" Indeed, following the 1991 ceasefire, part of the Ta'ang movement and new key political figures had rejected the agreement and had constituted a new political organization, the Palaung State Liberation Front (PSLF). They had established a base, first at the Thai-Burma border in Manerplaw and later at Mae Sot. From there, PSLF had looked at the blow of the 2005 PSLO/A disarmament ceremony in Manton as the apex of a process of sociopolitical and economic marginalization.[4] This scenario was further destabilized just a few years later by the proclamation of the 2008 Myanmar constitution. Sanctioning that "all the armed forces in the Union shall be under the command of the Defense Services," the new constitution provided a legal basis for Sit-Tat's aspirations to disarm EROs and integrate them as paramilitary or militia forces under its structures.[5] Amidst the faltering political landscape of northern Shan State, in 2009 the reconstitution of a Ta'ang army appeared to the PSLF as an inevitable step in light of the lack of credible peaceful solutions to the root causes that had underpinned almost sixty years of Ta'ang politico-military struggle.

K.'s father, now a man in his late forties, had also fled Myo Thit when he was young, as he recounted, joining the conversation in K.'s room. During the late 1980s, with the Communist Party of Burma (CPB) slowly crumbling and the regime achieving a series of military successes later accompanied by cease-fire agreements with rebel formations, Ta'ang regions of northern Shan were progressively targeted by the infamous four-cuts counterinsurgency strategy. Myo Thit and surrounding villages had historically been a hotspot of rebellion, being the headquarters of Battalion 5—one of the founding units of the Palaung National Force (PNF), the first Ta'ang rebel movement. Thus, entire villages on the mountains were forced to relocate into areas defined as "white" by the army, and many families had to flee to avoid the extreme waves of violence unfolding. K.'s relatives had been displaced to Kyaukme town and his father had attended school there. In the town he had become friends with the first secretary general of the Ta'ang movement, Brigadier General Tar Bong Kyaw. Also born in Myo Thit, Tar Bong Kyaw had likewise moved with his family to Kyaukme as a child in the late 1980s. But after the ceasefire the two friends had gone their separate ways. Tar Bong Kyaw becoming part of the newly established PSLF, while K.'s father moved back to Myo Thit. In the village, a community Pyi Thu Sit militia had been established by the military that included other minor villages in the area. Unlike larger militias on the rise in those years, which were allowed to operate in the narcotics trade and could thus have access to more powerful automatic weaponry, the Myo Thit militia had been provided with only a dozen old rifles. Its main purpose was to expand state authority and colonize Ta'ang areas, while preventing possible reignitions of hostilities.[6] Moving away from Kyaukme back to the village at the end of the 1990s, K.'s father had been forced to serve in the militia, later even acting as its head from 2002 to 2006.

While K.'s father continued recalling his experiences as a militiaman in Myo Thit, my eyes lingered on a portrait of a man in uniform hanging on the wall. Another native of Myo Thit village. Another Ta'ang. Another man of arms, as K.'s father would explain. "Maybe Captain Khun Thaung or Khun Aye, the leaders of PNF's Battalion 6 and 5?" I asked. A question I was addressing partly to him—in a clumsy attempt to impress them with a reference to the history of the Ta'ang resistance—and partly to myself, still trying as I was to make sense of the figure of K.'s father. At first glance, he appeared politically controversial: a Ta'ang tea plantation owner who had held leadership roles in a Sit-Tat community militia, but nonetheless maintained close personal and ideological links with the Ta'ang rebel movement. So close that he had even pointed out to his son enlistment with PSLF/TNLA as a life-changing opportunity. "No, Major Kham Lang," he replied, a major and fighter pilot of the Tatmadaw Lay (the Myanmar Air Force) who had been assassinated during an exercise allegedly in a purge of non-Bamar officers. While the military insignia of the Sit-Tat uniform Major Kham Lang was wearing had escaped me, they added a further layer of complexity, bringing into the picture the materiality of a martyrdom. For the image, the body it pictured, and the death that that body had endured all bore witness to both direct and structural forms of violence. A violence with which the Ta'ang polity had been confronted and oppressed by the military-state—be it through selective inclusion and exclusion from the tea markets, targeted killings, or the deployment and management of militias.

The stories of K., his father, and Major Kham Lang illuminate the realities of different governmental assemblages of control over the means of violence across several generations. In Myo Thit village one can see the stories of a rebel movement that experienced waves of ceasefire, disarmament, and rearmament. One can hear the story of a community village militia that is part of broader Sit-Tat strategies informed by the concept of People's War, and informing processes of nation and state building by attempting to territorialize the borderlands (such as the so-called four-cuts and other counterinsurgency techniques). These are the stories of a state army, the Sit-Tat, arranged along ethnic lines and aiming to monopolize the state security sector in environments where authority and order are fundamentally being questioned and constantly renegotiated. In such stories there are the weapons that K. left in the camp, as he planned to travel without uniform and through contested territories, as well as the weapons distributed by the Sit-Tat to militias throughout decades of war. There are different territories and territorialization processes that alternate and shift in time and space. There are different geographical scales of resolution in contexts of protracted armed conflict. In other words, these stories point to the implications of the politics of governing weapons and military means for the reproduction of Ta'ang areas as territories and frontiers of different political orders.

To be fully appreciated, however, such politics of governing military means must be situated against longer histories and broader scales—the longer histories of the politicization of weapon flows, in the broader *borderworld* scale of the Shan

State(s). This chapter focuses on weapons as important elements of the frontier assemblages that (un)make the borderlands. Recounting the historically and geographically heterogeneous processes throughout which these technical objects have circulated, it traces two main dimensions. First, the chapter traces the relevance of the main rationalities of governing weapons and military means—that is, narcotics eradication, ethnonationality, humanitarian security—for shaping rebel polities. Second, the chapter foregrounds how the production, flow, and acquisition of weapons (geared toward the formation of armed groups) shapes borderland areas into territories and frontiers.

Weapon Flows through Colonial Frontiers: From Free Trade Rights to Differential Arms Control

Throughout the nineteenth century the flow of munitions at the frontiers between southwest China, northeastern Burma, western Laos, and northern Thailand linked local traders and political rulers with the nodes of imperial networks. From the perspectives of European colonial apparatuses, movements of weapons turned these areas into frontier spaces in need of security and political stability in light of the economic potential they held. Besides representing a potential threat, though, weapons circulation was also considered part of the exercise of free trade rights and a valuable bargaining chip for political purposes (Tagliacozzo 2004, 361).

English and French colonial apparatuses had long maintained port areas or docks along the coastal and maritime frontiers of the strait of Malacca and Siamese dependencies that were used as trade operation bases and free trade areas for the sale of guns. Similar dynamics occurred on the coastal areas of Arakan, Tenasserim, and the lower reaches of the Burmese kingdom, which through the first two Anglo-Burmese conflicts had come under British Bengal. Rangoon, for example, developed into an important mercantile base involved in weapons smuggling upcountry and from there toward the Shan principalities (Tagliacozzo 2004, 365). Likewise, at the frontiers between the latter and the ill-defined northern borders of Siam, colonial interests involved the sale of muskets throughout the broader frontier regions. For British colonial rule, the territories under *sao hpa* rule were increasingly looked at as a buffer to be brought under control to contain French influence in the region and to build economic bridges into southwestern China.

During the central part of the nineteenth century, and up to the annexation of upper Burma by the British in 1885, the situation in the Shan plateau was one of a kaleidoscope of territorial claims, political authorities, and contestation. The region was organized into a series of principalities essentially under the indirect control of the Burmese kingdoms. These were held by so-called *sao hpa* princes (Yawnghwe 2010, 70).[7] Between the 1840s and 1870s a series of rebellions led by some *sao hpa* willing to declare independent authority had swept the Shan States (Yawnghwe 2010, 70–71).[8] Against this backdrop, the rebels managed to source firearms via smuggling by European and southwestern Chinese traders, while the Burmese kingdom was supplied by the British with Enfield rifles in exchange for commercial concessions (Tagliacozzo 2004, 369). Moreover, as commercial expeditions to identify viable routes into China intensified so did the flows of weaponry: these were accompanied by gifts of arms and ammunition to secure deals of passage with the *sao hpa* (Tagliacozzo 2004, 371).

In these decades, the frontiers of the golden triangle witnessed the commerce and smuggling of arms by British and French traders—arms, like muskets, that local merchants and authorities recirculated locally (Tagliacozzo 2004 372). On the one hand, Burmese and Siamese courts purchased weapons from Western governments and traders while, on the other, the Western governments exchanged weapons for economic concessions or goods with frontier commercial and political elites indirectly controlled by the Burmese and Siamese polities. British colonies and commercial bases in particular emerged as hubs for free trade and unchecked flows in Southeast Asia. British Burma was used as a third country trans-shipment springboard in order to avoid regulatory restrictions. Similarly Siam operated as a clearinghouse for regional weapons trafficking (Tagliacozzo 2005, 298). Weapons were also flowing through the Chinese frontiers for profit or as part of commercial operations. The huge revenues to be made with firearms sales—100% profit per purchase at times—diffused such technologies throughout Southeast Asia in the mid-nineteenth century (Tagliacozzo 2005, 262). Enfields and modern firearms, German Mausers, Beaumont rifles, American Winchesters, cannons, gunpowder,[9] and explosives, including dynamite:[10] munitions were highly valued as commodities and a wide variety of materiel of war was trafficked in the region (Tagliacozzo 2005, 292–294).

Weapon flows were furthermore linked to technological aspects. For example, weapons testing. The expedition led by Sir Harry Prendergast to lead the Burma Field Force into the third Anglo-Burmese war in November 1885 highlighted in part this aspect of the circulation of weapons in the Southeast Asian frontiers. Sailing up the Irrawaddy River, the flotilla of the Burma Field Force was accompanied by forty newly released Maxim guns. The Maxim gun, the world's first machine gun, had just been unveiled the previous year, in 1884, and was being fielded in Burma in order to be tested before its deployment in many more colonial campaigns and the battlefields of the First World War (Thant Myint U 2006,

5; Hicks 2020, 124–125).[11] In this sense, exploring and testing the technological frontiers of making violence and death proceeded hand in hand with the opening of political-geographical frontiers.

In an effort to govern—or to pretend to govern (O'Morchoe 2023)—such frontiers, the British, who up to the end of the nineteenth century had stuck to the primacy of free trade initiatives, started to adopt stricter regulations and to consolidate borders so has to stem the circulation of firearms. As colonial authorities attempted to project rule, both administrators and public opinion converged on the view that access to firearms should be limited for Southeast Asians. They moved to demarcate the frontiers between the Shan principalities, Siam, and China—key areas to the incorporation of both Burma and Siam within British influence—and to impose control over commodity flows (Yawnghwe 2010, 75; Tagliacozzo 2004, 365; Tagliacozzo 2005, 264, 298–9).

After the establishment of colonial rule over upper Burma in 1885–1886, the *sao hpa*'s principalities were individually placed under the governor of British India and administered indirectly as quasi-autonomous areas under British suzerainty and supervision by the 1888 Shan States Act (Sai Aung Tun 2009, 166; Han 2019, 48). In turn, a twofold system of governance was instituted, with the so-called Burma proper governed directly, while the so-called frontier areas were governed indirectly (Sai Aung Tun 2009, 194).[12] The impacts of the establishment of indirect colonial rule over the frontier areas were severe. Wading through the complexity of highland societies in an exercise of what has been termed "order without meaning" (Aung-Thwin 1985; Yawnghwe 2010, 76), the British attempted to crystalize highly mobile and fluid social groups by instrumentalizing ethnicity. Following European enlightenment rationalities and scientific racism, British authorities decided to organize and label the population of the colony by equating the language of each subject and group with its ethnicity (Ferguson 2021, 45). The administrative category of *lumyo*—which literally means "type of person" and was already in use in Burma but to designate social status/class distinction—was used to this end and gradually became coterminous with one's own language and "race" (Ferguson 2021). In turn, previous forms of political rule and sovereignty that had their own elites and sociocultural groups were also somehow made to coincide with ideas of ethnos and lumyo, thus establishing and bordering immovable authorities (Smith 1999, 47). For example, the Shan were mostly recognized as rulers to the detriment of heterogeneous arrangements of political authority and ethnic diversity to be found in the hills, although often to speak and dress like a Shan constituted more social status symbols rather than ethnic identity markers. Limited recognition was given to other groups by granting three principalities to Wa rulers in eastern Shan and recognizing the

Palaung/Ta'ang dynasty in Tawngpeng (Smith 1999, 47). Overall, bureaucratic and security mechanisms characterized by political and administrative simplifications informed by the equating of lumyo with language and ethnicity would later politicize anticolonial violence along ethnoterritorial and racial lines (Aung-Thwin 1985; Callahan 2003, 16).

Against a landscape of increased autonomy for the government of India and capitalist-motivated considerations framing Burma as politically immature (Callahan 2003, 22), the 1921 Government of Burma Act transformed Burma proper into a province of the British raj. The following year, the Federated Shan States were created as a purely bureaucratic body to administratively locate the frontier areas of the Shan States under the governor of Burma. The gradual separation process of Burma from India required a reconsideration of security-related matters in the frontier areas. Of particular concern were two issues: the regulation of firearms and the setup of the local security sector.

As part of such processes in 1924 the governor of Burma promulgated the so-called Shan States Arms Order, which was deemed to modify the customary norms characterizing the Shan States in matters pertaining to the manufacture, sale, trade, possession and transportation of firearms and ammunition. After the inclusion of Burma proper into the British Raj, the 1878 Indian Arms Act, which regulated arms control, was extended to Burma as part of the direct rule system. Nonetheless, the Shan principalities under indirect rule had been excluded from it. Different from the 1878 Indian Arms Act, the 1924 order authorized the possession of firearms and ammunition in the Federated Shan States (as opposed to Burma proper) but mandated a number of control measures. Weapons already present on the ground had to be stamped and registered, while the manufacture of new ones was allowed only with written permission, registration, and a system of fees. According to the law, the imposed permit and registration system fell within the purview of village and circle headmen or officials while the chief of the Federated Shan States had ultimate authority over firearms-related matters.[13]

Such geographical differentiation between Burma and the Shan States in terms of firearms control was in part linked to an overall consideration of the latter as a region of key relevance: a commercial crossroads and potential strategic threat. Thus, besides firearms regulations, throughout the inter–world wars period, the British also increased troops deployments in the frontier areas and developed a more encompassing territorial defense of the hills by allowing the sao hpa to form their own levies (Callahan 2003, 27). This was part of a broader reorganization of the security sector in the colony were military forces in Burma were revamped and previous local leaders and structures were eliminated in the midst of evictions and dislocations due to the expansion of extractive capitalism (Callahan 2003, 22; Smith 1999, 55).

In 1935–1937 Burma was eventually separated from India and became an autonomous colony, but the status of the Federated Shan States did not change. In the years leading to World War II, the question of the armed forces of Burma became increasingly politicized. As Mary Callahan and Mandy Sadan have noted, the exclusion of the Bamar ethnic majority from the Indian Army developed into a spinous political issue on which Burmese nationalists leveraged in order to denounce discriminatory racist policies of the British (Callahan 2003, 32–33; Sadan 2013b). Although the number of soldiers recruited from among ethnic minority people was extremely low and the exclusion of locals, rather than being based on a thought-out divide-and-rule tactic, was motivated by a preference for relying on an established, better-trained and oiled British-Indian security apparatus (Callahan 2003; Sadan 2013b), on the eve of World War II a series of political rationalities and discourses had developed that revolved around the depiction of ethnic minority recruits as part of the foreign colonial armed forces occupying Burmese territory (Callahan 2003, 35–36). Such rationalities and discourses were part of a broader politics of opposition to foreign oppression by European, Indian, and in part Chinese rulers and economic elites playing divide-and-rule strategies in Burma and frontier areas by co-opting local (*taingyintha*; i.e., indigenous races) elites (Cheesman 2017, 3).

Coercion-intensive state apparatuses and repression, combined with the disarmament of Burma and the lack of Bamar inclusion in the armed forces, generated a politicization of the differential regulation of armed violence between Burma and the Shan States as frontier areas (Callahan 2003, 43–44). Moreover, British colonial policies were characterized by the lack of initiatives to benefit social forces demobilized after military service. In the ambit of Burmese political movements, this favored the formation of militias as a tool to mobilize political consent and support by leveraging state exclusion and ethnonational rationalities (Callahan 2003, 43–44; Smith 2007).

Tales of Floating Arms and Hidden Caches: Disarmament and Integration at the Margins (1942–1960)

Throughout World War II the Japanese enforced a strict separation of the Shan States from Burma proper. Burmese troops, militias, and any kind of weapon flows were prohibited to enter the Shan States (Sai Aung Tun 2009, 194; Yawng-hwe 2010, 81). Such spatial differentiation was underpinned by an understanding of the Shan States as a key frontier space in a twofold sense. On the one hand, firm control over flows of weapons into the area was aimed at preserving the status of bargaining chip that the Shan States maintained in the eyes of Thai and Burmese nationalists and that the Japanese authorities exploited to push them to cooperate. On the other, the Shan States continued to play a militarily strategic role in the war as allied supply lines passed through the Mandalay-Lashio corridor to move arms into China (Sai Aung Tun 2009, 194). Thus, when Burma proper was granted independence in 1943, the Shan States remained under Japanese occupation. In the process, the Thakins (Burmese nationalists) maneuvered to convince the *sao hpa* to mobilize in favor of the incorporation of the Shan States into Burma. Many saw the resistance opposed by the hereditary rulers as reactionary, generating the first Shan youth movements opposing *sao hpa* rule as a counterreactionary position (Yawnghwe 2010, 81; Ferguson 2021). Such political stances were cross-fertilized by Marxist and socialist ideologies as well as aspirations of autonomy and self-government inspired by the violent and repressive experiences of foreign occupation (Ferguson 2021).

In the immediate postwar years, the whole of Burma was considered to be awash with weapons and the Shan States, especially the areas straddling the Burma Road, had been particularly affected by the hostilities. Callahan estimated that after the war as many as fifty thousand weapons were in private circulation alone (2003, 91; Selth 2000b, 5). These were mostly arms supplied by the Allies to resistance fighters in the frontier areas, or arms left over by retreating armed forces at different stages[14] (Callahan 2003, 99, 109; Yawnghwe 2010, 96; Smith 1999, 97; Sadan 2013a, 326–327n87). Such abundance of weapons outside the hands of political and military authorities intersected with three main social and political questions concerning the control of arms and military means.

First, the spread of weaponry raised the issue of how to not only disarm but also integrate the armed formations that, in various capacities, had fought the war. The high availability of firearms was in fact connected to the disarmament and reintegration mechanisms established at the 1945 Kandy Conference. At Kandy, British authorities and the Burmese leaders of the anti-Japanese resistance had negotiated the structure of the new army of Burma. They had decided to constitute class-based (read: ethnic-based) Burma Army units that would be formed through class-based (read: ethnic-based) recruitment. Nonetheless, the conference had left some spinous issues unaddressed, one being the very important point of *unit* versus *individual* reintegration of the myriad of resistance combatants and formations across the country. Many soldiers who rejected individual recruitment, as opposed to unit-based integration, simply left with their guns (Callahan 2003, 99).

Second, the problems linked to disarmament and integration posed the question of the influence that these men-at-arms may have played in the future political landscape. As the formation of militias by political parties, movements, or charismatic figures had long been a key political dynamic, the reinstated colonial government saw disarmament as a crucial endeavor to be completed before independence (Sadan 2013a, 326–327 and footnote 87). In 1947 the legislature passed a temporary amendment to the 1878 Indian Arms Act—still regulating firearms in Burma—which tightened up existing penalties targeting in particular the possession, carry, or storage of specific WWII weapon-types.[15] The amendment acted in conjunction with collection and disarmament efforts, undertaken especially in key conflict zones, such as Kachin and northern Shan (Sadan 2013a, 326–327n87).

Last, reluctance and opposition to disarmament, as well as political mobilization through militia formations, were closely linked to broader questions and contestations vis-à-vis the restoration of British rule in the postwar era and the interferences of colonial authorities in matters related to the very constitution of Burma as a postcolonial polity, such as how to structure the new armed forces of the country. From the political struggles concerning the governing of military means characterizing these years, a rationality of ethnicity-based structural divisions of the armed forces emerged. Such logic was tied to a "two Burmas" vision for political independence that distinguished between Burma proper and frontier areas (Callahan 2003, 112). With the 1947 Panglong Agreement the question of the political configurations that the relations between Burma proper and ethnic minorities' authorities should have assumed after independence was relegated to the background (Walton 2008). While autonomy and self-government for the frontier areas was recognized, with the Shan and Karenni States being constitutionally entitled to activate a secession clause in ten years' time, Panglong's foremost objective was the swift attainment of independence from the British. Thus, problems pertaining to the management of political, administrative, and military arrangements among the entities of the newly born Union of Burma were left unaddressed (Sadan 2013a; Yawnghwe 2010, 104; Walton 2008).

These three sets of issues became ever more relevant amidst the military turbulence unfolding between the late 1940s and early 1950s in the Shan States. During the parliamentary multiparty era that followed independence (1948–1962), the prevalent political vision of Bamar Union governments—and even more so of the Sit-Tat—did not reconcile with the possibility of a strongly decentralized union. Union governments and the Sit-Tat sidelined open discussions concerning federalism and autonomy for the single states. Such vision was in part underpinned by a conflation of the logic of (Bamar) nation building and the idea of state building (on this conflation see Callahan 2003, 13, 95; Sakhong 2014, 2–8). The merging of nation and state building from Bamar perspectives also accompanied the expansion and politicization of the role of the Sit-Tat as a kalei-doscope of insurgencies quickly expanded in the Shan States,[16] and in 1949–50, divisions of the retreating Chinese Nationalist army, the Kuomintang (KMT), invaded the eastern borderland areas of Mong Yang and Keng Tung.

In this context, the management of the single Shan States' security and armed forces, as well as the relations between the Union and Shan States' gov-ernments in this ambit, became particularly problematic. Amidst the expan-sion of armed rebellion, Shan States politicians had requested the Union government to increase military police posts to secure the frontiers, and to integrate a Shan battalion into the Union armed forces (Sai Aung Tun 2009, 335). Confronted with central government and Sit-Tat's denials, when the government declared martial law in Shan State in 1949–1951, in many areas the *sao hpa* started raising levies of troops on their own (Sai Aung Tun 2009, 338–339; Yawnghwe 2010, 99). The Shan States government proposed the transformation of these levies into a Shan State battalion for self-defense orga-nized along the lines of the Sit-Tat, but the Union authorities refused. Cen-tral political and military establishments were against the idea that the Union government should have led the demobilization and/or reintegration of such troops at the end of hostilities and counterproposed to create a civil defense force to be managed by the Shan State government instead (Sai Aung Tun 2009). On top of this, the mobilization of their own armed formations by the *sao hpa* highlighted a political divide at another scale: a divide between sup-porters of the hereditary rulers and anticolonial, Marxist, and socialist forces openly against the perduring feudalist system of the *sao hpa* and their privi-leges (Sai Aung Tun 2009, 332; Yawnghwe 2010).

In light of the deteriorating security context, particularly in connection with the KMT occupation, a State–level conference was organized in 1951 by the Shan State government to discuss the role of the police, the army, and other com-ponents of the security apparatus (like the *sao hpa* and their police or levies) (Sai Aung Tun 2009, 343). In debating the institutionalization of levies for local defense some raised the issue of training: the levies should have been properly disciplined and trained in the use of armed force, especially concerning latest firearms technology use and handling. Such concerns were justified by firearms misuse and misconduct during the war that fed civilian grievances, as well as by the evolving guerrilla warfare strategies and tactics (Sai Aung Tun 2009). Crucial in this regard, besides being considered a security threat in itself, the military buildup of the KMT forces represented both a key channel for the acquisition of technologically more sophisticated munitions and a training and advisory springboard for rebel movements gravitating around the Shan States borderlands (Taylor 1973, 16–17; Tun 2009, 308; Yawnghwe; Gibson and Chen 2011).

In the decade 1951–61 in particular, KMT armed formations based at the Burma-Thai-Lao borderlands catalyzed weapon flows in various ways. In 1950–1951 the KMT secured agreements with the Thai National Police and the US CIA for the provision of munitions, military supplies, and training. These were arranged via the covert mediation of a front company operating through a scheme the latter two had finalized the year before for the creation of a Thai border police force (Taylor 1973, 33; Tun 2009, 310; Gibson and Chen 2011, 37, 52, 62). Supplies were shipped or flown to Thailand (and Laos later) and from there transported into Shan State, or sometimes directly airlifted on KMT base areas (Gibson and Chen 2011, 62). The weaponry delivered in successive waves throughout the 1950s included US-manufactured arms stockpiled in Okinawa or supplied to Taiwan and Thailand through military assistance programs; European-manufactured weapons legally transferred to Taiwan; and Taiwan's own military stockpiles (Gibson and Chen 2011, 126). In 1953, empirical evidence of ex-KMT forces' weapons in Shan State revealed they held arms that had just been recently issued and deployed in Korea by the US Army (Taylor 1973, 38).[17]

Chinese Nationalist armed organizations soon started to build roads, military instalments, airfields, and runways (Taylor 1973, 38). KMT forces also exploited dissent among ethnic people in Shan State through psychological and propaganda warfare programs (see Tun 2009, 312–315), practiced the formation of satellite militias among ethnic minorities, and attempted to form a sort of rebel front coordinating Karen and Mon rebels in particular (Taylor 1973, 64; Gibson and Chen 2011). Militias and guerrilla detachments were often formed through the provision of weapons to local formations, although borderlands inhabitants looked at weapons as useful currency to be bartered in exchange for other goods (Gibson and Chen 2011, 71–2). Similarly, while acquiring weapons through the Thai forces, Karen and Mon rebel movements would barter rice for KMT weapons at times of decreasing food supplies for the KMT (Gibson and Chen 2011, 122). In Shan State, US contractors in Thailand and French Foreign Legion instructors embedded with the KMT spread military knowledge concerning organizational techniques, military training and discipline, command-and-control structures, guerrilla warfare, and weapons handling and use (Taylor 1973, 38; Tun 2009, 318; Gibson and Chen 2011, 133). For example, in the context of Sit-Tat operations to dislodge the KMT and dismantle the rebels, a technique of porterage started to be consolidated and systematized, the porter becoming a military category through which human bodies were turned into military vehicles (Tun 2009, 316).

Disarming the Shan State population had become a key priority for the Sit-Tat and the Union government. When in 1952 the military administration of Shan State was declared, the first decree issued by the interim government concerned the suspension of the 1924 Shan States Arms Order. The decree mandated that civilian firearms possession be controlled through a system of licenses released by the army (Tun 2009, 346). In parallel, the years of military administration, when ended in 1954, were characterized by the enhancement of chauvinism and Myanmarization of the borderlands (Tun 2009, 325). This was in part connected to different intersected layers of propaganda wars revolving around the main political issues of the moment: land redistribution requests, increased democratic representation, and the abolition of the *sao hpa*'s rule system. For example, the Sit-Tat and related political parties supported antifeudalists positions, and communists fanned anti–*sao hpa* sentiments (Tun 2009, 346). Meanwhile, the KMT encouraged Shan State politicians and the *sao hpa* to secede from the Union on the one hand and, on the other, encouraged ethnic minorities to set up their own armed forces and claim independence.

Through a mixture of Sit-Tat military operations and interstate diplomacy, by 1954 KMT forces had been partially evacuated.[18] As the deadline of the ten-year trial period granted to Shan and Karenni State under chapter 10 of the 1947 Union Constitution approached, the issue of secession gained center stage from Bamar perspectives. The possibility for the two states to activate the secession clause was politicized by the central government and the Sit-Tat as a threat to the integrity of the Bamar nation-state (Yawnghwe 2010, 106). The role of weapons and control over the means of violence in such environment was key from several points of view.

Anti–*sao hpa* political campaigns supported by the Sit-Tat and the Union government depicted the hereditary rulers as promoters of secession and feudalist powerholders. In this sense, from Bamar nationalist perspectives, the fact that the *sao hpa* held firearms and police forces was understood as highly problematic in light of their potential role as organizers of armed rebellions (Ferguson 2021, 64). Those in the military and government viewed the widespread availability of weapons throughout Shan State, and especially the eastern and northern borderlands, to be a potential future enabler of armed insurrections (Tun 2009). They argued that weapons availability among the population was connected to long-standing differential regulation of firearms possession and traditions of weapons carrying in the Shan States, but also to World War II weapons surpluses and KMT arms stashes. In particular, various tales circulated concerning hidden leftover arms caches from World War II that were said to be dispersed throughout the countryside. These stories—based in part on communal memories of airlifted military supplies floating down from the sky (Yawnghwe 2010, 107; Sadan 2013a; Ferguson 2018), and in part on state propaganda concerning the threat of foreign military support to rebellion—sustained narratives about the ubiquity of military supplies and stockpiles at the borders (Yawnghwe 2010, 107–8). Moreover, from the perspectives of some Shan State politicians and officials, the armed rebellions of the first half of the 1950s, in combination with violent indiscriminate Sit-Tat operations and Union government encroachments, posed the question of the social and political consequences of the complete lack of control over the means of violence by the Shan States authorities.

In 1956, the Sit-Tat and Union government decided to deploy military operations to the Shan States with the aim to locate and seize hidden arms stocks and disarm the civilian population (Yawnghwe 2010, 107). The following years were characterized by a marked militarization of the Shan States. The disarmament campaigns entailed violent repression of the civilian population, land and property confiscations, and progressive construction of military installations (Yawnghwe 1989, 2010, 108; Tun 2009, 391). This situation sparked political instances to negotiate with the Sit-Tat and the Union government about the control of armed violence. However, the Union authorities' monopolization of defense-related matters through the 1948 constitution prevented any dialogue on the issue and made political contestation a matter of "with" or "against" us (Yawnghwe 2010). At the same time, the violent disarmament campaigns and militarization processes unfolding further polarized communities and in many (especially peripheral) areas pushed people to strengthen links with local ex-KMT armed actors, ethnic minorities' nationalist leaders, or elites open to armed rebellion (Yawnghwe 2010).

The first waves of armed rebellion, which then triggered the military caretaker government at the Union level throughout 1958–1960, were initially very much a matter of fragmented and bottom-up formations coalescing around local figures and authorities such as peasant leaders, influential traders, monks, village heads, or police defectors (Yawnghwe 2010, 109). Such armed insurrections and the weapon flows that they intercepted were strongly decentralized, being mostly armed with weapons held in households and communities— such as swords and spears, hunting muskets, and bolt-action rifles of different assorted manufacture. Contrary to the disarmament efforts and the propaganda of the Sit-Tat suggesting the imminence of foreign support to rebels, the organization of arms acquisition and armed groups formation was extremely poor. People joining were often misled by stories of alleged munitions warehouses and military training camps already set up and waiting at the border with Thailand (Yawnghwe 2010, 108).

Access to Weapons and Fragmented Political Geographies of Resistance (1960–1970)

The armed rebellions sweeping Shan State had emerged amidst a combination of violent repressive military administration and the inability of the Shan States institutions to provide protection and catalyze the instances of middle-ground political positions (Yawnghwe 2010, 110; Ferguson 2021, 72–73). During the military caretaker government of 1958–1960 the eventual removal of the *sao hpa* signaled the need to rebuild Shan State political civilian institutions to regain legitimacy among the population (Tun 2009, 355, 391; Yawnghwe 2010, 110). To attain this goal, the provision of security and protection from increasingly assertive Bamar-led military-state authorities was seen as a paramount priority. Shan State political leaders initiated a federal movement for the modification of the constitutional structure of the Union, which was considered unequal and far from the principles of federalism.

Not by chance, a major element of the federal reform proposal focused on rethinking the arrangements regulating the management of defense-related matters and internal peace and security, whose purviews, it was hoped, should at least partially come under the competence of the federated states' governments (Tun 2009, 391; Yawnghwe 1989, 110–111). After the end of the periods of martial law (1949–1951), military administration (1952–1954), and caretaker government (1958–60), the Sit-Tat had consolidated administrative and law-and-order prerogatives that should have been reentrusted on the Shan State government (Tun 2009, 391). Thus, the proposal focused on a clearer demarcation between civilian and military competencies, as well as a rebalancing of the relationship between the Shan State government and the Sit-Tat.

Overall, the complete devolution of defense matters to the Union, and hence to the political influence of the Sit-Tat, generated grave concerns—especially in connection to the demobilization and reintegration of levies in case of armed conflicts, and concerning the lack of overview on Sit-Tat-affiliated militias. In an attempt to counter the rebellions, in 1955 the Sit-Tat had started to organize villagers, providing them with arms and constituting the so-called Pyu Saw Hti (Smith 1999, 95; Buchanan 2016).[19] This militia program entailed the distribution of some (most often very basic) armaments to local villagers. Arms distributed remained under the control of local garrisons, which had managed them with a system of record-keeping and stockpile management measures. In this regard, the distribution of surplus decommissioned weapons to local defense militias by the Sit-Tat was particularly problematic in the "postconflict" period. Similarly, military cantons, garrisons, camps, and bases constituted a highly contentious topic, for their establishment, delimitation, and self-government fell completely outside Shan State local administration (Tun 2009, 391–392). Through successive waves of militarization, several areas had been expropriated to create cantons and bases. But no compensation had been provided to the evicted owners or communities, no agreements or consultations before or after had been carried out, either with local populations or with the Shan State government. And the regulation of housing and land management inside cantons remained opaque, to say the least.[20]

In the federal reform proposal, claims were advanced concerning the Shan State government's primary role in the maintenance of peace and security. These proceeded hand in hand with complaints about a lack of modern weaponry for Shan State forces—like the levies and civil defense—especially in comparison to rebel movements (Tun 2009, 391–392). The lack of weaponry was connected as well to a lack of control by Shan State authorities over arms procurement processes and/or production of arms and ammunition for police forces. For example, in the first half of the 1950s the Sit-Tat had started supplying surplus weapons, materiel, and support to local *sao hpa*'s police forces (Lintner 1999, 184–186). But the weaponry supplied to these local forces often consisted of firearms decommissioned by the army as a result of successful attempts to obtain foreign military assistance since the beginning of the decade. Besides direct government-to-government training provision and conventional weapon transfers from countries like Australia, Sweden, and Tito's Federal People's Republic of Yugoslavia, the Sit-Tat was able to partner with the German engineer Fritz Werner in order to set up a factory and start manufacturing firearms under licensed production of the arms producer Heckler & Koch (Lintner 1999, 155–156; Bourne 2007, 72; Abel 2000).[21] After the *sao hpa* resigned in 1959, the police force had increased in size, but this created a shortage of arms. *Sao hpa*'s arms had been considered private properties of the princes under their relinquishment agreements, and such weaponry had been taken away from the police force.

The federal reform movement and its proposal were crushed by the 1962 coup d'état. The Sit-Tat's crackdown on the milieus connected to the *sao hpa*, as well as regional authorities and political leaders, engendered a sociopolitical vacuum throughout Shan State. The first rebellions were organized around ethnonational elements and relied strongly on local leaders and the *sao hpa*'s networks (Ferguson 2021, 64). Illustrative of this fact was the trajectory of the first Shan armed forces formed in 1958, Noom Suk Harn (Young Brave Warriors). Noom Suk Harn was organized by Sao Noi, a Shan who had been able to coordinate support from various *sao hpa* and reshape their local police forces (Lintner 1999, 184–186). Similarly, in the Ta'ang areas of Tawngpeng (today Namhsan township), an armed movement was being organized by Chao Nor Far, member of the royal house of Tawngpeng (Smith 1999, 95; 2007, 14). The rebellions drew on patronage systems and firearms availability connected to them in order to transform popular political movements into rebel movements with their own armed structures (Smith 1999; 2007, 14). These armed movements were not ethnically homogeneous but comprised people of different backgrounds; in fact, a key political rationality in the formation of these groups revolved around the constitution of a Shanland, meaning a politico-administrative entity with its own sovereign territoriality that coincided with its (Shan) racial subjects, but at the same time included a diverse tapestry of ethnic communities (Ferguson 2021, 72–79).

Taingyintha

A key conjuncture for the development of an ethnonational rationality was represented by the 1963 Enterprise Nationalization Law passed by the Ne Win military government. As Nick Cheesman has noted, such economic nationalizations were part of a larger program for the political domination of Burma launched by the socialist dictatorship that was informed by the concept of *taingyintha* (national races) (2017, 5–7). Military-dominated statehood and politics were characterized by the aspiration to gift a socialist economy and nation-state unity to all the national races, which—so the regime official narratives recited—had historically stood united against capitalists and imperialists, with the Bamar majority front and center. The nationalization of all productive and commercial activities provided the conditions for the emergence of thriving extralegal economic practices, with the borderlands—especially those of Shan State—representing the lungs of such informal economy. Given the extremely poor status of the Burmese industry, nearly all products requiring industrial processing would be sourced from neighboring economies and smuggled in, while raw (or very basically processed) agricultural products, natural resources, and particular handicrafts would flow out.

Similar trends also characterized in-country weapons production and informal weapon flows. In fact, throughout the late 1950s and early 1960s the Sit-Tat had consolidated the state's weapons manufacturing capacity. It had built several production plants in the central plains on the western bank of the Irrawaddy River, near Pyay, starting licensed production of small arms, grenades, landmines, mortars, and ammunition under the Directorate of Defense Industries, also known as Ka-Pa-Sa due to its Burmese abbreviation (Maung Aung Myoe 2009, 105–107). Part of a weapons supplies self-sufficiency policy and import-substitution scheme, domestic production was integrated with tightly controlled imports.[22]

Sit-Tat's approach to weapons production and control in these decades was part of a broader set of reforms in which defense-related institutions were understood as vehicles of Bamar nation-building-cum-state-building (Callahan 2003, 177–179). In this ambit the military government drew heavily on the example of Yugoslavia through study visits for weapons and know-how purchases or practices and tactics/strategies development (Callahan 2003). Yet it is also interesting to note what the military government did not replicate. In Yugoslavia the federal states not only maintained considerable stockpiles of arms but also managed decentralized weapons stocks of their so-called Teritorijalna Odbrana (Territorial Defense) forces and maintained stashes of arms in public venues throughout the territory (Griffith 2010, 184; Hajdinjak 2002, 9). In addition, federated states were part of an overall industrial complex, and some hosted segments of the arms industry. The Sit-Tat instead was very careful to avoid any conspicuous decentralization of arms production and control. Even general industrial production capacity that could have potentially been converted toward military purposes was practically absent throughout the borderland states. For the emerging rebel movements in Shan State, this meant that munitions could be accessed only via leaks from Sit-Tat forward bases in the borderlands and corrupt officials. Similar to the political-economic dynamics of other sectors, the homeostatic adaptation of the borderlands' economies to the 1963 nationalization measures constituted a major possibility to intercept weapon flows and industrial capacities to rely on for the refurbishment, and to a certain degree the manufacturing, of weaponry.

In the post-1963 Shan State borderlands, opium trade emerged as a key element both facilitating and at times hampering the acquisition of weapons and the consolidation of armed movements (Gibson and Chen 2011, 107, 242–246; Lintner 1999, 232–234). In fact, by the mid-1960s ex-KMT and Taiwanese intelligence armed groups had established several bases for opium and other trade flows that created a sort of north-south corridor into the southern Shan-Thai and Thai-Burma border and the eastern tri-border area of Laos, China, and Burma (Yawnghwe 2010, 115–116). Local armed couriers and factions of ethnonational rebel movements gravitating closer to the orbits of ex-KMTs' bases would often become their auxiliaries, while more structured armed actors would act as front operators (Yawnghwe 2010, 117; Gibson and Chen 2011, 107). The armed actors would work for ex-KMT forces in their areas of presence, and the ex-KMT would provide arms and ammunition in exchange. Collaboration with the ex-KMT and trading militias in the midsixties allowed rebel movements to access weapons sources and locate military training often performed in KMT bases at the Thai-Burma border (Gibson and Chen 2011, 254; Yawnghwe 115–116).[23] At the same time, local armed traders and rebel formations conducted their own activities, escorting caravans to the borders and accessing weapons mainly through the mediation of the Thai police and army (Smith 1999, 332–333). In these years some militias in particular were consolidating their role in the shifting opium and consumer goods trade networks, such as Khun Sa's Loimaw, Lo Hsing Han's Kokang militia, and Maha San's Wa militia.[24]

The more fragmented rebel movements tended to be those gravitating closer to the orbits of ex-KMT base areas and the most important "trading" militias that were spread along the borders with China and Thailand, overlapping and interspersed with the ex-KMT. In the areas where the two were mostly present, their centralization of weapons acquisition ran in parallel with the fragmentation of ethnonational rebellions. Rebel movements in Shan State had thus to cope with the problem of molding highly fragmented and disparate armed formations that had emerged since the outbreak of violence in 1958–1959 and especially in the aftermath of the 1962 coup. In the following years different minor armed political forces would consolidate around larger ones that allowed them to access weapon sources and organizational resources.[25] Among others, this would have been the case for Ta'ang ethnonational rebel movements that consolidated in the orbits of Shan and Kachin ones.

Armed Minorities of the Armed Minorities

In 1964, under the leadership of the Mahadevi of Sao Shwe Thaike—the defunct *sao hpa* of Yawnghwe and first president of the Union of Burma—for the first time a pan–Shan State armed force was formed, the Shan State Army (SSA) (Lintner 1999, 225).[26] Of the five brigades around which the SSA structured itself, one was commanded by a half-Ta'ang from Tawngpeng, Sai Hla Aung, and incorporated two Ta'ang battalions (Smith 1999, 220).

Just one year before, on the January 12, 1963, a first Ta'ang national armed movement had been formed under the name Palaung National Force.[27] The PNF had emerged out of different components that in a sense were illustrative of the main lines of the sociopolitical turmoil of the 1950s and 1960s. These political dividing lines were maintained when PNF decided to take part in the SSA's project. Inside SSA's brigade structure, the Ta'ang force was reorganized in two battalions: the Fifth Battalion under Chao Nor Far favored the reinstalment of *sao hpa* rule, while the Sixth Battalion under Captain Kham Thaung favored federal democratic positions.

Part of the PNF was led by three relatives of Tawngpeng's last *sao hpa* (Hkun Pan Sing)—Tar Khun Li, Khun Aye, and Chao Nor Far—who had mobilized against the uprooting of the feudal system by the Burmese Union government and had joined previous Shan rebels (Smith 1999, 220). The other main faction, under the leadership of Tar Khun Thaung, had developed in close connection with the Kachin revolutionary movement and aspired to a democratic federal system, thus opposing the feudal rule of the *sao hpa* (see also Meehan 2016a; Smith 1999, 220).[28] Active in the northernmost parts of Ta'ang areas, like today's northern Namhsan and Namkham township, such factions overlapped with the so-called Kachin substate of Kutkai in northern Shan State. Here the Kachin Independence Organization (KIO) had rapidly evolved from a guerrilla movement to a structured army (Kachin Independence Army; KIA) (Smith 1999, 219–220). The KIO/A had been isolated by impervious mountains to the west and the absence of connections with communist China to the east, thus encountering difficulties in acquiring munitions (Sadan 2013a, 328; Lintner 1999, 234). Through linkages with the KMT, however, in the mid-1960s KIO/A managed to set up a base in Tam Ngob, at the Thai-Burma border.[29] This allowed it to secure an outlet for its main potential revenue sources (jade and opium), obtain military training from the Chinese Nationalists, and acquire arms—directly from the KMT or through their enmeshment in the securitization of the borderland areas by Thai forces. In addition, since the mid-1960s the KIO/A quickly transitioned into an organic military structure providing a hub for smaller armed movements to build their military capability, establish safe presence areas, and in general link up with regional dynamics (Smith 1999, 332–333).

Confronted with the expansion of a panoply of armed actors, the Sit-Tat revamped a series of militia programs deployed since the 1950s. Two in particular, the so-called Ka Kwe Ye (KKY; "Defense") and Pyi Thu Sit ("People's War") militias, assumed particular importance.[30] The latter entailed the creation of village community militias through training and weapons provision,[31] while the former entailed the official recognition of different kinds of armed actors as Sit-Tat-affiliated counterinsurgency units in exchange for the tacit approval of their lucrative activities, *in primis*, the smuggling of opium (Buchanan 2016, 8–9). From the point of view of weapon flows and arms acquisition, the unfolding of the KKY strategy throughout Shan State entailed an "explosion" in arms proliferation (Smith 1999, 221). Militia commanders were able to use opium revenues to acquire military equipment linked to the Vietnam and Lao wars and transport it from across the borders thanks to the military government's collusion (Lintner 1999, 232).[32]

Nonetheless, rebel movements' access to weapons became regulated through what Patrick Meehan, in another ambit, has defined as a "limited access order" (2011). The KKY strategy created a system of rents providing militias with preferential access to the extralegal trade and forms of taxation linked to it (Lintner 1999, 231; Meehan 2011, 379). While KKY militias maintained ties to the rebels and at times facilitated their arms acquisitions, overall the arrangement financially undermined the rebel formations and "institutionalized" the KKY as weapons acquisition gatekeepers. The geographies of weapons acquisition were reconfigured. For not only could militias pass through military-government roads and areas but they were also in a better position to manage flows in remote areas, access the borders, and mediate with authorities at the interface of the China-Lao-Thai borderlands.

The KKY assumed particular importance after the expansion of the CPB's North Eastern Command (NEC) in Shan State. Throughout the 1960s the Chinese Communist Party (CCP) supported communist organizations in Laos, Vietnam, Thailand, and Malaysia. In 1967 the CPB and CCP agreed to a ten-year aid program consisting of full-scale support in terms of training, logistics, medical services, and weaponry (Smith 1999, 248).[33] The CPB consolidated bases and weapons stockpiles along the Sino-Burmese borderlands, from the Wa areas of Panghsang to the west of Namkham, and managed to co-opt ethnic local leaders, militias, and rebel factions often through the provisions of arms (Yawnghwe 1989, 30; Buchanan 2016, 9). While often met with resistance by ethnonational rebel movements perceiving it as an ideological opponent and threat to their very existence, the CPB quickly developed into a major source of weapons and a channel for acquisition (Yawnghwe 2010, 120). Its infiltration into southern Kachin and northern Shan State, for example, generated tensions with KIO/A. An alliance agreement signed in early 1968 to set up a united front allowed KIO/A to obtain a consignment of several hundred weapons, otherwise always difficult to obtain, but the agreement quickly broke down after a few months (Smith 1999, 252).

The case of the PNF is illustrative of these processes of weapon flows, acquisition, and armed forces formation. In the midst of such intricate conflict trajectories, in 1966 the two Ta'ang battalions of SSA broke away to form their own separate front again, the PNF (Smith 1999, 220). Previous internal lines were maintained: the Sixth Battalion remained in charge of the areas to the north, from Namhsan to Namkham and Muse; while the Fifth Battalion covered the southern and western parts of northern Shan State (lower Namhsan, Kyaukme, Hsipaw, Mogok, Maymyo).[34] Soon after, in 1968–1969 the PNF split into two Ta'ang forces, one under the command of Kham Thaung and Kyaw Hla, who decided to ally with the KIO/A, and the other under the three sons of the *sao hpa* of Tawngpeng, which maintained ties to the KMT and SSA.[35] In 1970 the Fifth Battalion was forced to retreat from its base at Myo Thit village (from which this chapter opened) by a joint operation of the PNF and KIO/A.[36] As a TNLA officer recalls, "At that time the two battalions of Ta'ang were all together but actually at a certain point Chao Nor Far had been sent to the Thai border to obtain some weapons and when the internal split broke out he was not able to go back anymore. At that time, it was very difficult for us to acquire arms, as we had to travel to the Thai border to try and get."[37]

As a result of these developments the PNF under Kham Thaung and Kyaw Hla reasserted its autonomy vis-à-vis the SSA and carved out a strategic place for itself in the areas of Namhsan and Namkham. The escalation of armed hostilities linked to the expansion of the CPB and an increase in opium cultivation, which by the early 1970s had spread across Shan State, heavily affected Ta'ang communities (Howard and Wattanapun 2001, 75; Lintner 1999, 233). Thus, the Ta'ang rebel movement secured a military pact with the KIO/A and the SSA in 1974. In an attempt to pursue a more effective political approach, in 1976 it changed names, from PNF to Palaung State Liberation Organization/Army (PSLO/A).[38] This name change reflected the aim to align with the stances of the Kachin movement, distance itself from the SSA, and advocate for the autonomy of Ta'ang areas within the Shan State (Meehan 2016). As chapter 4 will analyze more in depth, at this stage the movement continued to use the exonym Palaung with which the Bamar-dominated central authorities referred to Ta'ang minorities, often pigeonholing them as a subpopulation group within the Shan.

From the perspective of the Ta'ang movement, KIO/A and SSA monopolized weapons acquisition sources and processes.[39] At times they would obtain Chinese-manufactured munitions by negotiating arms consignments with the CPB (Meehan 2016a). PSLO/A never got close to the CPB but would access arms via the KIO/A and—as the current commander in chief of the TNLA, Tar Ho Plan, formerly a member of the PSLO/A, notes—rumors circulated that KIO/A was "telling CPB they did not need to provide [arms] directly [to PSLO/A]. It remained difficult to acquire arms because they [PSLO/A] had to travel to the Thai border."[40]

Thus, in 1976 PSLO/A took part to the creation of the National Democratic Front (NDF)—a military alliance of ethnic politico-armed movements. This granted the movement with important connections and the possibility to set up a contingent at the Thai-Burma border. Moreover, in the late 1970s, PSLO/A stipulated the so-called Ngon Savit agreement, through which the commanders of the SSPP/SSA, PSLO/A, and KIA's Fourth Brigade demarcated their respective territories in northern Shan State.[41] The main logic on the basis of which territory was demarcated was what one SSPP/SSA-N officer once referred to as "compound communities": that is, the idea that territory was to be demarcated on the basis of the distribution of the ethnic population.[42] These strategic alliances, in particular with assistance from the KIO/A, allowed the PSLO/A to consolidate and grow into a one-thousand-strong armed force mostly characterized by its guerrilla nature but solidly based in the mountainous areas of Namhsan (Smith 1999, 334).

War(s) on Drugs (1970–1980)

By the early 1970s the complex tangle of armed forces in Shan was mirrored by a bewildering overlap of guerrilla war zones and more stable, so-called liberated areas. The consolidation of frontier spaces and areas of territorialized rebel control in the borderlands during this period was connected to the entrenchment of KKY militias in the opium and consumer goods trade, as well as to their role as weapon flows "brokers" (Buchanan 2016, 11). Militias had the resources, capacities, and connections to access weaponry. At the same time, they maintained close ties to rebel movements. In an attempt to curb insurgencies and reterritorialize the frontiers where "liberated areas" were emerging, beginning in 1966 the Sit-Tat started to unfold a strategy named the four cuts (Pya Lay Pya, which was officially adopted in 1968).

The strategy aimed to cut off the linkages between "civilian" populations and rebel forces by hampering supplies of four resources: food, money, information and intelligence, and weapons and recruits (Maung Aung Myoe 2009, 26; MacLean 2022). It entailed the disconnection of rebels from their sociospatial background by involving the very environment surrounding the insurgent in fighting the insurgent. But while it has been noted how the four cuts took various shapes—for example, from the aggressive application of a "national language policy" and the prohibition to publish or disseminate any information in minorities' languages in ethnic areas, to the implementation of exploitative military-economic partnerships (Sakhong 2014, 16; Woods 2016)—I argue a key dimension of the four cuts concerned the governing of weapons.

The strategy's consistent adoption has been linked to the use and control of weapons in at least two ways. First, the four cuts rationalized and institutionalized the links between weapons, armed violence, and ethnicity. In fact, through forced relocations or blockage of supplies, the coercive division of the "population" from the "rebel" created an artificial insurgent body (a collective body that in reality was much more blended than civilian-combatant divides assumed by counterinsurgency perspectives). In the logic and practice of the four cuts, ethnic minority people's bodies and entire ethnic areas could become enemies to be included or excluded from the polity just due to their proximity to the "rebel." People could thus become enemy weapons to be targeted and/or weapons to be harnessed. This was the case, for example, of the systematic use of villagers as porters, human minesweepers, human shields, or improvised militiamen. This division—between an ethnic minority population to be re-territorialized into the Myanmar polity and an ethnic polity to be erased—embodied broader policies of Bamar nation building aiming at the constitution of "one voice, one blood, one nation" (Sakhong 2014, 2–8).[43]

Second, the strategy entailed a depoliticization of weapons and armed violence in given space-time settings and a multiplication of the entities that could be understood to act as a weapon. As part of the four cuts, the Sit-Tat created a vast network of military bases occupying large swathes of land along the main roads, in key junctures and towns, while posts were built in rural areas, close to villages, or in remote areas. This occurred especially in borderland areas designated as "black" and "brown" frontier zones. In fact, the four cuts was strictly intertwined with the delineation of areas of control along a white-brown-and-black chromatic spectrum. "White" designated military-government-controlled areas; "brown" designated areas in which both rebel movements and government forces operated; and "black" designated areas outside military government control. Interestingly, the designation and enforcement of such areas was connected to specific techniques of weapons use and management. Black areas were free-fire and scorched-earth zones, while brown areas were spaces subject to a regime of curfew hours and no-go zones, in which the same free-fire and scorched-earth practices may be applied.

The four-cuts strategy started to gain momentum in the early 1970s, in conjunction with the dismantlement of the KKY program. While both counterinsurgency techniques—the KKY program and the four cuts—had been pursued within the same people's war logic, the KKY came under strong criticism for being counterproductive to four-cuts efforts. The KKY was thus abolished in 1973. Its disbandment, and the crackdown on influential militia leaders such as Khun Sa (Shan United Army), Maha San (Wa), and Lo Hsing Han (Kokang), was carried out on the premise of a radical (albeit cosmetic and hollow) condemnation of the production and smuggling of opium and narcotics. It has been argued that the military government was moved by the intent to regulate the more structured militias, which had relied heavily on narcotics smuggling and had maintained close connections with rebel movements, but also by advantages in aligning with the US "war on drugs" and antinarcotic elements of its foreign policy in the golden triangle (Yawnghwe 1989, 127; Lintner 1999, 312[44]; Buchanan 2016, 11).[45] Adopting a "war on drugs" stance granted the possibility to justify military actions against recalcitrant and cumbersome militia actors in the borderlands while legitimizing state authority as a gatekeeper and regulator of the illicit narcotics markets.

Although the termination of the KKY did not mark the end of other Sit-Tat militia strategies, it represented an important moment for the development of an antinarcotics rationality of control over military means. From the perspective of rebel movements and militias, the war on drugs would provide a platform, an arena, to construct and consolidate legitimacy—either through affiliation with state authorities, in opposition to them, or in opposition to other armed actors. In the decades to come, and especially after the fall of the CPB, for armed actors the fight against narcotics slowly became a technique to shape themselves and their political-geographical spaces as legitimate embodiments of the means of violence (see also Ferguson 2021, 127–128). Up to that point, involvement in the circuits of narcotics smuggling had been framed as a necessary evil and a means to locate the financial resources needed to acquire arms and pursue the higher goals of self-determination and federal democracy. But, with the condemnation of "drug lords," antinarcotics started to emerge as a rationality for control of arms and armed violence. It became a field for building nonstate armed forces' authority and spaces. A "war on drugs" rationality became more and more intertwined with the stated goal of protecting ethnonational populations from the spread of narcotics as a weapon deployed by the state and communist entities to weaken resistance movements.

In the aftermath of the termination of the KKY program, some of the militias teamed up with rebel movements, ensuring the rebels had continuous weapons access.[46] Not only did militias bring along with them weaponry, military specialists, and knowledge (Smith 1999, 257; Kramer 2007, 11, 22; Gibson and Chen 2011, 274; Buchanan 2016, 11). They also provided connections for arms acquisition. Events in Thailand were key in this sense. In fact, from the Thai state perspective, the expansion of the CPB in Shan State was a worrisome development given the simultaneous threat posed by the Communist Party of Thailand (CPT) on the Thai-Laos border (Lintner 1999, 299). The CPT, similar to the communist party in Burma, was able to obtain supplies from the People's Republic of China and became a further node of arms proliferation in the northern areas of the Thai-Laos borderlands. In turn, Thai authorities became keener to actively support or facilitate the stabilization of ethnic armed movements at the Thai-Burma-Laos borders. They allowed them to establish bases, to liaise with the Thai police and army, or to access black and gray weapon transfers with the help of former KKY militias. All in order to forestall any possibility that the two communist entities (CPB and CPT) could link up.

Moreover, in the same years, Thailand became a privileged weapons transit and retransfer third country due to its stances vis-à-vis the conflicts in Cambodia. As Cambodia had quickly become a crossroads for arms proliferation due to intricate proxy relationships, the US government favored the strengthening of Thailand's security apparatuses as a buffer against the expansion of Vietnamese influence and agreed to a militarization plan with the Thai army (Pongpaichit et al. 1998, 153).[47] The latter increased its budget and entered agreements with the US to create weapons reserves on Thai territory (Pongpaichit et al. 1998).[48] General Chavalit Yongchaiyudh's So po ko to bo 315 unit was tasked with supervising the delivery of covert supplies into Cambodia through Thailand (Pongpaichit et al. 1998, 134). Along the Thai-Cambodian extended borderland a series of warehouses were set up, where covert supplies were initially directed (Pongpaichit et al. 1998, 137, 147).[49] Police and military officers were then tasked to transfer them to the Cambodian border. Parts of such shipments were diverted to the Thai-Burma border, where they were further moved into Shan State and southeast Burma via links with the actors present there—especially ex-KKY militias, ex-KMT armed traders, and some ethnic rebel movements (Karen, Mon, Wa in particular).

These rebel movements and militias became the most immediate recipients of weapons smuggled through Thailand. For example, industrially manufactured landmines were mainly acquired by Karen and Mon rebel movements, Khun Sa's Shan United Army and later Mong Tai Army (MTA), besides the CPB.[50] Khun Sa's MTA was particularly important in this sense due to the considerable stockpiles maintained at the headquarters of Ho Mong and to the lathes and furnaces used for the production of factory-grade landmines, among other items (Selth 2001, 34).[51] Here weapon specialists and technicians produced copies of the POMZ-2 stake-mounted landmine initially designed in the Soviet Union and widely used in Vietnam and Cambodia.[52] Furthermore, the main nodes of gunsmithing, weapons refurbishment workshops, and basic factory-grade production sites developed at the southern Shan State border with Thailand and in the CPB strongholds from Panghsang to Kokang in the north.[53]

Together with weapons, techniques and practices circulated as well—like the construction of village complexes and community resettlements; the arming of community or larger partner militias; and the construction of roads, military instalments, and fortifications. The 1970s and 1980s were a period of intense territorialization processes in the Thai-Burma-Laos borderlands, and these were definitely not limited to the ambit of state-led counterinsurgency (Peluso and Vandergeest 1995, 2011; Gibson and Chen 2011, 296–298).

Post-1988 Politics of Weapon Flows (1990–2000s)

By the beginning of the 1980s the intertwining of a series of political and economic shifts was gesturing toward future trends. In 1980 KIO and CPB entered peace negotiations with the government, but these fell apart very soon. Despite the failure to reach any agreement, the talks signaled the intention of Rangoon to halt hostilities in the north while scaling up military operations in the southeastern borderlands.

Moreover, since the early 1980s Deng Xiaoping's China had started to identify Burma as a primary economic partner to resuscitate the economies of its borderlands, Yunnan *in primis*, and as a potential prospective outlet on the Bay of Bengal and the Indian Ocean (Smith 1999, 360). With Deng Xiaoping, the CCP decided to gradually scale down its support to the CPB (Kramer 2007, 16). In turn, more open to forging partnerships, the CPB entered into a military pact with the NDF and especially the KIO/A, SSPP/SSA, and PSLO/A. This virtually consigned the Kachin-Shan-Yunnan borders to the control of KIO/A and CPB, except for Pansay and major army checkpoints, such as Muse (Smith 1999, 334, 360).

Shortly thereafter, though, in 1988–1989, the CPB broke up, generating a panoply of politico-armed actors and in part fracturing along ethnic lines (see Lintner 1990, 47).[54] Five EROs emerged: (1) the Kokang force of CPB under Pheung Kyashin, renamed Myanmar National Democratic Alliance Army (MNDAA); (2) the Wa force, which became today's UWSA;[55] (3) a Shan brigade; (4) the Mekong River division, renamed National Democratic Alliance Army, tucked up against the Chinese border in northeaster Shan State; and (5) CPB's military region 101 in Panwa, Kachin, under Ting Ying, which was renamed New Democratic Army Kachin (NDA-K) (Smith 1999, 378; Min Zaw Oo and Win Min 2007, 16).[56] As a result, these armed actors controlled vast stockpiles of weapons, mainly received as direct CCP support to CPB—especially in the decade 1967–1977—or acquired across the border with China via the involvement of CPB commanders in the opium and consumer goods trade (Kramer 2007, 16–18).

The military government quickly negotiated bilateral ceasefire agreements with the former factions of CPB. In parallel, Sit-Tat offensives unfolded against KIO/A in Kutkai and along the Kachin borders with China. These waves of attacks led to the split of some one thousand troops of KIA's Fourth Brigade, which agreed to a bilateral ceasefire in January 1991 and turned into the Kachin Defense Army (KDA) (Saw Oo and Win Min 2007, 18). Reshaping the configurations of rebellion in the Shan State borderlands, the ceasefires were accompanied by the delineation of special regions in which the rebel organizations would have been allowed to maintain some degree of political and economic autonomy. Overall, these political, economic, and military developments had important implications for the consolidation of the main rationalities of governing weapons and military means, as well as for the politics of weapon flows.

Governmental Rationalities: Ethnonationality, Drug Eradication, and Humanitarian Security

The creation of the special regions assigned to armed groups was underpinned by the political rationality of *taingyintha*. The organizations that emerged from the crumbling of the CPB were granted zones of special autonomy as representatives of "national races"—national races either considered to be included under the Shan as a sub-*taingyintha* or to be a minority group within the territory of Shan State. Refraining from tackling any significant political issue, the waves of ceasefire agreements launched at this point in the north retained some common ingredients. Armed groups were offered to maintain their arsenals and were encouraged to trade in opium and narcotics. At the same time, they were pushed to transform into militia forces outside the structure of, but controlled by, the Sit-Tat—something that was especially true for smaller groups and splinter factions (like the KDA) (Buchanan 2016; Min Zaw Oo and Win Min 2007).

Meanwhile, in the landscape of the 1988 pro-democracy protests (also known as the 8888 uprisings,[57] since the early 1990s the SLORC interim government initiated a National Convention process that aimed at drafting a new constitution after the abolition of the 1974 constitution promulgated by the dissolved Burma Socialist Programme Party (Smith 1999, 421). With the Myanmar army setting its "three main national causes"—nondisintegration of the Union, nondisintegration of national unity, and perpetuation of national sovereignty—at the center of the constitutional process, it soon appeared clear that negotiations with pro-democracy forces would be long and contentious.

Such political developments could not be disentangled from the moment of renaissance experienced by the rationality of *taingyintha* (Cheesman 2017). The new SLORC regime in fact was quick to fall back on the ideology and regime of truth of the national races. It justified its rule through the need to once again bring together all *taingyintha*, to bring them back to their "natural" status of unity in a single political community struggling against internal rebels and external pressures and enemies (Cheesman 2017, 7). However, the national races were prominently characterized as subsections of the national political community living in a condition of social and political backwardness due to civil war (Cheesman 2017). These subsections needed the Bamar-led state and the Sit-Tat to be brought back into civilization and into their "natural" condition of *taingyintha*.

From the politico-armed rebel movements' perspectives, the National Convention process marked a renewed emphasis on the question of autonomy for ethnonational minorities and the need to define the role of their armed forces vis-à-vis the Sit-Tat in the future state security apparatus. The question of how to redistribute control over the means of violence became progressively intertwined with the logics of ethnonational stateness and territory in a federal union, whereby the different ethnic movements aspired to their own territorial entity. Throughout the late 1990s, and especially in the 2000s, rebel movements in

Shan State came under recursive waves of military and political pressure behind requests to disarm, disband, and/or integrate under the structure of the Bamar-led Defense Services.

At the same time, the waves of ceasefire agreements concluded by the military government in the first half of the 1990s was accompanied by an increase in opium production and trafficking in Shan State (Kramer 2007, 50). For virtually all the rebel movements in Shan State, the possibility to maintain some sort of involvement in the narcotics trade represented a key element for the continuation of their political struggle.[58] Unlike the EROs in Burma's southeast, such as the KNU, adopting antinarcotics policies was not feasible (Kramer 2007). Yet, the increase in narcotics production and smuggling observed until around 1997 triggered a series of sociopolitical implications that eroded the legitimacy of rebel movements.[59] This erosion, combined with the recalibration of the drug-control approaches of neighboring countries (China, Thailand) and the US (via the United Nations Office for Drugs and Crime), started to favor the adoption of antinarcotics political stances (Kramer 2007, 24). Beginning in 1997, different armed actors in Kachin and Shan State adopted opium bans and put a curb on drugs.[60] In this landscape, the idea of a "war on drugs" and opium bans could become, in the decades to follow, a material and discursive domain to reassert and reproduce rebel movements' legitimacy and authority. It provided a rationale to acquire weaponry, build armed forces, forge connections and alliances, demarcate us/them boundaries, assert the need to maintain control over the means of violence, or territorial rebel movement's authority in opposition to or coordination with state apparatuses (as we will see more in depth in the next chapters).

Meanwhile, in the southeastern borderlands the Sit-Tat scaled up military operations, in particular in Karen areas and southern Shan State. As the military government progressively encroached on EROs' influence areas in the borderlands—both in the southeast, through aggressive military operations, and in the northeast, through military-private business partnerships (Woods 2011, 2016; Meehan 2011)—consistent reliance on landmines was becoming ever more a leitmotif of the armed conflicts. Since the mid-1990s rebel movements as well as the army increased their use of landmines and explosive items in a shift toward much more mobile guerrilla tactics connected to the progressive erosion of the so-called liberated areas (Selth 2001, 23). This was also an important juncture for the craft manufacture of explosive items and the consolidation of workshops and gunsmithing skills. Particularly iconic became the explosive devices assembled using chopped bamboo trunks and empty bottles of a renowned energy drink.[61] Being used both defensively and offensively, landmines terraformed certain regimes of mobility and life in the land. They prevented the use of given routes, forcing deviations to or staggering the movements and presence of opponents.

They increased the costs of logistics, instilled fear and thus diminished voluntary recruitment, while also delegitimizing the military. These regimes of mobility were further enhanced by the use of landmines and explosive items to regulate access to specific areas, enforce village relocations, clear the way for prospective or already built infrastructures (such as dams, electricity lines, pipelines, logging concession areas, or agribusiness estates), or making local agricultural fields and plantations unserviceable for local populations and rebels.[62] In this sense one can see how landmines would come to fully embody the logic of the four cuts, if not move beyond that logic due to haphazard (or completely absent) techniques of landmines record-keeping and mapping.

Linked to processes of economic restructuring through military-private partnerships in development projects and natural resources extraction (Brenner 2017a),[63] these offensives were accompanied by apparently contradictory trends. For, on the one hand, the army had embarked on a process of expansion, "modernization," and restructuring[64] while, on the other, the new SLORC military government initiated a gradual partial opening toward civil society and nongovernmental organizations (Smith 2016, 67, 71–73). Standards, norms, and practices of humanitarian and human rights INGOs (international nongovernmental organizations) permeated (and were reformulated by) civil society and community-based organizations in their peace-related activities, social activism, and refugee communities' self-organization, particularly in the southeastern borderlands.[65] Here many EROs had maintained a presence as well, as the KNU headquarters at Manerplaw was also the headquarters of the NDF alliance. Strong links and overlaps often existed between EROs and civil society organizations, the lines between the two being particularly blurred in the milieus of ethnonational political movements (see also Brenner 2019). Thus—through notions and practices of human rights violations monitoring and recording, training in peace action and advocacy, storytelling and recording of ethnic minorities' conflict histories—the strengthening of civil society activism developed into both a vehicle of political contestation vis-à-vis the military government and a form of checks and balances on rebel movements.[66]

These practices and logics operated also in the ambit of humanitarian arms control, especially in relation to the use of landmines by armed actors, and the distinction between "civilians" and "combatants" or "conflict spaces" and "humanitarian/relief spaces."[67] Although we will delve more into this in the following chapters, one should underline that the humanitarian rationalities of control over the means of violence that consolidated at this stage concerned the governing of weapons and armed forces in relation to a social body of "civilians," and a "civilian" space to be preserved.

Weapon Flows and Acquisition

Throughout the 1990s, weapon flows in the borderlands were characterized by three main processes: the recirculation of surpluses stashed in Thailand and Cambodia; the crumbling of Khun Sa's MTA; and the rise of the UWSA. After the 1993 elections, which established a new coalition government in Cambodia, parts of the arms stashes destined to conflict parties that had been accumulated in Thailand in previous decades remained there (Pongpaichit et al. 1998, 137–138). When the conflicts in Cambodia scaled down, this became a source of surplus weapons moved to the Thai-Myanmar border.[68] In addition, weapons would flow from the Cambodian borderlands, sold by former or active Khmer Rouge or leaking from the Cambodian government army stockpiles and then sold in Thailand (Pongpaichit et al. 1998). Similarly, Thai military and police stocks were not exempt from leaks (Pongpaichit et al. 1998, 138–139). A further significant weapons-smuggling source was the diversion of imports from licensed companies with close links to police and army officers. These flows operated through broader links between business, political, and security apparatus figures, which allowed weapons to transfer to various actors at the Thai-Myanmar border. In particular, logging companies (as well as border military and police officers) covered a pivotal role, being directly enmeshed in the southern Shan State and Karen borderlands. Here they may get involved in weapon transfers in exchange for fee waivers vis-à-vis militias and rebel movements taxation.

Quite relevant in terms of weapons and military expertise proliferation in these years was also the crumbling of the MTA. Khun Sa's armed group based at the southern Shan State–Thai border had long been a pivotal actor in the acquisition of weapons via regional flows.[69] In 1995–1996 the Sit-Tat carried out a joint military operation together with UWSA, which in return obtained verbal agreements recognizing new strategic base areas taken over from MTA at the Thai border (Kramer 2007, 22–23). The armed forces that emerged from the disbandment of MTA capitalized on stockpiles as well as on social and physical military architectures of the armed group.[70] In particular, while some other disbanded MTA units turned into military-government-recognized militias, Yawd Serk—one of the most prominent commanders in MTA—formed a new Shan nationalist rebel movement.[71]

The involvement of the UWSA in the offensive against MTA was part of a broader consolidation of its military and political role in these borderlands. Throughout the 1990s and into the early 2000s the UWSA continued to manage firearms acquisitions from Yunnan-based PLA's personnel (Capie 2013).[72] New weapon acquisitions by the Wa rebel movement allowed for the recirculation of older stocks being substituted by UWSA to other formations (in particular the SSPP/SSA-N). In addition, in parallel with a strengthening of the relations between Beijing and Yangon, the UWSA managed to establish small-arms and ammunition production capacity in the second half of the 2000s (Weng 2008; Capie 2013).[73]

By the end of the 2000s these developments marked a shift in the geographies of weapons acquisition in the Shan State borderlands. For rebel movements in Myanmar it became more difficult to acquire weapons via the southeastern border areas.[74] Three different waves of arms control efforts in Cambodia and the slow depletion of Thai-based stocks bound for the Cambodian conflicts led to a reduction in weapons availability. Meanwhile, tougher counterterrorism and border control approaches by Thai authorities combined with a strengthening of Sit-Tat-Thai diplomatic relationships and processes of state co-optation of KNLA's brigade commanders. While these developments made it more difficult to access weapon sources in the southeast (Brenner 2017b), UWSA's regions instead consolidated their role in weapons acquisition processes. Between 2010 and 2013 the UWSA undertook a rapid renovation and expansion through a series of shipments originating from China (Davis 2019).[75] Apparently, these supplies were directed to the Laotian defense ministry with end user certificates produced by North China Industries Group Corporation and China National Precision Machinery Import-Export Corporation (Davis 2019).[76] The weapons were transported over land to Laos' northwest, moved through the Mekong to the river port of Sop Lui (in Myanmar), and on to the Mongla area (Shan State Special Region 4). From here they were shipped to the Wa region (Davis 2019). The Wa became a fulcrum for weapons, uniforms, and training acquisition for a number of EROs.[77] This also marked a significant change in the types of munitions circulating. As the acquisitions of the UWSA included small arms and light weapons previously uncommon throughout the borderlands, in recent years such weaponry became available to some of the rebel movements politically tied to the Wa polity.[78]

As the following chapter will explain, for the Ta'ang movement this overall shift in weapon flows from the southeastern to the northeastern borderlands occurred at a particularly sensitive time. Like other groups, the PSLO/A faced increased pressure to disarm and integrate under the state defense services' architecture. But at the same time, being a minor and recalcitrant resistance movement of a "minority of the minorities," the PSLO/A remained at the margins of the waves of commercial and development projects involving Chinese economic actors in partnership with Sit-Tat-affiliated businesses that were transforming the borderlands (Woods 2016). This political and economic marginality translated into a lack of access to otherwise available weapon flows. Throughout the 1960s–1980s, alignment with ethnonationalist pro-democracy and anticommunist stances (for example, the KIO/A and SSPP/SSA) had allowed the PSLO/A to resist the frontierization of Ta'ang land. In that scenario, to mold a political community, it had become key to link with the networks of military and political alliances that emerged around the UWSA in the Shan State borderlands (Ong 2018).

From the "uncivilized" frontiers of colonial order to the fringes of the ethnocentric Bamar polity envisaged by the Sit-Tat, or the ethnonationalist federal projects of EROs, the flow of weapons was never a smooth, apolitical problem. Annihilating forms of colonial and state military violence had made it virtually impossible to resist without weapons. In turn, resistance movements—including the Ta'ang—politicized the acquisition and control of weapons, as well as the formation of their armed branches, as an essential and integral dimension for the formation of their political entities. Weapons have been an important element of the frontier assemblages that made the borderlands into state and nonstate political territories and frontiers.

FIGURE 3. PSLF/TNLA narcotics-burning ceremony at the 2017 International Day against Drug Abuse and Illicit Trafficking. Source: Author's private correspondence during fieldwork.

DISARMING AND REARMING TA'ANG LAND

When I first met TNLA's commander-in-chief I knew his name—Tar Ho Plan—but not his face. With a trip mirroring the one to Myo Thit and then Kyaukme that opened the previous chapter, K. had brought me from Lashio to Namtu, and from there up to what I had been told was the rebel movement's "temporary mobile" headquarters, somewhere on the mountains of Namhsan. On the bamboo porch of the command center, towering over the drill ground located at the center of the base, sat four high-ranking officers sipping tea. From afar, the low neon lamps hanging from the ceiling that illuminated the veranda—the only lit place in an otherwise pitch-black military camp at seven o'clock on an early November evening—drew yellow halos blurring their sturdy figures. As a fifth officer was about to introduce me to Major General Tar Ho Plan, I realized I would not be able to guess which of the four he was. I had managed to find his name in the news but, contrary to other ERO commanders-in-chief, sourcing a picture of the TNLA leader had proved troublesome.

Quite a few days had to pass before the major general could spare me some of his free time for a conversation. Revealing how he had seen me "busy doing research" and thus wished to avoid "disturbing" me, one evening Tar Ho Plan called me back to the veranda—the same high-ranking officers again sitting there, bathed in the same, yellow-shaded light illuminating the same end-of-day tea ritual. "Would the researcher spare some time for me?" he asked with an eloquent profusion of respect that nonetheless reinstated the not-so-subtle displays of authority with which he had disregarded the interview requests

I had addressed to him through his subordinates in the previous days. "Where in Italy are you from?" he asked, while a gesture by one of the other officers kindly invited me to help myself to tea. "Bologna, in northern Italy," I replied, never imagining that the name of my hometown would lead him to reveal his passion for Roberto Baggio, the world-famous Italian football player who, among other teams, had played in Bologna too.

My simple answer to his simple conversation starter unexpectedly brought us straight back into the ceasefire era of the PSLO/A. For the sporting trajectory of "the Divine Ponytail"—as he referred to the attacking midfielder, without failing to mention his conversion to Nichiren Buddhism in 1985—had somehow paralleled the commander's personal experiences in the Ta'ang armed force. Baggio's exploits, which he had followed on TV during three consecutive World Cups in the 1990s, had accompanied Tar Ho Plan's years as personal assistant to the PSLO/A's general staff officer (GSO). The end of the footballer's career had coincided with the demobilization process, which culminated in the disarmament of the rebel movement in 2005, after which Tar Ho Plan had decided to go underground (Buscemi 2022):

> From 2005 to 2008 I was not doing anything, but in 2008 I started to travel around our Ta'ang areas, and during those trips I recall I was noticing *there was something wrong in our Ta'ang land. The land was full of drugs, something is missing here to control the problem,* I thought. It became a motivation for me and at the same time I had never agreed, since the beginning, to the ceasefire. This antinarcotics goal, in particular, became another very strong motivation together with the *refusal of the ceasefire first and the delusions of disarmament later*; two *very strong motivations to do something.*[1] [Emphasis mine]

Visiting Ta'ang communities in the borderlands in the aftermath of the PSLO/A disarmament, Tar Ho Plan experienced firsthand what he, as his recollection proceeded, diagnosed as the gangrene of a polity: an anthropomorphized geo-body with its vital space—a "land full of drugs"—in which something was missing to "control the problem." What he had witnessed was the political and geographical discrepancy between Ta'ang areas, as areas where Ta'ang populations live, and Ta'ang Land. Better still, between "our Ta'ang areas" and "our Ta'ang land," as he repeatedly put it, addressing an imagined Ta'ang audience. A discrepancy, a gap, whose delineation has been the object of the Ta'ang rebel movements since at least the 1962 coup d'état, as we saw in the previous chapter. In fact, the biologization of the political community that the TNLA's general deployed by framing a fully sociopolitical phenomenon in terms of an organic geo-body has been accompanied by the politicization of a number of geomorphological traits. Both the governmental apparatuses of the PSLO/A's first, and those of the PSLF/TNLA later, have presented these traits as constituting the objective boundaries of a Ta'ang polity with its biological and geographical body.

The PSLO/A politicized certain physical features of the so-called Shan Hills, arguing that the geomorphological characteristics of Ta'ang areas could define Ta'ang Land beyond the confines drawn by the British with the recognition of the principality of Tawngpeng, and subsequently the Namhsan *sao hpa*. Mountain ranges and rivers in particular offered some coordinates. In Shan State a fault line runs from the Goteik pass east toward Kunlong and further into China. The Namtu River winds through this fault line and separates the undulatory Shan plateau to the south from a mass of different mountain ranges descending almost parallel in a north-south direction. For the PSLO/A the Ta'ang regions laid in the western portions of such massifs. Stretching from the border towns of Muse and Namkham all along the Shweli River valley southwest to Mongmit, Mogok, and northern Kyaukme townships, Ta'ang Land was construed as a land traversed by a series of high mountains separated by narrow gorges and highland valleys, dotted by the green of tea plants, and bounded by the waters of the Shweli.

And yet, not only were such geographic delimitations inadequate to encompass Ta'ang areas. They also represented a slippery slope leading to potential frictions over the reification of boundaries and borders that other (at times allied) rebel polities framed from a completely different perspective. Many Ta'ang had long populated areas of the central, southern, and eastern Shan State (such as Mongyai, Tangyan, Kalaw, Keng Tung), or lived in diaspora-like communities in China (where they are named De'ang), northern Thailand, Mandalay, and Yangon. Areas right at the edges of Ta'ang Land, such as Kutkai, have been considered part of a so-called Kachin substate under the KIO/A, or of a Shanland under Shan State armies' political geographies (Ferguson 2021). Since decolonization, the political and civilizational projects of disparate rebel movements, militias, and military-state regimes intersected Ta'ang areas. As the previous chapter explained, the politics of access to available weapon flows, so as to constitute armed formations, contributed to molding the political space of the borderlands. Each of these political orders generated its own political geographies with its own frontiers, meaning the political spaces where their putative "centers" of power happened to be unevenly constituted and distributed, and where multiple armed authorities would overlap. In this context the very idea of a Ta'ang Land, as the territorialization of a Ta'ang polity, remained a highly contentious topic.

Throughout the decades following the ceasefire between the PSLO/A and the Sit-Tat, Ta'ang Land was the object of intense frontierization and reterritorialization processes involving different armed actors and rebel polities. Processes that, as Tar Ho Plan was eager to remind me, were linked to narcotics production and circulation—("the land was full of drugs," as he stated) (figure 3)—as well as to extractive capitalism and accumulation by dispossession under different guises—("there was something wrong in our Ta'ang land"). The areas of Pansay and Namkham to the north, and of Kutkai and Hsenwi to the east, consolidated not only as important narcotics production and smuggling hubs, but also as crucial crossroads into Yunnan, Kokang, and the Wa regions. Similarly, the bustling town of Muse was growing more and more as a key economic hub due to its role as an intensely trafficked Sino-Burmese border crossing, home to lucrative casino businesses and later to the Muse 105th Mile Trade Zone inaugurated in 2004. Right to the south, the old Burma Road connecting the Burmese plains and Mandalay to Lashio, and eventually across the Chinese border farther north, became the key commercial artery of the post-1988 capitalist conjuncture. Meanwhile, to the west along the Shweli River valley, plans were being made in the same years to build the twin pipeline that today connects the offshore fields of Kyauk Phyu, off the coast of Arakan, with Nanning in China. Thus, against the backdrop of these complex frontier assemblages, the relinquishment of weapons and actual disbandment of the PSLO/A in 2005 did shine a bright light on the political problematique concerning the need of a new Ta'ang armed collective in order to (re)shape Ta'ang Land and the Ta'ang polity. ("Something is missing here to control the problem," the major general argued, referring to the "antinarcotics goal" and the state of Ta'ang Land in those years in conjunction with his personal "refusal of the ceasefire first and the delusions of disarmament later" as "very strong motivations to do something.")

So, reflecting on his own postdisarmament experiences, and alluding to the need of forming a new armed force, Tar Ho Plan depicted a certain governmentality of weapons and military means as coterminous with the shaping of a Ta'ang polity with its population and political geography of territory. But before turning to the production of territory (chapter 4), let us pause on the more explicit aspects of Tar Ho Plan's reflections in this third chapter: the role of political rationalities in the governmental assemblages that shape armed assemblages. Conveying the fully political nature of weapons acquisition and the formation of an armed force, this chapter explores the ways in which the rationalities and techniques of governing the relations between humans and weapons that we encountered in chapter 2 circulated in the Ta'ang polity. It looks at their consolidation during the ceasefire and disarmament process, and at how they have contributed to shape the rearmament of the resistance movement and the formation of the TNLA in 2009. Ultimately, the chapter shows how the feasibility of access to munition supplies, the processes, and practices of weapons acquisition, and eventually the very creation of an armed collective have been shaped by rationalities and techniques of ethnonationality, narcotics eradication, and humanitarian security that traveled in time and in space.

The first three sections outline a micro-genealogy of the emergence and consolidation of these rationalities in the Ta'ang polity during the ceasefire, disarmament and disbandment years that unfolded against the landscape of larger frontierization and territorialization projects, and moments characterizing the last three decades. The last section analyzes the trajectory of rearmament detailing how rationalities and practices have been harnessed to shape the acquisition of weapons and the re-creation of an armed force.

Ceasefire: Lack of Access to Weapons and Ethno-depopulation

April 21, 1991. Approaching almost three decades of armed struggle for the establishment of a Ta'ang autonomous political entity within a democratic federal union, a component of PSLO/A's leadership decided to come to terms with the military government of the State Law and Order Restoration Council. Several elements underpinned their decision. Under the pressure of counterinsurgency offensives, as Patrick Meehan has noted (2016a), the Ta'ang movement had become politically and militarily isolated with no prospects in terms of weapons and supplies acquisition.

In what could be described as one of the most underreported cases of the infamous four-cuts counterinsurgency strategy, in the previous years Ta'ang communities had been targeted by extreme violence and forced displacement.[2] This was especially the case for those areas in which PSLO/A had maintained a strong influence, such as Manton, Namhsan, Namtu, Kyaukme, Hsipaw, and Namkham. As part of the four cuts, the Sit-Tat delineated so-called black and brown areas that were subject to harsh depopulation practices. In black areas everybody was considered an insurgent and everything was a potential target, while in brown areas the same praxis was limited to curfew hours after sunset and specific no-go zones (like pathways or strategically positioned tea fields). As these techniques of violence hampered livelihood activities and disrupted food and supply chains, whole communities were profiled and targeted on the basis of ethnicity and assumed links between the "civilian population" and the "ethnic rebels." Scorched-earth tactics and shooting on sight displaced entire communities away from villages on mountainous areas and hilltops toward temporary "hamlets"—white zones—located in lower areas closer to main roads and towns.[3]

The massification logic inherent and codified in such modalities of making violence and killing had the effect of shaping an "ethnic" rebel polity with its population body. Those who experienced such violence recall this aspect in retrospect: "[W]e [Ta'ang] have to rely on tea for living, and at that time it was the time to pick the tea [leaves] [i.e. late March until mid-April] but our people could not pick up the tea. So . . . it became impossible to make a living. That is why our people came to us and proposed us to make a ceasefire with the Myanmar Tatmadaw."[4] Memories, emotions, and effects of the ethnically profiled violence experienced due to the four cuts in the years leading to the ceasefire were to leave a deep mark. Current leaders of the Ta'ang rebel movement, today in their late thirties and forties, still recall such experiences, emphasizing the inherent spatial dimensions of the techniques of violence they encountered:

> Me myself I faced that situation at that time, as I was about six years old. I . . . and the whole of my family, and the villagers, were being forced to move, to move to Manton, by the Myanmar Tatmadaw. But as my grandfather was afraid and scared of the Burmese army . . . we decided to move to Namtu instead. As far as I can remember, we had to run about three days, but we did not dare to go to the town of Namtu. . . . [W]e remained instead just in one of the Shan villages near Namtu. . . . [T]he Burmese army forced the people not to stay in their village . . . nobody could stay in the village, and nobody could return back to their own villages. . . . [T]he Burmese army . . . warned that if they (villagers) were seen somewhere they would shoot people dead. . . . This was also part of their strategy to identify areas as white, brown, or black areas.[5]

The technical possibility of deploying automatic fire massified human and non-human elements of the space of a given black or brown area into the space of the rebel polity that Sit-Tat violence at the Ta'ang frontier had to reterritorialize. This engendered a distinction between Palaung (the Bamar exonym for the Ta'ang) civilian populations to be included into the Myanmar military-state polity and a rebel polity to be eliminated.

The impact of military offensives was further aggravated by the fact that, all around Ta'ang areas, EROs were entering ceasefire agreements.[6] In the final years of the 1980s, under General Khin Nyunt as head of the intelligence services, the Sit-Tat had negotiated ceasefire agreements with armed actors in northern Shan State, starting on May 5, 1989, with the Kokang component of the CPB. With sixteen ceasefire agreements concluded in the period 1989–1995 (Jolliffe 2015, 18–19), this was to turn into a strategy characterized by an important spatial element. In fact, the military state proceeded to legally and administratively territorialize a number of rebel "ceasefire" areas as "Special Regions." These state reterritorializations mushroomed in key nodes adjoining the Chinese border and in general surrounding Ta'ang land: Special Region 1 was granted to the Kokang MNDAA; Special Region 2 to the UWSA; Special Region 3 to SSPP/SSA; Special Region 4 to the Mongla. Meanwhile, the Sit-Tat was intensifying military offensives against other rebel movements, especially PSLO/A's longtime ally, the KIO/A.

Overall, these developments modified the political geographies of weapons access, making the PSLO/A arms procurement processes even more difficult than before. As TNLA's commanders previously fighting in PSLO/A recall, it had been "always difficult to acquire weapons."[7] The Ta'ang armed force had relied heavily on connections with KIO/A to source weaponry, as shown in the previous chapter. In particular, the Fourth Brigade of KIA, located in Kutkai township, just to the north of the main Ta'ang operational areas, used to broker arms acquisitions for PSLO/A. When the Fourth Brigade broke up with the Kachin rebel army and decided to transform into a pro-regime militia before agreeing to a ceasefire in January 1991,[8] the terms of the accord obliged it to sever ties with EROs. The territory of the special region granted to the new pro-regime militia came to divide PSLO/A from both the Sino-Myanmar border and the KIO/A in Kachin. Although KIO/A continued its armed struggle,[9] the Kachin army remained geographically apart and the breakup of KIA's Fourth Brigade entailed the shutdown of the main munition pipeline of the Ta'ang movement. Furthermore, the combination of Sit-Tat offensives and ceasefire agreements made the only alternative left unviable. For PSLO/A it became almost impossible to logistically connect with its own contingent of troops dispatched at the National Democratic Front (NDF)'s headquarters in Karen, which in the past had facilitated the acquisition of military supplies across the Thai-Myanmar borderlands.[10]

Threatened with economic sanctions that foreshadowed the possibility of an internal "embargo" on the tea leaf produce originating in Ta'ang areas, and continuously confronted with the violent realities of an unfolding four cuts, Ta'ang communities mobilized to request the PSLO/A to come to terms with the Sit-Tat.[11] Under the leadership of Ta'ang Buddhist monks and village heads, a series of national consultations was initiated, to which communities from Manton, Namhsan, Namkham, Mandalay, and other areas were invited.[12] Consultations resulted in the formulation of a petition requesting PSLO/A's leadership to halt hostilities and agree to the ceasefire proposed by the military.[13] The petition was consigned to the PSLO/A leadership by Ta'ang monks as the four cuts had isolated the movement. At the same time, PSLO/A was confronted with military-government promises of enhanced economic and development opportunities, including the possibility to keep its weapons and be granted with territorial and administrative autonomy.[14]

In this context, discourses of communal suffering and civilian protection understanding the ceasefire agreement not only as a tool to give respite to PSLO/A but especially as one to stop violence vis-à-vis Ta'ang communities assumed particular relevance.[15] These discourses, which elaborated on the negative effects of being armed and the possibilities inherent in putting guns on hold, are still recalled by the PSLF/TNLA's leadership when recounting the whys of the ceasefire, often drawing on larger historical trajectories of violence: "[A]t that time . . . there were some advantages . . .: like, in this country [Ta'ang Land] before we [Burma] got independence we had to suffer the conflict by the English and by Japan, after independence we suffered from the conflict made by Kuomintang and then other ethnic armed groups, but as PSLA made a ceasefire with the government for our civilians they did not have to suffer from that conflict anymore."[16]

Eventually, on April 21, 1991, PSLO/A entered a ceasefire agreement with the military regime. In the negotiation phases of what would have been a verbal ceasefire without any piece of paper signed,[17] the Ta'ang movement was promised a territorial autonomy comprising a minimum of eight and a maximum of ten townships to be established through a peace accord.[18] Sidelining any meaningful political discussion, eventually PSLO/A was granted with a so-called Palaung Special Region—Shan State Special Region 7—which included limited parts of Manton, Namhsan, Namtu, and Namkham townships (see also Meehan 2016a). The agreement—a verbal agreement that today is recalled as a unilateral imposition by the military state—was accompanied by a series of obligations. In the future the movement would have to refrain from engaging with any political party or politico-armed organization opposing the military government, neither abroad nor in Myanmar. It had to sever connections with any alliance, inform authorities in advance about movements in its Special Region, and halt any military recruitment, training, and operations. In addition, PSLO/A would stop collecting taxes or providing services (Meehan 2016a, 369–370).

Yet, different views circulated among PSLO/A's leadership concerning the decision to agree to a ceasefire. Strong disagreement characterized the phases preceding and immediately following the April 1991 accord.[19] Less than a year later, on January 12, 1992, some members of PSLO/A and new political figures regrouped into a new political front branded Palaung State Liberation Front, which set up base at Manerplaw, the KNU headquarters at the Thai-Burma border that were also home to the NDF headquarters.

The formation of the new front at this stage was nested in the relations between PSLO/A and NDF. As PSLF's third secretary general Tar Parn La explains, the rebel group had maintained a contingent in Manerplaw: "As PSLO/A was a member [of the NDF], every two or three years we had to send representatives there according to the strength of the armed force memberships. At that time, as PSLA probably had around five hundred or six hundred soldiers, . . . we had to send about thirty or fifty soldiers as representatives to NDF."[20] In the aftermath of the 1988 uprisings, the PSLO/A contingent at Manerplaw had been joined by a number of Ta'ang students and activists fleeing Yangon and Myanmar's central areas. During the years leading to the ceasefire, among the Ta'ang community at the Thai-Myanmar border it was believed that "it was not reliable to come to terms with . . . the Tatmadaw as they were forcing us to make a ceasefire with power and there was no contract at that time and the agreement was only on word. So that promise could have been broken at any time."[21]

The four main founding figures of PSLF in 1992 mirrored the main political and social bases from which the front emerged. Two were relatively young PSLO/A officers in charge of the Ta'ang troops at NDF: Tar Joke Jar and Tar Aik Bong, who would later become deputy chairman and chairman of the organization, respectively. And two were political characters linked to the 1988 prodemocracy protests: the political activist Mai Aik Pan and Dr. Mai Tin Moung.[22] With the creation of PSLF, the larger landscape of the Ta'ang resistance came to be broadly articulated around three main components. PSLO/A continued to operate in northern Shan State, albeit under the terms of the ceasefire agreement. The troops and officers dispatched at Manerplaw remained active at the Thai-Myanmar border, although PSLO/A had to interrupt any official involvement in NDF. And the mixture of former PSLO/A members and new political activists constituting the front of PSLF was to continue as an ethnonational political resistance movement.

Constituting a political front without forming a new armed branch, but maintaining instead only one Ta'ang armed resistance force (PSLA), reshaped the contours of the Ta'ang polity with its population body and its spatial extent. Two coordinates—one temporal, the other spatial—illuminate this point and the ethnonational rationality of governing military means that infused the process. Temporally speaking, the decision to found the new political front on January 12, 1992—the anniversary of the formation of the first Ta'ang resistance movement (PNF), which continues to be celebrated as Ta'ang Revolutionary Day—aimed to shape the PSLF in continuity with, and as a metamorphosis of, previous rebel formations representing the Ta'ang population. Along the same lines, but on a different scale, the organization was named using the exonym *Palaung* rather than the endonym *Ta'ang*. While *Ta'ang* has been connected with the political projects of specific elites of the Ta'ang polity that developed in particular after the ceasefire, *Palaung* has been the label employed by the *taingyintha* regime of truth/ignorance. The term *Palaung* has long been used by the military state to categorize the Ta'ang as a population subgroup under the Shan national race that is situated in a specific frontier—mostly Namhsan and Manton—within the confines of the Shan State. For PSLF, reemploying the exonym *Palaung* was a way to reaffirm the historic political geographical space of the Ta'ang polity vis-à-vis the Myanmar military-state polity and the other ethnonational rebel polities in the borderlands, thus holding on tightly to the political achievements of the previous decades.

Nonetheless, spatially speaking, the creation of the PSLF brought together different visions of the Ta'ang polity and its Ta'ang Land. The three components of the Ta'ang resistance movement were operating across different scales. Although the armed force of PSLO maintained a presence in northern Shan State that had been basically downsized to small portions of Namhsan and Manton townships, at the Thai-Myanmar border its troops intermingled with Ta'ang political activists fleeing Myanmar and taking part in the PSLF, which maintained a broader vision of the Ta'ang polity. Across such political and geographical scales, the rationality of ethnonationality informed the shared agreement among these three components of the broader movement that the Ta'ang polity should have had only one armed formation: PSLO/A. In part also due to the previously mentioned intricacies of access to weapons that made it difficult to acquire arms, PSLO/A would have continued the politico-armed struggle in northern Shan, and PSLF would pursue a broader, purely political resistance exceeding the political and geographical terms imposed by the ceasefire. The creation of the new front was the expression of a balancing exercise that kept together unity and dissent. Producing a coherent and united polity was a precondition for generating political claims and to liberate a "Palaung" state within the matrix of the *taingyintha* regime, but at the same time the front had to accommodate different conceptions of the polity spatially exceeding the "Palaung" one.

The issue of the internal fractures around the decision to agree to the ceasefire and the formation of PSLF became a delicate one, especially in light of a long tradition of suppressing dissent in the name of unity (*nyi-nyunt-yay*) that has long characterized Myanmar politics (IRN 2022; Walton 2015). Sit-Tat's strategies entailed two important aspects: on the one hand the designation and designing of ethnic minority groups as population units; and, on the other hand, the creation of divisions within and among rebel movements and alliance organizations, as well as between them and the main Bamar antimilitary party, the National League for Democracy (NLD). Since PSLO/A was supposed to sever any connection with political and armed movements according to the terms of the ceasefire, only the newly born PSLF remained inside the

NDF: "At that time [PSLO/A and PSLF] had an understanding according to which the PSLO/A would have continued to work for the development of Ta'ang areas, while PSLF would have carried out the nationality cause in foreign countries or areas far away from us . . . the Thai border especially . . . together with KNU and KNPP."[23]

Here one can see an important thread that characterized the rearming of PSLF down the line: the mobilization of logics linked to ethnonational identity as a rationality to shape politico-armed movements and spaces of struggle. Ethnic unity, fractures, and contestation among Ta'ang movements are carefully treated as a sensitive issue that could be easily exploited by the Sit-Tat or Union governments. Tar Aik Mone, PSLO/A chairman at the time of the ceasefire, looks back at the divisions concerning the agreement, discarding disagreement as a possible explanation:

> it [the formation of PSLF] did not mean that we did not agree to sign the ceasefire, but it meant that we could not communicate and contact anymore with those who were at the border and also, at that time, the leaders thought that, although the PSLA signed the ceasefire, they still had PSLF, because at that time the PSLA had sent people to NDF . . . they still had the feeling that PSLF was a part of NDF so maybe they [PSLF] might continue to work.[24]

In a similar manner, the current PSLF/TNLA leadership interprets the relations between those who agreed to the ceasefire and those who promoted the continuation of struggle underlining unity and agreement rather than the nuances of different stances: "Later we genuinely understood each other. Because in our Ta'ang values, every one of us agrees that we only want to have one: one army, one party . . . only one, we do not want to separate—even now, for example, we have one political party, one army, only one, we do not want to separate each other."[25] The ethnonational rationality through which these arguments are articulated has to be understood not only as a discursive strategy deployed to counter the *taingyintha* regime of knowledge/ignorance of state authorities (Cheesman 2017), but also as a way to navigate its ramifications at the scale of the rebel polity.

Approaching Disarmament: Tea Crisis and the Rise of Narcotics

The inaccessibility of weapon flows previously available to the Ta'ang rebel movement and its isolation during the ceasefire years were nested into wider frontierization projects unfolding throughout the 1990s and 2000s in Shan State's northern borderlands. Throughout the 1990s, under Sit-Tat's strategies of "ceasefire capitalism" (Woods 2011, 2016), EROs were granted business concessions in the ambit of natural resources exploitation or large-scale agriculture. In the case of PSLO/A, the ceasefire agreement brought with it only small economic opportunities and minor benefits for the leadership. Limited economic zones to log and trade teak wood were assigned to PSLO/A members in southern Namkham, Namhsan, and Manton, and Kyaukme, while plots of land were donated to build houses or small businesses in Lashio.[26] For example, traveling on the bumpy road from Kyaukme to Mogok, after a shiny nursing home and just in front of a Sit-Tat fixed roadblock located at the entrance of the town, one can still see a compound, a sort of small village hosting the houses and Buddhist monasteries of a Ta'ang community. Red brick walls delimit the perimeter, while the main entrance gate is towered by a gigantic Buddha statue named *nyein-chan-yae-taung* (Peace Mountain).[27] Here PSLO/A was allowed to deal in teak, drawing from a small forest inside the perimeter of the compound. Besides minor concessions, the regime also promised the authorization to participate in the national convention to draft a new constitution and issued reassurances concerning the possibility of creating a Ta'ang political party in the future.

Nonetheless, different from what was happening for larger EROs and breakaway factions with more direct connections to the borders—such as Kokang, Wa, or Kachin—Ta'ang areas (and in particular those of PSLO/A) were becoming increasingly disconnected from potential opportunities generated by Sino-Burmese border economic networks. In fact, ceasefire EROs around Ta'ang areas became progressively enmeshed in lucrative economic partnerships involving military-state authorizations and cross-border Chinese investments, which were reconfiguring the landscape of authority and expanding the influence of the Sit-Tat (Woods 2019a). Similarly, pro-regime militias were part of these schemes. In exchange for their quiescence and cooperation in quelling the rebels, they were allowed to manage the production and smuggling of narcotics, in particular opium at this stage.[28] Illustrative in this sense is the example of the Pansay militia led by Pansay Kyaw Myint.[29] Linking up entrepreneurs and local farmers, the Pansay militia encouraged the cultivation of opium not only in the rural areas surrounding Namkham but also farther south into Ta'ang areas of Maiwee, east of Namhpakar and Kutkai, as well as north of Manton and Namtu.[30]

The progressive expansion of poppy cultivation in Ta'ang areas in these years was perceived as being strongly connected to economic changes accompanying the ceasefire agreement. Notwithstanding the historical role of tea as the main cash crop in Ta'ang areas, the tea industry had slipped along a persistent decline that would culminate with a marked drop in market prices in the mid-2000s.[31] The crisis of the tea industry converged with the expansion of opium production and narcotics smuggling whereby—especially in the triangle between Namkham, Manton, and Kutkai—farmers were increasingly turning to cultivation of or labor in the local opium economy. Institutionalized exclusion was perceived also in relation to other economic opportunities in the regional landscape of resource extraction, like the lead and silver "Bawdwin" mines in Namtu:[32]

> In the areas of Namtu there are also some mines of minerals and raw materials, the mine is very big. . . . [T]he Ta'ang villagers were prevented from working in the mines or obtaining benefits from it. There was a big company that was granted the concession to work in the mine, but the people around the area were not involved in the activities nor benefited from any opportunity. The villagers' lives were greatly impacted by the presence of the mine and the extraction activities—especially because of holes in the ground due to the excavation activities. And yet none of the civilians could sue the company in front of justice.[33]

Further strengthening these views, many tea and other crops plantations, as well as communal lands, were being confiscated by armed actors in the context of large-scale land grabs and infrastructural projects (Meehan 2016a; see Kramer and Woods 2012; Woods 2019a). Since the late 1990s and early 2000s, the military regime had authorized preparatory works for two main megaprojects affecting Ta'ang areas of northern Shan: the Shweli dams and the Shwe Gas and Oil Pipeline.[34] Land grabbing was becoming a constant experience for communities linked to these infrastructure projects but also for those who found themselves on so-called wasteland—land designated according to a 1991 law that granted the government the right to freely dispose of all the land without official legal titles, regardless of it being cultivated or not, communally used or not.[35]

Land confiscations were strictly intertwined with Sit-Tat redeployments, the rising of new local militias, and a general militarization of Ta'ang areas. Local estimates reported an increase in military battalions and units in Ta'ang areas, passing from four battalions in 1991 to fifteen battalions and thirty mobile infantry units in 2007 (Meehan 2016a, 373; TSYO 2011, 41; TSYO 2012, 21). Infrastructure projects and land grabs were often accompanied by the construction of large Sit-Tat military bases and camps. Entire villages and hills could be adapted to the needs of the army bases under construction. To these the army most often annexed neighboring fields and agricultural areas, which were delimited with fences, walls, or barbed wire for the sustenance of the military units deployed there. The landscape was being modified by fixed concrete checkpoints on roads but also at the entrance of some villages, as well as open-air shooting ranges and training areas built by terracing hilly slopes whose soil debris was randomly dumped in the surroundings (PYNG 2007, 21). Along with such infrastructural expansions, mobility restrictions were being introduced, particularly in nearby villages and communities. The Sit-Tat and militias introduced population lists and mandatory checks with forms to be compiled and consigned when moving in or out of those areas, while the presence of armed actors was impacting people's (especially women's) free movement and general feelings of safety. The militarization of Ta'ang areas was extended along the networks related to such infrastructure. It extended through the roads to be built or renovated in order to transport the materials required for the construction of installations and high-voltage transmission grid lines with their pylons and wires. This irremediably altered many locals' tea fields and paddy farms by changing the course of irrigation streams or impeding access to them. The labor related to both this militarization process and (militarized) "civil" infrastructures was sustained by the realities of forced conscription for porterage, forced labor, displacements, confiscation of tools, food, or vehicles for construction-related purposes.

Overall, this generated a buffer area around the PSLO/A. For the PSLO/A, a relatively small ERO, these processes progressively shrank the armed and territorial influence of the movement.[36] After the 1991 accord PSLA had withdrawn to its bases, camps, and positions and was now practically bounded mostly to the town of Manton and, only limitedly, Namhsan.[37] Instead, administrative and governing activities continued but remained episodic and significantly constrained in space, covering only portions of Manton and Namhsan.[38] PSLO/A's self-support mechanisms related to local communities had shrunk in connection with the reterritorialization of Ta'ang areas. The organization managed to build a few road connections to try to facilitate trade, and launched some new schools in Manton and Namhsan, but with the confinement of its army, the movement now lacked a real operational branch.

With the diffusion of narcotics trade and consumption, the most prominent endeavors of the PSLO/A consisted in efforts to manage and contrast drug addiction.[39] The movement had issued a series of decrees instituting rehabilitation camps and enjoining farmers to stop the cultivation of opium poppy (PWO 2006, 27). Village heads were mandated to draft lists of drug addicts who were then gathered in Buddhist temples' compounds and hosted in designated areas with dormitory huts (PWO 2006). Drug addicts were then employed in community services while practicing forms of rehabilitation through forced abstention. By the mid-2000s, the circulation of narcotics had turned into a serious, multilayered social issue throughout Ta'ang societies. Drug use was increasingly diffused, especially among males (PWO 2006). It seriously affected many households, impairing the latter's workforce and income, while often pushing women to be the sole breadwinners (PWO 2006). At a larger scale, narcotics abuse triggered or aggravated a whole series of other social and health issues: from theft and household property losses to domestic violence, from mental disorders to HIV/AIDS, from child neglect to community stigma on women,[40] from school dropout (when formal education was accessible) to slave labor through opium as a wage and consequent drug addiction, or work-related migration and subsequent drug dependency (PWO 2006; Kramer and Woods 2012).

Among Ta'ang communities, a key defining feature of these frontierization processes was represented by a growing sense of marginalization at the hands of different armed formations—*in primis*, the Sit-Tat and affiliated militias—which were militarizing the territory of PSLO/A. By extension, the demilitarization and loss of an effective Ta'ang armed force was similarly becoming an issue.[41] A certain logic of narcotics eradication—understood as a step to solve the abovementioned societal problems—was progressively emerging and circulating throughout Ta'ang communities. Socialized into discursive and material practices, such as different forms of community stigma on drug addicts and their relatives (particularly on women, who could be stigmatized both as drug users and partners thereof), such rationality was front and center in research, publications, activism, and advocacy activities carried out by local Ta'ang civil society organizations (CSOs).

The meaning of *local* in this sense was networked and multiscalar, for these CSOs were not based in Ta'ang areas, or northern Shan either, but at the Thai-Myanmar border. A first organization, the Palaung Youth Network Group (PYNG), was rooted in the context of the KNU and NDF headquarters at Manerplaw.[42] As we have seen, younger generations of students and civil society figures had come here after the events of the 1988 pro-democracy demonstrations. Different lived experiences and backgrounds met here, not far from the humanitarian complex that European and US donors' relative attention (especially to the All Burma Students Democratic Front movement and the NLD cause) was contributing to create at the Thai-Myanmar border. Many CSOs and CBOs were headquartered at Manerplaw, where the Karen Human Rights Group (founded in 1992) was of particular importance as a model for others to come.

In such environment the main liaison figure for the PSLF was Mai Aik Pan. After the fall of Manerplaw and the displacement to the Thai-Myanmar border of all the community and civil society organizations based there, in 1996 Mai Aik Pan initiated the creation of PYNG, which was eventually founded in 1998. The founding of PYNG was closely influenced by the work of other ethnic nationalities' CSOs in terms of youth political training, human rights and humanitarian law violations documentation, environmental action, and more broadly the idea of shedding light on the bottom-up history of border-land communities.[43] Similarly in October 2000, the Palaung Women Organization (PWO) was founded in Mae Sot with the aim to develop Ta'ang women's educational opportunities, promote and document human rights, as well as Ta'ang history and culture.

The Ta'ang CSOs developed at the Thai-Myanmar border in the late 1990s and early 2000s were not part of PSLO/A, or the political front of PSLF per se, but were certainly involved in a larger resistance social movement. For example, a key figure at this stage, the current PSLF/TNLA's first secretary general, Tar Bong Kyaw, was active in PYNG, where he worked as a representative to the Network for Human Rights Documentation Burma (ND-Burma), but at the same time maintained close connections with PSLF's leadership.[44] As noted by David Brenner in the case of the Kachin and Karen rebel movements (2019), the lines between different forms of actors could not be reduced to an "armed organization" or a "political organization" and "civil society organizations," for the realities of all of these remained very much intertwined. Acting in the ambit of various social issues like health, family planning, birth spacing, education, political reform, environmental action, human rights, and humanitarian law violations documentation, Ta'ang CSOs were operating across borderworld scales bridging different Ta'ang communities in northern and southern Shan State with those at the Thai-Myanmar borderlands. The larger resistance movement was molding a Ta'ang polity (and a Ta'ang Land) transcending the politico-territorial horizons cemented by the ceasefire imposed on PSLO/A.

The work of PYNG and PWO was crucial in shaping and circulating a rationality of narcotics eradication into PSLF. Different conceptions of struggle were circulating throughout the Ta'ang movement, with the components involved in PYNG being more "hostile to any form of militarism, guns and the creation of armed groups."[45] Yet, in a context where PSLO/A appeared unable to take care of Ta'ang people and Ta'ang spaces exceeding Myanmar state-framed identifications and geographies, the eradication of narcotics was more and more intertwined with the formation of a new armed collective for the preservation of a Ta'ang ethnonational body population.

The threads of such logic were articulated by PYNG, PWO, and the movement's grassroots in light of the material realities of three interconnected frontier governmental assemblages enmeshed in the management of narcotics.

First, at the scale of the military-state Myanmar polity, around the turn of the 2000s international outcries concerning opium production in Shan State had led the regime to take some cosmetic measures. Such measures left unscathed the main militia protagonists, such as Pansay Kyaw Myint, and instead turned toward the criminalization of small producers and the eradication of some farmers' plantations (PWO 2006, 28). From Ta'ang perspectives this fed a sense of injustice and lack of accountable and effective institutions. Second, official statements and reports concerning a decrease in opium production and trade issued by the United Nations Office on Drugs and Crime, as part of its programs in the northeast, became contentious across both borderlands and international scales. The UN agency had been claiming that since around 1996–1997 opium cultivation in Shan State had undergone an encouraging decline (UNODC 2003). Yet, from Ta'ang perspectives this was difficult to reconcile with the realities of the borderlands. Such statements were criticized by other ethnic CSOs as yet another distortion based on state-framed data collections to legitimize military-state rule forcibly incorporating minorities at the frontier (PWO 2006; PWO 2011, 19). Speaking of southern Kachin, for example, a Baptist pastor highlights how:

> between 1994 and 2011 the production, trafficking, and use of drugs increased because it was allowed by the police and the Tatmadaw, even inside university compounds for example. The drug issue deteriorated, and this deeply affected the ethnic groups. Recently the antidrug agency of the UN, UNODC, published a report, but this was not clear and thorough because it did not report about this following issue: opium farms are very often to be found close to Tatmadaw bases; they surround the military posts.[46]

Third, with the growing political and economic influence of a narcotics proliferation assemblage made of landowners, entrepreneurs, brokers, and smugglers linked to Sit-Tat-affiliated militias or ceasefire EROs, the othered image of "non-Ta'ang"—"Chinese," "Shan," or "Kachin"—profiteers and capitalists came to incarnate a broader governable order oppressing Ta'ang communities and spaces along different center-margin trajectories. Ta'ang activists argued that these actors were undermining Ta'ang communities in similar ways as the military-state regime, in a situation where PSLO/A was not able to look after Ta'ang people and spaces. With an explicit reference to a frontierization of different parts of Ta'ang Land, one of the current PSLF/TNLA's secretary generals recalls how:

> even though there was an organization representing the Ta'ang, PSLA was just able to set up administration in a very little portion of our Ta'ang

area without the southern parts, and . . . even in the northern parts its area was very limited . . . like the areas in Mine Lon, Mine Gone, and Kyaukme were controlled by the SSPP and the areas of Namkham, like Mong Gyi and Muse . . . were controlled by the KIA and KDA [Kachin Defense Army]. Actually, those are regions where our people live. In Namkham, as there was the Pansay Pyi Htu Sit, militia, PSLO/A administration could not cover that area as there was the militia present. As the ethnic groups like Shan or Kachin militias were controlling most of our Ta'ang region, and as our ethnics (people) had to supply them with food or recruits, new soldiers, we suffered a lot from those militias . . . our people suffered, I mean."[47]

Opium cultivation and trade was progressively politicized: for the regime the eradication of narcotics continued to be a policy tool to legitimize itself in relation to international agencies, while for ethnic nationalities movements it was becoming a political logic to claim and negotiate autonomy, not only vis-à-vis the military regime but also in a messier scenario of fragmented sovereignty (Lund 2006, 2011, 2016; Woods 2019a). Narcotics were linked to questions of institutional autonomy and to the need to reshape the configurations of armed collectives and control over the means of violence in Ta'ang areas:

> [T]he root causes of the drug problem must be understood and addressed if the problems of drug addiction are to be effectively tackled. These root causes include increased military controls over the local economy that have undermined the Palaung tea industry and caused local farmers to turn to opium growing as well as the regime's policy of allowing local militias to grow opium in exchange for controlling local populations and sharing drug profits with the Burma Army. These root causes can only be addressed by means of genuine political reform in Burma." (PWO 2006, 54)

The work of PYNG and PWO foregrounded not only claims for recovery programs throughout society but also for more robust measures to tackle what were perceived as root (political) causes of the narcotics issue. Among these, in particular, there was also the need to reshuffle the configurations of the means of violence and their control.

During the ceasefire years, landgrabs related to agribusiness and infrastructure projects, the expansion of narcotics production and trade, the tea industry crisis, and militarization materialized core Ta'ang communities' grievances vis-à-vis not only Bamar-dominated state authorities but also other EROs and militias. Spatially one can see how these processes had started to reterritorialize

Ta'ang areas and were making people, resources, and land more legible from state authorities' perspectives. Along the Mandalay-Lashio-Muse road running to the east, a series of military bases had been constructed, with annexed agricultural lands to sustain them, fixed checkpoints, and camps. Muse and Namkham had been turned into a valuable border economic area, a hub for cross-border trading. To the south, the Pansay and Namphakar militias had been allowed to manage the trade of opium while, to the west, the actual and prospective dams along the Shweli River and connected infrastructure had come with further militarization. A general sense of absence of political and economic equality was circulating due to the realities of exclusion from jobs in the mining industry, unfair employment, forced labor, and market distortions in the tea industry. Coupled with drug addiction, displacement, and limited access to education and health care, these experiences fed feelings of discrimination and a sense of insecurity along ethnic lines, while the marginal role of PSLO/A was progressively highlighting the need for renewed institutions.

Disarmament: "The House Without a Wall"

The problem of access to weaponry and of forming a Ta'ang armed force became more neatly delineated around the mid-2000s, with the unfolding of specific political developments across borderworld scales. The Sit-Tat was accelerating the national convention process that would have led to the promulgation of a new semiauthoritarian Union constitution. After the ousting of General Khin Nyunt at the hands of more hardline military regime entourages (Selth 2019),[48] the military state started to take a more aggressive stance vis-à-vis armed actors throughout the country. It now demanded their disarmament and integration into the political system on the basis of the prospects of political inclusion allegedly offered by the so-called "road map to discipline-flourishing democracy" announced in 2003. The political logic of *taingyintha* and the three main national causes of the Sit-Tat—"non-disintegration of the union, non-disintegration of national solidarity, and perpetuation of national sovereignty"—informed this vision, whereby the military viewed itself as the sole guardian of the Bamar nation-state's integrity, and thus also the only legitimate arms bearer in the Union. Military pressure was being scaled up vis-à-vis larger ceasefire rebel movements, such as the KIO/A and UWSA, to push them to integrate under the Sit-Tat. Smaller EROs, meanwhile, such as PSLO/A, were targeted by "arms for peace" initiatives entailing their unconditional disarmament. Disarmament and disbandment were imposed on PSLO/A, in part through the very entrepreneurial relationships and commodified spaces generated by the ceasefire (MacLean 2008).

As we have seen, PSLO/A was highly disconnected at this point. The head-quarters in Manton had been isolated from the bases and camps in Ta'ang areas, while its presence among Ta'ang communities had been limited to Namhsan and Manton. A series of consultation processes began, both between PSLO/A lead-ership and rank-and-file members, and between the former and Ta'ang com-munities. PSLO/A organized consultations with village heads who had to refer to villagers and then back to local PSLO/A figures: "[T]he leaders asked the people what to do and then they asked the battalions, their militaries, whether they wanted to disarm or not."[49] Various ideas circulated about disarming and disbanding PSLO/A, with some battalion commanders and other leaders think-ing it would be better not to bow to the military regime's demands.[50] Yet, when looking back on these events, emphasis is placed more on the lack of alterna-tives and the reality of oppression: "Actually all of us did agree that we did not want to disarm, but PSLO/A was forced to disarm and finally we decided to do so."[51] Similarly, other voices prefer to talk about the disconnection between the "center" and "margins" of the Ta'ang polity, between the headquarters in Man-ton and "the jungle": "The soldiers, and even the commanders of the battalions in the jungle, did not know the situation . . . they just received the order from the headquarters that they had to come back to the headquarters with all their weapons."[52]

Eventually PSLO/A agreed to the "arms for peace" initiative, and an official disarmament ceremony was held on April 21, 2005, in Manton.[53] Around 568 assault rifles, rocket-propelled grenade (RPG) launchers, mortars, grenades, two landmines, handguns, and ammunition[54]—apparently only half of which was still functioning—were showcased in expository racks towered by State Peace and Development Council flags, as prominent military-government figures paraded in front of them, followed closely by PSLO/A's leaders and national TV cameras.[55] On the opposite end of the Manton sports ground hosting the event stood 150 rice bags piled up together with one hundred viss of edible oil and clothes that had been brought in by the SPDC delegation.[56] While the opposite position of the two on the ceremonial ground symbolized the exchange of armed struggle for peace and development that—in the jargon of the Sit-Tat—the event embodied, the opposite journey undertaken by such technical objects and essential goods after the ceremony provides an allegory of the developments that would have fol-lowed the Ta'ang disarmament. After ritual photo-ops for the media, the SPDC supplies were swiftly returned to the Manton- and Namhsan-based wholesalers from whom the regime delegation had borrowed them only a few hours ear-lier. The weaponry was not destroyed or burned, but rather collected and car-ried to Sit-Tat military bases in Manton and Kyaukme. In the days preceding the event, PSLO/A's leadership had been invited by the SPDC to draft a text of their ceremonial speech and share it with the military-government officials. Once at the ceremony, PSLO/A leaders were to find out that the SPDC officials had censored large parts of the text, particularly the sections in which PSLO/A touched upon Ta'ang areas' trajectory during the ceasefire years.[57] For its part the SPDC delivered a note promising that the military regime would have taken good care of demobilized PSLA soldiers. It was guaranteed that development would have reached Ta'ang areas, and that the Sit-Tat would have provided security for the region from that moment on.[58] PSLO/A was applauded as a pioneering example of the way in which EROs throughout the country should have embarked on the "arms for peace" initiative on the path toward democracy. In the end, though, as had been the case with the ceasefire, no written agreement officialized the disarmament of PSLO/A.

In this scenario disarmament was primarily understood by the Ta'ang resistance as a technique to manage violence and protect ethnonational civilian communities,[59] emphasizing a link between humanitarian and ethnonational logics:

> If PSLA had not disarmed, we would have had to fight the government again. . . . it would have been very, very difficult for us to fight again, so it would also destroy the whole of the people, it would affect the people, so instead we [thought we would] disarm and then build the . . . promote the community development and build the development for the sake of the civilians, Ta'ang people, Ta'ang civilians. Not only for the organization.[60]

But the years following the disarmament of PSLO/A and its transformation in a purely political organization[61] were characterized by the materialization of lingering sentiments of lack of trust connected to the unkept promises made by the military regime concerning socioeconomic development and security for Ta'ang communities. The disarmament ceremony, with the physical removal of weaponry and dismantlement of the armed force, in a sense became the epitome of such experiences and feelings. Strong individual and collective emotions emerged in relation to the removal of weapons and dismantlement of PSLA, both among leaders and civilians.[62] The theme of the functioning and age conditions of the weapons being handed in at the ceremony is often recalled today either to highlight the farcical character and draconian nature of the disarmament (when the consigned material is remembered as "old and malfunctioning"); or to underline how giving in "good weapons" laid the foundations for the problems to follow. Recalling how he experienced the events of the ceremony, current TNLA commander-in-chief Tar Ho Plan brings up the quality of PSLO/A weapons and illuminates dissident positions against the decision to disarm in 2005:

> When I served in PSLA, I was just a personal assistant to PSLA's GSO [general staff officer], I was only a sergeant at that time. In 2005, when the disarmament occurred, since I was a personal assistant to the general staff officer, he had ordered me to work in Kachin state as he had gone there. So, at the ceremony I was not there [in Manton], I was in Kachin and I only received partial information about the situation, I knew only approximately what was happening. . . . At that time, after the news arrived through the media, I remember that I asked my GSO if we had all decided to give back our own weapons or whether instead we had kept some good arms. He replied that we had already given away all our weaponry. After our leader told me so, . . . I remember that I felt so heartbroken that I did not know anymore what to do. After the disarmament ceremony, since there was a permanent alliance contract with KIO/A, PSLO/A had a duty travel to the Kachin to explain to them about what had occurred and the new situation. The person tasked with this embassy task was Tar Nyi Lone. At that time the trip of Tar Nyi Lone was misrepresented by the media, who had circulated the false information that he was traveling to the KIO/A areas inasmuch as he had rejected the disarmament and wanted to continue the struggle. So, coming to know this false news I remember they made me very happy and . . . I had decided I would join Tar Nyi Lone in his rejection. So, I called my GSO again to ask about this, was it true or not? No, he replied, Tar Nyi Lone was not traveling there to look for allies but merely to inform the old allies of PSLO/A about what had happened with the disarmament. But the answer of my GSO at that time did not matter much to me as I had already decided what I would have wanted to happen . . . I had really strong feelings and felt very bad about both these news, the disarmament first and then the subsequent fake news.[63]

After the 2005 disarmament ceremony the Sit-Tat began to intensify its militia strategy in Ta'ang areas. The army increased the number of local community militias by dictating that villages must send "volunteers" to local Tatmadaw bases where they would receive military training and weapons before going back to their villages.[64] In Namhsan, for example, every Ta'ang village was mandated to provide ten people. Between October 2005 and January 2006 alone around one hundred youth had been inserted into the militia program (PWO 2006, 23).[65]

Part of the arms that the Sit-Tat distributed at this stage originated from the stashes just handed over by PSLA and stored at Manton and Kyaukme's military bases, while some of the militiamen were previously among the ranks of the dismantled Ta'ang force.

The militia strategy was accompanied by strong Burmanization policies, especially through the management of the education system. After the PSLO/A disarmament, Burmese teachers could be dispatched to rural areas. The teaching of native language, culture, and history was prohibited. Students' Ta'ang names were often changed into Burmese ones, for teachers' ease and institutional purposes (TSYO 2010).[66] At the same time, the dispatching of both government teaching staff and SPDC officials or Sit-Tat units brought about interethnic marriages— sometimes forced, sometimes not—whose offspring would not be recognized as Ta'ang but registered as Bamar at the registry of the Department of Population in the Ministry of Immigration and Population.[67]

The expansion of the influence of militias runs in parallel with the expansion of narcotics. Drug production, trade, and consumption continued to rise, tending to shift toward rural areas under the influence of regime-affiliated militias (Kramer 2007, 23–24, concerning southern Shan State). As Tom Kramer has noted, after a decade of decline between 1996 and 2006, opium production more than doubled in the subsequent decade in Myanmar, while that of amphetamine-type stimulants also increased (Kramer and Woods 2012; Kramer 2016). Since 2006 an opium-substitution investment plan for Myanmar and Laos adopted by the Yunnan province triggered a huge increase in large-scale agribusiness plantations—especially rubber and maize—along the portions of the Mandalay-Lashio-Muse highway more adjacent to the border (Kramer and Woods 2012). Commercial plantations squeezed out small farmers from land and markets while offering few job opportunities to local populations. Apart from relocations of dispossessed local inhabitants who would be hired for cheap labor, work in such estates was predominantly contracted out to outside migrant workers from central regions of Myanmar (Kramer and Woods 2012, 4). This contributed to pushing dispossessed farmers and workers toward labor in narcotics production.

Frontier areas of the Ta'ang polity in which regime-affiliated militias operated were particularly affected. Local research by the PWO documented a deterioration of opium and narcotics production and use in the years following the PSLO/A disarmament, with particular reference to the areas of Maiwe, Pansay, southern Namkham, and Namhpakar (PWO 2010, 2011). Moreover, military-government antinarcotics policies focusing on the arrest of drug addicts and small-scale farmers (often Ta'ang), or the eradication of single fields, were once more understood as symbolic and highly discriminatory measures vis-à-vis

Ta'ang communities. The regime failed to target militias' areas and to provide any awareness-raising, rehabilitation or social welfare program. Narcotics diffusion was also increasingly seen as a social problem generating a crisis of the family and having harsh consequences on households in terms of living conditions, education, livelihoods, and job opportunities.[68]

In parallel, Ta'ang areas were witnessing additional preparatory works being carried out for the construction of two more dams—the Shweli Dam 2 and 3—with all the annexed assemblage of roads, high-voltage transmission lines, military camps, and militias necessary to connect the hydropower plants to the Namtu-Bawdwin mines.[69] Grassroots organizations argued that hydroelectric projects backed by international investment should be seen as a dimension of the growing militarization of Ta'ang land, as they were generating revenue with which the junta could "purchase more weapons" across Myanmar (PYNG 2007, 43). Similarly, the extraction of energy from the frontier through its militarization epitomized a broader disproportionate allocation of funds to the military and security sectors' budget by the SPDC, which perpetuated the oppression of Ta'ang populations via lack of, or discriminatory access to, education and health services and infrastructures (TSYO 2010).

Militarization was not limited to Sit-Tat bases, regime-affiliated militias, and local village militias, though. In 2006, the RCSS/SSA-S attempted to move to Ta'ang areas of northern Shan State, in part with a view to link with Shanni communities in southern Kachin. Meanwhile, the SSPP/SSA-N and KIO/A too could expand their influence.[70] Connected to the economic restructuring of northern Shan, the expanded presence of multiple EROs often translated into expropriations, interethnic recruitment, and the use of Ta'ang villagers as porters.[71] Before disarmament, repression by other EROs had been more common in areas of southern, central, and eastern Shan State under their influence. Here Ta'ang communities have often been denied access to formal education beyond primary school, and the promotion of language and history through Ta'ang CSO education initiatives, or ethnic festivals, has been prohibited or hindered.[72]

When placed in perspective with the crisis of the tea industry, especially since 2008 (Meehan 2016a; TSYO 2010), rising drug addiction, consistent smuggling, and land grabbing, such postdisarmament developments were interpreted as a threatening reshaping of the Ta'ang polity's ethnonational community and territory proceeding hand in hand with militarization. In the aftermath of disarmament, the landscapes of authority in Ta'ang areas were reconfigured and a sense of exclusion and oppression that fueled autonomist sentiments grew progressively.[73] Militias had consolidated their presence as indirect projections of central military-state authorities, other EROs had expanded their influence in Ta'ang areas, while households and rural communities were undergoing a socioeco-nomic crisis. Most important, after the disbandment of the PSLA what remained of the former PSLO political and governance apparatuses had lost legitimacy and concrete presence on the ground.

Thus, there appeared to be no valid alternative channel for Ta'ang sociopoliti-cal demands. In this sense, the lack of an armed force was being understood as a problem of ethnicity and institutional vacuum: "[D]uring the PSLO/A adminis-tration people in the areas under the administration of the group were experienc-ing slightly more comfortable conditions than those living in other Ta'ang areas. They were enjoying some advantages like security or control of narcotics, but at that time our people under the administration of other ethnic organizations suffered a lot."[74]

The problem of controlling weapons and military means came to be linked to larger processes and practices of incorporation of individual and collective Ta'ang bodies and spaces through their reterritorialization into Sit-Tat or other EROs' polities. The governmentality of weapons and military means overlapped with the shaping of the geographical vital space of the polity: "[A]t that point we did not have an army that would look after our people . . . when we stopped to have an armed force, when we did not have the arms anymore . . . it seemed as if we did not have a wall . . . the house without a wall, people could enter easily . . . to grow the opium in our territory, like the Shan . . . arresting people and forcing them to work in their army."[75]

Rearmament: Reshaping the Polity amidst Frontier Assemblages

The actual rearmament process of the PSLF in the second half of the 2000s was shaped by an entanglement of actors, people flows, political logics, and dynamics of weapons production and circulation. Navigating these frontier assemblages, the PSLF harnessed the political rationalities of ethnonationality, narcotics eradi-cation, and humanitarian security to govern processes and practices of weapons acquisition. Through this governmentality of military means and the forma-tion of the TNLA, the Ta'ang polity's population and political space were being reshaped.

In October 2009 PSLF convened its third national congress and announced the constitution of a new armed branch of the movement, renamed Ta'ang National Liberation Army (TNLA). The use of the endonym *Ta'ang* rather than the exonym *Palaung* in the name was part of a broader reshaping of the eth-nonational collective identity of the polity that had become ever more urgent since the disarmament in 2005. For the Ta'ang political front based at the Thai-Myanmar border, the blow of disarmament signaled the need to take "full" responsibility for a Ta'ang polity transcending Shan State Special Region 7—the region assigned to PSLO/A after the ceasefire. In order to do so, majoritarian stances among congress participants argued, the PSLF needed to recalibrate its struggle via the constitution of an armed branch. Different factions of the resistance politicized the events of the ceasefire era, arguing that taking away the Ta'ang armed collective had been tantamount to removing an organ from the Ta'ang body. It had destroyed Ta'ang lives and Ta'ang Land (see TSYO 2011, 58). Similarly, the disruptive developments that had punctuated the previous years were all understood as part of a genocidal strategy by the military-state authorities and other armed actors against Ta'ang people and spaces.

For example, the transformation of PSLO/A into a pro-regime militia after disarmament was managed from this perspective. After just a year, in 2006, the military had requested the leadership of PSLO to form a militia in Manton. Such a request was potentially divisive. PSLO members initiated lengthy internal discussions and a public consultative process. Eventually they decided to form a militia under the understanding that—although "most of them [PSLO] . . . were not happy to form a militia because they did not want to listen [to] orders from the Tatmadaw"—if they "did not come to be a militia, others would come to be a militia in Manton. And if others would come to be a militia in Manton, PSLO would have to listen to them. So instead of listening to Chinese militias . . . better to be Ta'ang."[76] With a logic of preserving the Ta'ang polity from its outside, creating a militia would have prevented the Sit-Tat from institutionalizing other, non-Ta'ang, armed actors in the area. Armed actors that here, in retrospect, are construed through, and evocative of, the image of the othered "Chinese." At the same time, the creation of a state-affiliated militia by the PSLO remained a politically sensitive move within Ta'ang communities. As noted, the PSLO and PSLF had agreed to jointly pursue the ethnonational struggle across different scales. Moreover, negative feelings circulated throughout societies vis-à-vis militia formations, due to their controversial role in Ta'ang areas during the ceasefire years, as well as to the life trajectories of many PSLO/A's soldiers who, after disarmament, had been quick in joining the ranks of local village militias.

Weapons and armed forces—or the lack thereof—had become a dimension of a larger sense of oppression and a lack of political, social, and physical space for the Ta'ang polity. They had become intertwined with questions of individual and community management to preserve the body population. Some former PSLA officers in northern Shan shared the view that PSLF "should not . . . stay without weapons."[77] Along similar lines, from the perspectives of those at the Thai border, with the disbandment of PSLA, and PSLO relegated to Manton as a militia, the PSLF had now become "fully responsible for . . . the Ta'ang people."[78] It could not focus only on promoting "the nationality cause in foreign countries or areas far away" from Ta'ang land.[79] Thus, in 2007 the PSLF leadership and former PSLA officers organized a national congress at the Thai-Myanmar border to discuss the issue of rearmament together with Ta'ang civil society representatives. At this congress it was decided to cooperate and re-create an armed force under the name of Palaung National Liberation Army (PNLA). The project was placed under the supervision of Tar Nyi Lone, the former PSLA commander-in-chief. Interestingly, in the name of the embryonal armed collective, the term *Palaung* was maintained, but reference to a "nation" rather than "state" was introduced with a view to bridge and include national communities across and beyond the political geographies settled by the military state.

Yet, just a few months after the formation of the PNLA, the promulgation of the 2008 Myanmar Union constitution curbed PSLF's political aspirations by enshrining a Palaung Self-Administered Zone (P-SAZ). The P-SAZ fell short of the liberation front's claims in spatial and juridical terms.[80] On the one hand, the constitutional provisions concerning the new administrative entity enshrined the political conditions imposed via the 1991 ceasefire agreement and the 2005 disarmament. This limited the SAZ to Manton and Namhsan townships, thus codifying a geography of Ta'ang areas that did not resonate with members of the rebel movement.[81] On the other hand, it was lamented that in the zone there was "absolutely no self-determination." People had to "follow the rules of the central government. So, only in the name self-administration was granted, but no power had actually been transferred."[82]

The shift of the new Ta'ang army's name away from "Palaung" to "Ta'ang" National Liberation Army constituted a dual political move—a move of both contestation and reproduction of the *taingyintha* governmental regime.[83] The name *Ta'ang* expressed the idea that the population body, and political geography, that the armed force should free from all forms of oppression (not only state, but also other rebel polities oppression) were different from, and broader than, the social and geographical limitations established by external identifications and impositions. In this attempt at rearticulating the political geography of the Ta'ang polity there was one specific concern: the inclusion of those communities living far away from the P-SAZ. Throughout decades of armed conflict, Ta'ang communities in southern Shan, for example, had been particularly affected by waves of violence and were now taken as the epitome of the denial of Ta'ang self-determination. The term *Ta'ang* contested the exonym that, as mentioned earlier, Bamar authorities have used, not only to identify a Palaung ethnic population but also to construct it as composed of three main subethnic groups—Shwe ("gold" in Burmese) Palaung; Ngwe ("silver" in Burmese) Palaung; and Rumai (alternatively called "black Palaung" in Burmese).[84] The use of the term *Palaung* has become increasingly contentious due to the exploitation of such exogenous identifications by Bamar state governments in order to manage ethnic populations. *Palaung* has been used to locate the exogenously identified ethnic group under other "major" so-called *taingyintha lumyo*, such as the Shan, for the purposes of incorporation into the Myanmar polity (TNI 2014, 16; Cheesman 2017). The choice of *Ta'ang* as a name was connected to the creation of the army as part of broader nation-building projects of Ta'ang political and economic elites. These involved the activities of the association Ta'ang Literature and Culture Committee concerning the creation of a common script and the possibility to unify different Ta'ang dialects.[85] In the same years, after the 2008 constitution, the main CSOs changed their names as well, using Ta'ang rather than Palaung.[86] Once again, the change of name represented an attempt to reshape collective identities that were lamented to be manipulated by state authorities, as a TWO member notes: "We have only one single name, that is Ta'ang, and it is the Burmese that divided us into Shwe and Ngwe, golden and silver. . . . So there are not three main subgroups of Ta'ang, those are Burmese identifications. If we say Palaung that is not representative of all Palaung. If you say Palaung the people will ask you, are you Shwe or Ngwe? But these are Burmese identifications.[87]

Bodies

After the 2009 formal establishment of the TNLA occurred at the Thai-Myanmar border provinces where the PSLF was active (Mae Sot, Mae Hong Son, and Chiang Mai), a complex combination of the rationalities of ethnonationality, narcotics eradication, and humanitarian security inflected the recruitment process to boost the ranks of the armed force in the Shan borderlands. Throughout 2010, the front initiated a series of recruitment drives that brought the secretary general and the chairman of PSLF, Tar Bong Kyaw and Tar Aik Bong, to carry out advocacy activities and meetings with local communities in northern Shan's Ta'ang areas. Recruitment revolved around four main techniques to collect and manage the bodies necessary to form the TNLA. These techniques also reproduced Ta'ang land as a vital space for the self-determination of the polity.

One. The PSLF aimed to dismantle village community militias created by the Sit-Tat in the previous two decades, reinsert Ta'ang militiamen into the TNLA, and secure civilian communities. In this sense, recruitment embodied the reterritorialization of zones conceived as the frontiers of Ta'ang land—particularly the areas of southern Namkham, Maiwee/Minewee, Manton, Namhsan, Namtu, and the western portions of Kutkai and Hsenwi—where the Sit-Tat had expanded village militias. Larger pro-regime militias and their poppy fields were also heavily targeted with a view to recruit among the Ta'ang farming communities working in these frontier spaces.

Two. Disrupting poppy production was not the only way in which recruitment was informed by the idea of eradicating narcotics. Recruitment drives in northern Shan relied heavily on the press-ganging of drug addicts into military service with the TNLA. Deploying an antinarcotics version of the Sit-Tat's four-cuts strategy, TNLA troops rounded up villagers in order to cut off means of communication and sustenance to drug addicts in the area. Once forced into withdrawal, addicts could be identified and then forcibly recruited into the PSLF's drug rehabilitation program. Besides providing a labor force that could be immediately deployed in military support work, such as construction and maintenance of military camps, press-ganging drug addicts served to identify "deviant" subjects deemed to affect the Ta'ang polity's body. These subjects, from the TNLA's perspective, needed to be rehabilitated and then reinserted into society.

Three. The logic of forming the TNLA to shape the social base of the movement informed a "one Ta'ang household, one Ta'ang body" recruitment policy. Ta'ang families were turned into the recruitment base of the armed force. A system of exceptions was established so as to allow households to continue sustaining themselves after the removal of valuable workers. At the same time, this recruitment practice embodied the PSLF's conception of the Ta'ang polity. Recruitment was strictly limited to Ta'ang individuals, a guideline that PSLF/TNLA's apparatuses have continued to follow, and one that the leadership prides itself on. Although the act of recruitment per se would not reify the ethnic identity of recruits that remained a fluid construct, this policy distanced PSLF's idea of an ethnonational polity from the vision usually proposed by other ethnonational rebel movements. From the perspective of the PSLF, while Ta'ang land includes other ethnonational communities too, the governing apparatuses should not unite them and subjugate them under the Ta'ang polity—for example, via recruitment into the armed force. Instead, different Shan or Kachin ethnonational postures have often understood Shanland or Kachinland as made of a tapestry of communities that belong to the Shan or the Kachin and are to achieve unity, autonomy, and harmony within the Shan or Kachin struggle (Ferguson 2021). Yet ethnonationalism remained at the core of the recruitment practices. The PSLF/TNLA did not refrain from targeting non-Ta'ang individuals in its antinarcotics programs and exploiting them in some form of labor for the TNLA. Yet, at the end of the "rehabilitation" process non-Ta'ang individuals were not recruited into the army. While the project of freeing the Ta'ang geo-body from drug addiction and making it a drug-free space for different communities was carried out by exploiting social and class distinctions, recruitment and combat service were harnessed to build the Ta'ang population body at this stage.

Four. To authorize and boost the household-based recruitment endeavor, the PSLF geared its enlistment support campaigns toward women. In a socioeconomic environment in which narcotics consumption had particularly impacted the male population, in many villages women had become the actual pillars of the household and could be leveraged to authorize recruitment drives. Rebel discourses argued that Ta'ang women personified key traits of the Ta'ang people more broadly. By crafting the figure of the responsible, money-saving wife and the gentle, child-rearing mother, they mobilized women's consent to household-based recruitment to legitimize the formation of a Ta'ang armed force. The voices of women were instrumentalized and mobilized to rally support for the rearmament effort as a response to the narcotics and ethnonational issue, while their unpaid labor and contributions were mobilized in service-delivery roles.[88] Ultimately this technique presented recruitment as a practice to liberate Ta'ang land from drugs by freeing Ta'ang women, households, and village communities from drug-related impacts. Thus, governing recruitment embodied the polity as an entity with a specific political geography by linking different localities to each other and relating them to a single Ta'ang territory.

Weapons

Rationalities and techniques of governing recruitment were strictly intertwined with specific modalities of access to weapons. At the time of the official constitution of the TNLA, the PSLF was struggling to acquire weaponry. Throughout 2008 several official letters were sent to the PSLO/A's former allies—the KIO/A and the SSPP/SSA-N—requesting support, but these remained unanswered. The political relations between the Sit-Tat and EROs in Shan State and Kachin were growing increasingly tense, and the politics of rebel movements was in agitation (Ong 2018; Brenner 2019), but for the time being they still maintained ceasefire agreements. Even though PSLF was willing to re-create an armed force, it lacked solid connections in northern Shan State and the financial means to intercept potentially available weapon sources, training, and supplies.[89] Yet, connections and money were not the only dimensions. As touched on in the previous chapter, the geographies of weapons and support acquisition had shifted in the second half of the 2000s. Access had become more difficult at the Thai-Karen and Thai-Shan borderlands, while the UWSA had managed to acquire stocks decommissioned in Yunnan and to progressively set up manufacturing and assembling capacity. In addition, the whole process suffered a setback when Tar Nyi Lone, the commander tasked with the formation of a new army, died in 2008.

For the very first steps of the TNLA, the organization relied on established connections through which PSLF had been able to maintain a small underground armed branch. The TNLA project in fact had not really been the first initiative to shape an armed force for the Ta'ang front. After its formation in 1992, the PSLF was able to source military training from Karen and Wa rebel movements. While the PSLF was based at Manerplaw first and KNU areas later, a small armed component—at that time referred to as PSLF-Force—was hosted in bases of the Wa National Organization/Army (WNO/A), along the border between Myanmar and Thailand's Mae Hong Son province.[90] Such connection was forged in the ambit of the NDF's Manerplaw headquarters, as WNO/A was also a member of the umbrella organization. In 2005, after PSLA's disarmament, the Ta'ang front arranged to bring some recruits from northern Shan on foot to the WNO/A's Thai-border base for training.[91] These were hectic decades, with the movement focusing on shaping a new generation of leaders and members, by trying to capitalize on the social context in which it was embedded. Flows of Ta'ang on the move had continued to arrive at the Thai border since around the second half of the 1990s. Here possibilities were emerging for exchanging different kinds of resources with different movements. Tar Bong Kyaw, for example, in Mae Sot and Chiang Mai, could attend English classes and take part to NGO-driven programs on transitional justice,[92] foreign affairs and diplomacy, and human rights and humanitarian law, with exchange visits that would bring him also to South Africa.[93] Nonetheless, it remained crucial but extremely difficult for the PSLF to establish access to weapon sources in Ta'ang areas of the Shan State borderlands.

During the first year of TNLA operations the PSLF's leadership politicized the rationalities that informed the recruitment drives just discussed in order to build consensus among disgruntled PSLO/A members and tap into weapon sources they controlled in northern Shan. The roughly five hundred men disarmed and disbanded in April 2005 had experienced three main life trajectories. Some of the PSLA's rank-and-file and commanders had joined or had been recruited by other EROs. Others had enlisted in militias constituted by the army, either village community or nonintegrated ones.[94] And yet others had decided to stash weapons away, remaining in northern Shan and avoiding involvement with armed groups. The articulation of a Ta'ang armed force as a project to delineate a Ta'ang polity spatially exceeding the status quo enshrined by the 2008 constitution reawakened the political aspirations of PSLO/A's former officers that had been deeply frustrated since the 1991 ceasefire. To support the TNLA project, they granted access to weapon stocks they had stashed away during the 2005 disarmament process. Moreover, they facilitated attempts to remobilize former PSLA soldiers, some currently inactive and others defecting from militias and other EROs, who would bring their weapons along.[95]

A key turning point for the acquisition of weaponry, though, was represented by a shift in the political and military posture of different EROs linked to a series of broader developments. The sanctioning of the new constitution in 2008 had provided the Sit-Tat with a legal and political argument for justifying its more aggressive policies vis-à-vis EROs throughout the borderlands in the following years. The military regime aimed to try to disarm EROs to then integrate them into the structure of the armed forces through its so-called Border Guard Forces (BGF) and People's Militia Forces (PMF) programs. The military regime argued that the promulgation of the 2008 constitution, as a result of the 2003 Roadmap to Discipline-flourishing Democracy, had led the country to democratic federal rule. Thus, this new constitutional arrangement rendered rebel movements redundant (Huang 2016, for a view of the military's version of democracy; Jones 2014). Linked to constitutional provisions, the BGF and PMF programs entailed the transformation of all rebel movements and armed forces in the country into paramilitary and militia groups integrated "under the command of the Defense Services."[96]

The long-standing issue of "one state, one army" and the Myanmarization of the armed forces of the Union, as well as weapons production and acquisition, gained central stage once again. The BGF and PMF programs remained particularly problematic for many EROs, due to a de facto imposition of disarmament and forced integration before any peace negotiation. With the general elections looming at the end of 2010, tensions increased between the Sit-Tat and rebel movements all around Ta'ang areas of northern Shan. Especially targeted at this point were KIO/A, the Kokang MNDAA, and the UWSA.

KIO/A formulated some counterproposals based on the logics of ethnic-based military units that would be incorporated within the Union armed forces rather than complying with the scheme proposed by the Sit-Tat, which entailed the insertion of military officers in BGFs' battalions, their coming under the logistic and command-and-control structure of the Sit-Tat, and the individual incorporation of soldiers (Sakhong 2014, 129–132).[97] Such proposals were justified on the basis of the unmet promises of Panglong 1947 and mimicked the structure of the armed forces established with the Kandy Conference that had subsequently been manipulated by Ne Win's regime (Sakhong 2014). Meanwhile, in August 2009, the Sit-Tat targeted the MNDAA in Kokang on the pretext of searching what was believed to be a drug refinery in Lao Gai (Kokang). Although the alleged refinery turned out to be a weapons repair factory maintained by the MNDAA, the Myanmar police issued arrest warrants for different MNDAA members (Keenan 2014b, 147). Days later, with the Sit-Tat about to carry out the 2009 Kokang offensive, a coalition of EROs in Shan State issued a statement stressing the illegality of the warrant and the liceity of the MNDAA's weapons refurbishment activities.[98] It was argued that ceasefire EROs, as entities officially recognized by the Union government, have a right to maintain workshops and factories to repair weaponry.

With Sit-Tat military offensives unfolding against the SSPP/SSA-N in central Shan State right after the 2010 general election, and KIO/A also coming under attack in northern Ta'ang areas in 2011, the Kachin and Shan rebel movements eventually agreed to support the formation of the TNLA.[99] The SSPP/SSA-N and KIO/A now looked at the Ta'ang front as a potential buffer and ally.[100] The Shan rebel movement provided financial support and donated a few pieces of weaponry. The SSPP/SSA-N had stockpiled weapons in the headquarter area of Wan Hai, just across the Salween River in central Shan State, with caches replenished over time through linkages with the UWSA.[101] UWSA was also becoming more inclined to support TNLA. In fact, the Sit-Tat was progressively encroaching on Wa areas and at the same time offensives against the SSPP/SSA-N were highlighting the precarious position of the Shan ERO as a buffer to the western banks of the Salween River for UWSA (Keenan 2013, 90, 135–137). Furthermore, the other main Shan group, the RCSS/SSA-S, with which the UWSA has historically maintained conflictual relations, was undertaking ceasefire negotiations with the Tatmadaw.

Meanwhile, in 2011 a first batch of fifty prospective officers was sent for training at KIO/A Third Brigade's bases, while others were trained with KIO/A's Fourth Brigade.[102] At the end of their training, TNLA soldiers traveled back with some older KIO/A surplus weapons. Since 2010 KIO/A had started to manufacture and assemble assault rifles and other weaponry at a production site located in Laiza, and these were now supplied to the TNLA.[103] As an interlocutor recalls talking about the breakdown of the 1994 Kachin ceasefire in 2011:

> This production site/factory started to manufacture arms only in 2010, that is why many civilians and some parts of the KIO leadership were furious with previous leaders, since they had neither planned to build the factory before, nor had purchased and stocked ammunition in advance. Once, we had the chance to meet and talk to the KIO chairperson who told us that in 2011, by the time the conflict had re-escalated, the KIA had only three thousand bullets in their military stocks. That is why the KIA started to consider the possibility to set up a weapons factory.[104]

The environment of extractive capitalism and infrastructural investment that characterized Ta'ang areas of northern Shan in these years offered various opportunities for the PSLF to obtain revenue through which it financed the formation of the TNLA. The TNLA was now in a position to request payments for granting access to business actors (or assurances of noninterference) (Meehan 2016a, 380). In particular, since late 2010 Chinese companies initiated the construction of the portion of the Shwe Gas and Oil Pipeline traversing Shan State, which fully intersects Ta'ang areas (TSYO 2012).[105] Eventually completed in 2014, this and other projects of road and railway network expansion were being carried out during the years of initial growth of the TNLA.

Between the end of 2011 and the beginning of 2012, the movement organized to bring back to northern Shan the troops and weaponry that had remained at the WNO/A bases at the Thai-Myanmar border. PSLF/TNLA soon launched a "war on drugs" in Ta'ang areas that led to clashes with militias involved in the narcotics trade, during which weapons, narcotics, and money were seized. The ERO enjoyed widespread local support and grew rapidly in this period, passing from a few hundred to 1,300–1,500 troops in 2013 (Keenan 2013, 124–125).[106] As the TNLA's capabilities evolved in the subsequent years, the movement was increasingly able to exert influence over the Myanmar-China overland trade routes and levying taxes on traders, transporters, and businesses. Nevertheless, attempts to expand its reach by establishing taxation and administrative nexuses beyond the Namhsan-Manton-Namtu triangle encountered the opposition of Kachin People's Militias and were met with more caution by KIO/A. By the end of 2014, the latter—up to that point the Ta'ang rebel movement's main source of support—scaled back its military assistance. This, however, did not imply a misalignment of the two EROs,[107] as a former TNLA soldier recruited at this stage noted: "We [TNLA] would always coordinate with them [KIA], yet the main source of arms was the UWSA, as arms were purchased mostly from them."[108]

While the relations with the SSPP/SSA-N and KIO/A were certainly key to acquiring weapons and training in the early years, the linkages with the UWSP/A proved decisive in a second phase. The Wa polity increasingly favored the strengthening of the TNLA when the RCSS/SSA-S started to expand its influence in Shan State, after signing a ceasefire agreement with the Sit-Tat at the end of 2011. The ceasefire with the Shan ethnonationalist armed group was part of a larger peace initiative by the Thein Sein's government (2011–2016).[109] Since late 2013 various rebel movements initiated a coordination process to negotiate the draft of a Nationwide Ceasefire Agreement (NCA) with the Sit-Tat and the government (TNI 2018, 41).[110] Nonetheless, such consultative processes were accompanied by military offensives and fighting throughout the country, from Rakhine to Kachin and Ta'ang areas. In 2015 the Sit-Tat refused to accept the TNLA as well as the MNDAA and the AA, on the grounds that the three EROs had been constituted after the promulgation of the 2008 constitution. Such refusal not only divided EROs throughout the country on whether to sign or not a non-all-inclusive "nationwide" ceasefire. It also pushed the TNLA, MNDAA, and AA closer to the UWSA and the Mongla NDAA. The latter two, since the very beginning of the negotiations, had expressed no interest in joining the process while acting as observers.[111] In late 2015, right after the signature of the so-called nationwide ceasefire by only eight EROs concentrated mainly in the southeast of Myanmar, the Sit-Tat launched harsh attacks on the SSPP/SSA-N's headquarters in Wan Hai, central Shan State. The SSPP/SSA-N moved its troops stationed north of the Mandalay-Lashio road back to Wan Hai. Yet, as the offensive began to wane, the Sit-Tat facilitated the RCSS/SSA-S, one of the NCA signatories, to move its troops from southern Shan to areas north of the Union Highway in Kyaukme, Hsipaw, and Namtu. Areas that, up to then, had been under the influence of the PSLF/TNLA and the SSPP/SSA-N.[112]

From Ta'ang perspectives, the expansion of the RCSS/SSA-S triggered various issues. Between 2016 and 2021, the presence of the Shan ERO generated further conflict with annexed displacement and violence, adding to fighting with the Sit-Tat. It also heightened pressure on local communities, at times creating friction among Ta'ang and Shan populations, and fed discourses about the "Shannization" of Ta'ang areas in part related to different forms of oppression experienced by Ta'ang minorities in southern Shan, often at the hands of the RCSS/SSA-S and the MTA before it.[113] The expansion of the Shan ERO was understood as part of older attempts to settle into Ta'ang areas and of a larger project entailing the imposition of Shan language and traditions on other ethnic populations.[114] In addition, for the RCSS/SSA-S this move allegedly aimed to establish a base area in the Shweli River valley, close to the China-Myanmar border. This represented a foothold from which the group could obtain economic concessions and cooperate with military-private business partnerships as security contractor.[115] From here, it was said, the RCSS/SSA-S would also expand its narcotics production and trade toward China's and India's border areas as well as the Hpakant mines, while also supporting the arming of the Shanni populations in southern Kachin and Sagaing.[116] For the Ta'ang rebel movement, such a scenario offered a further element to build up its political connections with the UWSA and strengthen the TNLA. In the following years the TNLA negotiated access to heavier military equipment (mortars, heavy machine guns, 107 mm rockets, RPGs, and eventually MANPADS). In 2018 it openly established a liaison office in the UWSA's capital Panghsang, sent more troops for training to Wa areas,[117] and cemented its relationships with EROs close to the UWSA.[118]

Access to weaponry produced and smuggled by Myanmar's largest ERO was shaped by the rationalities of ethnonationality and narcotics eradication.[119]

The PSLF progressively politicized the long-standing political connections between Ta'ang and Wa rebel organizations as the natural result of ethnic kinship between the two populations belonging to a larger Mon-Khmer ethno-linguistic nexus shared by Ta'ang, Wa, and Mon nationalities. Concerning the UWSA in particular, these relations dated back at least to 1995, when the Wa ERO had joined the Tatmadaw crackdown on Khun Sa's MTA.[120] Answering a request for support, PSLO/A had sent a contingent of troops to UWSA. After the disbandment of the MTA, this delegation of PSLA soldiers had settled down in what would become the UWSA Southern Command in Shan State.[121] Most important, the Ta'ang front could leverage a shared understanding of the ethnonational cause as a struggle unfolding at different scales: not only a struggle vis-à-vis central Bamar military authorities and state institutions, but also against the attempted assimilation of so-called minority within the minority ethnic polities via a Shannization process enacted by Shan armed groups (see also Ong and Prasse-Freeman 2021). Moreover, the two EROs maintained a shared political view concerning the necessity to carry out a "war on drugs" in the Shan borderlands. Between the late 1990s and early 2000s the UWSA had been one of the first EROs to adopt a ban on narcotics production and trafficking (Kramer 2007, 24).

Ultimately, as we will explore more in depth in the next chapter, the governmentality of military means—informed by the rationalities that run throughout the whole process of ceasefire-disarmament-rearmament—not only made it possible to access weapons and form the TNLA. It also constituted a political arena to produce a political geography of territory in the borderlands. The "war on drugs" of the PSLF/TNLA and the UWSP/A, for example, was linked to a shared vision of the nexus between territory and ethnicity. Both rebel movements promoted the sensitive idea that the political geography of Shan State should be redrawn according to a criterion of proportionality between ethnic minority population density and territory. Commenting on the issue of the territorial extension of a future Ta'ang state, PSLF's first secretary general, Tar Bong Kyaw, briefly explained this point during an interview with the *Frontier* in 2021, noting: "We [PSLF/TNLA] are not greedy—we are not demanding control of the cities, that will just lead to endless clashes. We are just demanding the regions where the majority of the people are Ta'ang" (Frontier 2021). But in what ways was/is the territory of the rebel polity reproduced by governing military means at the edge of the state?

FIGURE 4. "War on Opium," courtesy of Vincenzo Floramo.

TERRITORIES IN TA'ANG LAND

Every evening around 4 p.m., at the end of a working day that would usually start more or less eleven hours earlier, the offices of the PSLF/TNLA's "temporary mobile" headquarters emptied out. Soldiers and officers would take off military uniforms, put on their sportswear, and run in the direction of the soccer field and volleyball courts featuring bamboo goalposts and nets. Dug into the rugged terrain of the military base, the sportsgrounds were one of the main reasons why the PSLF/TNLA's headquarters cyclically returned to set up base here, Sit-Tat military offensive after Sit-Tat military offensive, displacement after displacement.

And as in other borderland rebel military bases, the courts dictated not just their own rules but their style codes too. Adidas- and Nike-branded jerseys of famous Thai and European professional soccer teams dominated the scene, customized on the back with iron-on vinyl decals bearing slogans rather than names: a navy blue Buriram United shirt with "Tea Land" on the back, "Free Tea Land" on top of some number tens; a bright red Liverpool jersey reading "Ta'ang Land" or "Free Ta'ang Land" on the backs of others.

Interpreting it as a metaphor for a self-identified ethnonational geography, due to the reference to the Ta'ang heritage of tea production (Dunford 2024), during an interview with a high-ranking officer following one of those after-hours sporting rituals, I had used that very term—"Tea Land"—to ask about the PSLF/TNLA's territory, only to be promptly scolded: "We do not call it Tea Land, just Ta'ang Land."[1] Although tea is politicized as a social and territorial element of the land that symbolizes the geographies of vital space of the Ta'ang communities, normalizing the expression "Tea Land" to refer to the territory of the PSLF/TNLA would be slippery. Saying "Tea Land" would shrink the ethnonational territory to just the areas where the plant is cultivated in what is instead seen as a larger "Ta'ang Land." The soccer jerseys, with their slogans, embodied all the fluidity of the politicization of biological elements (e.g. tea) to define a political-geographical space. In the discrepancy between "Tea Land" and "Ta'ang Land" lied a sea of territorialization practices that the PSLF performed by governing its armed force, the TNLA.

If in the political process of acquiring weapons and rearming the Ta'ang resistance movement there was also the delineation of a Ta'ang polity, with its biological population body and geography of vital space, as we saw in the previous chapter, then in what ways was (and is) the territory of the rebel polity produced? Territory is inflected in and reproduced by technologies and techniques of governing the relations between weapons and humans (Buscemi 2021a, 2022). The military uniform—for example, that uniform that TNLA soldiers and officers would take off to change into their jerseys at 4 p.m. every day—embodies a form of territorialization. The uniform, with the techniques to wear the garment that are part of it, is a technology that regulates and governs the relationship between people (individuals and collectives) and weapons (be it the very firearms that soldiers hold, or the soldiers as weapon-human technologies themselves). And by governing such relationship, the military uniform of the TNLA and related techniques to wear it contribute to reproducing a Ta'ang polity materially and symbolically. Different from soccer jerseys, the uniform turns the body with a stick and a gun immersed in a sloping field in figure 4 into a TNLA soldier busy slashing poppy flowers to eradicate the genocidal plague of narcotics from Ta'ang Land. Via the soldier, a Ta'ang territory of the Ta'ang polity is produced. In constituting and maintaining the TNLA, the PSLF did not only shape a polity with its ideal vital space, it produced its territory too.

Yet, in Ta'ang Land several territories of several rebel polities overlap, intersect, and at times, nest. Let us now see how. This chapter explores the ways in which the governmental assemblages managing military means produce territories with their boundaries and frontiers. It looks at how the technical objects, technologies, and techniques of managing the relations between weapons and the polity with its biological population and territory have been inflected by rationalities of acquiring weapons and forming the TNLA, which were exposed in chapter 3. To do so, the chapter explores disparate dimensions of governing, looking at military uniforms; strategies of narcotics eradication; recruitment and service; military infrastructures; weapons stock-piling; village militias; explosive devices control; and the management of the social effects of military means.

Uniforming Ta'ang Land

Upon rearmament in 2009, the PSLF adopted a military uniform that included a specific hat and badges for the new military force (figure 5). Military manuals and training routines formulated a specific doctrine that recruits would be instructed about concerning the uniforms' components and their meaning—including the hat and its details, as well as the badges and their details.

Hats

The PSLF's uniform hat was almost identical to that in use among the KIO/A, a detail that suggested the existence of common supply chains. Not only weapons but also militaria more broadly circulate among rebel movements in the

FIGURE 5.   UWSA and TNLA officers, November 2013. Courtesy of Thierry Falise.

borderlands. Yet, the cap clearly differed from that of other rebel forces due to a band crossing its upper panel transversally (figure 5). The TNLA's new hat drew heavily from the model that had been in service with the PSLO/A[2]: some pieces came straight out of leftover stocks, while others had been newly manufactured based on the same model.

The band connecting the left and right sides of the hat through the upper panel symbolized the symbiosis between the Ta'ang people and the armed force (figure 5): joining hands together, the people constitute the TNLA as much as the TNLA constitutes and preserves the people. Running along the circumference of the crown (where the soldiers would fit their heads), another band reinforced this role of the army as a machine that delimits and secures Ta'ang spaces and populations, like a band surrounding soldiers' heads (figure 5).

Besides these two bands, the hat's crown was made of a single cotton panel cloth folded back on itself six times. Giving coherent shape to the cloth and structuring it into a hat, this set of six folds was taught of as a metaphor for the arduous process that had led to the formation of the TNLA. The breakaway of the PNF, the first Ta'ang armed force, from the Shan State Army in 1966, and later its restructuring into PSLO/A under the leadership of Captain Kham Thaung and its Brigade 6, have been constructed as key passages not just in the formation of the TNLA but more broadly in the ethnogenesis of a Ta'ang nation and polity. At the end of 2016 the PSLF/TNLA General Staff Office renovated the gravestone of Tar Khun Aye—one of the three Tawngpeng *sao hpa*'s relatives who had participated to the founding of PNF and its Battalion 5—and transformed it into a war memorial in Pansali village.[3] Recounting the story of Battalions 5 and 6 at length, K.—the TNLA officer we met through the visit to Pansali village the book opened with—connects the memorial grounds with the hat of the TNLA combat uniform: "The story of Khun Li and Khun Aye provides us with direct experience that you cannot stand alone: Battalion 6 in fact was allied with the Kachin and therefore they were able to stand. Today, in our hat . . . we have six folds: this is used as a symbol to show our connections to the old Battalion 6 of the PNF and PSLO/A. We are now the direct descendants of Battalion 6."[4]

The TNLA combat uniform's hat, together with the training manuals and techniques as integral elements of the uniform, is part of a technology that connects and bonds the armed collective with Ta'ang populations and previous resistance movements across space and time. The act of wearing the uniform produces a Ta'ang polity and its territory. But this occurs in participation with related techniques and performances. For example, discursive and material practices of ethnonational security in Ta'ang Land are reiterated each year during the PSLF/TNLA celebration of National Resistance Day on January 12, the date marking the 1963 creation of the PNF. Similar to many other EROs in these borderworlds, the PSLF/TNLA holds parades on that day. The public display of weapons and uniformed troops plays a key constitutive role in light of the organization's objectives and relations with Ta'ang communities throughout a Ta'ang Land. This performance has not only a symbolic and deterrence value but also materializes the ways in which governing the means of violence makes a territory governable. As a civil society representative stated, speaking in a Ta'ang "liberated area" on the occasion of the movement's fiftieth anniversary in 2013:

> I . . . warmly welcome the reentering of TNLA into the track of national freedom movements. The Ta'ang army and the Ta'ang people resemble water and fish—the river without fish will be worthless. It means that the army cannot stand by itself if there is lack of people supporting, and the people will be insecure without the army. I, the Ta-ang people, have a clear picture that the army is needed in the current time of insecurity and threats posed by narcotic drug smuggling and consuming.[5]

The hat, the uniform as a technology, and the materialization of its scale-changing properties do not remain confined to the ambit of the PSLF/TNLA but circulate into villages and households via various media and artifacts and the interpretations that different audiences give of them. The circulation and diffusion of these techniques, though, is not to be understood in absolute terms either. During fieldwork in Kyaukme and Namtu townships the houses of Ta'ang village heads and villagers would often display copies of the PSLF/TNLA's promotional calendars—some up to date, others going back a few years. Entering the house and seeing a calendar on the wall depicting TNLA troops in full combat gear carrying out military operations or military medical staff absorbed in their tasks, I would initially take that as a paper-printed display of sympathy vis-à-vis the TNLA and its political vision. Yet, the calendars would often also be just presents distributed by the ERO to all the households in the area, presents that people did not necessarily care much about but which had to be accepted.[6] Cared for or not, the calendars were hanging there on the wall, sometimes even a few years down the line, as a potential reminder of the work needed to sustain the reproduction of an ethnonational territorial scale through uniform hats, war memorials and village entry gates, military parades, or promotional calendars that all act as markers on and of a Ta'ang polity's territory.

Badges

FIGURE 6A AND 6B.   PSLF/TNLA uniform's left-arm badge (above); Tar Aik Bong (PSLF's chairman) and Tar Bong Kyaw (PSLF's first secretary general) (below). Author's fieldwork, courtesy of the PSLF/TNLA.

But what is the territory of Ta'ang Land? What are the polity's Ta'ang spaces and people? The badges of the uniform that the PSLF adopted at the time of rearmament offered some coordinates (figure 6). The left-sleeve badge depicted a three-peaked mountain range in shades of green against a light blue background, nested between two tea-leaf branches joined by a tea flower at the bottom center of the image. Towered over by a stylized red sun, the mountains, tea leaves, and tea flower in the badge represent the hilly regions of northern Shan State were Ta'ang communities have long been commercializing tea. They depict the production of tea as a historical socioeconomic legacy defining the Ta'ang polity. Yet the three peaks and the stylized sun towering over the composition enlarged the scale of the PSLF/TNLA's territory and rejected state-framed scales of Ta'ang areas as restricted to the Palaung SAZ confined to Manton and Namhsan townships. The three-peaked mountain range stood in for the three macroregions of the borderlands that are inhabited by Ta'ang populations: northern and southern Shan State, the Myanmar-China border areas, and the Myanmar-Thai borderlands. The red sun instead aimed to provide an element of cultural coherence among these borderland archipelagoes. In fact, it acted as a reference to a Ta'ang cosmology shared by different communities across those three macroregions, according to which the Ta'ang were originally born out of the encounter between father sun and mother dragon, who eventually produced the egg containing the very first human beings.

The idea of a biological symbiosis between the armed force as a body, in its individual and collective forms, and the polity's population and territory was also embodied by the other badge. The right-sleeve badge displayed two Ta'ang traditional swords, named Boh,[7] against a red background with the acronym TNLA underneath (figure 6). The training manuals detailed how the two Boh swords represented the Uknown Soldier of the Ta'ang polity. By conveying the meaning of sacrifice for the higher goal of the defense of the polity, the weapons in the badge materialized the refusal to lay down the polity's weapons and disarm at any cost—even if this would mean to fight with traditional swords only. As key components of the camouflage service uniform, the two badges helped to mold the single bodies of recruits coming from disparate parts of Ta'ang land into a single ethnonational body with its territory space.

The materiality of TNLA's uniform should be read in terms of the properties, qualities, and "powers" that are activated by the act of wearing it. Through a purposefully stretched analogy with the insights produced by the anthropologist Viveiros de Castro about the role of ceremonial suits in rituals and practices performed by Amazonian populations, it can be pointed out how the act of wearing a uniform is not a matter of concealing the body—or at least not only—but one of "wearing the dress," meaning an activation of the properties of the suit/uniform (de Castro 2012). This analogy allows a change of vision to look at what the uniform does, rather than what it conceals. A military uniform does not only camouflage in order to conceal—to conceal and push aside the subjectivity of the person wearing it, as well as to conceal the body during service and combat. The uniform also uniforms bodies in specific manners. It uniforms bodies into a new subjectivity and new collective that upholds specific sociospatial coordinates. Wearing the uniform activates potential identities or identity-shaping emotions and affections that are articulated in induction procedures and codes of conduct instructing recruits about the uniform and its historical and spatial qualities and meanings vis-à-vis the Ta'ang polity. The TNLA's hat and badges, and the different components of the uniform, in combination with military doctrines and disciplinary techniques, function as a technology to shape the bodies of single recruits coming from different areas of Ta'ang Land. The hat, badges, and uniform are activated by the act of learning about and wearing the uniform. They contribute to reshape bodies, households, and villages into a Ta'ang ethnonational body, territory, and territorial scale distinct from state-framed categorizations of "Palaung" populations and state-mapped "Palaung" territories (within Shan State, within Myanmar). Via the uniform, this bio-geo-political body could be depicted and performed as both distinct from and broader than military state delimitations, beyond any zone of autonomy or hierarchical definition of the Ta'ang as a bio-geo-political body within Shan State.

If the act of wearing the uniform—as a technology that regulates the individual and collective body at its interface with weapons—has meaningful effects, the opposite is also true: *not* wearing it also contributes to reproduce the ethnonational polity with its territory. The hat and badges are not part of the "uniform" worn by drug addicts pressed into military service with or by the TNLA. Different from any other PSLF/TNLA member, drug users are shaved and provided with different (sometimes improvized) uniforms made up of civilian tops and decommissioned dress- or field-uniform pants. And differential clothing matches with differential accommodation, as they are housed into so-called Drug Fighting Centers—oftentimes consisting of a separate, downhill section of TNLA's military camps. "Rehabilitation" in these Drug Fighting Centers entails a mixture of military discipline, military-like drills, training (rigorously without weapons), and "community" social services structuring the days of the guests/inmates. While "community" here essentially ends up being the rebel movement itself, more than Ta'ang civilian communities, the logic behind this "rehabilitation" has to do with the idea of "cleaning up" drug addicts from Ta'ang society and reshaping Ta'ang Land into drug-free bodies, populations, and territories in opposition to the genocidal/oppressive territorialization strategies of the Tatmadaw and related militias spreading narcotics.

Violence is codified into military techniques to discipline the body of the addict that function through (and in combination with) the shaved head and the uniform that is worn (or not worn), the material and social architecture of the camp, as well as the acquisition and management of weapons. The uniform, as a technical object made of such logic and techniques, contributes to shape bodies as drug-addicted bodies that should be transformed into drug-free bodies—and eventually drug-free households, villages, communities, and territory. As people of different ethnicities are subject to the "rehabilitation" program and issued (or one should say *not* issued with) the uniform, by shaping those bodies the uniform-as-assemblage reproduces Ta'ang Land as a territory comprising various ethnonational communities, households, and families that have to be cleaned up from narcotics. In part, such rationalities and techniques come from and recirculate into families and households, communities and CSOs, as we saw in the previous chapter. Thus, not wearing the uniform during the "rehabilitation" process functions, by contrast, as a material practice that singles out the "addict" from the polity and its territory to gradually reinsert it into them, and to gradually recompose the polity and the territory (in part via the reinsertion of some of the "rehabilitated" bodies as soldiers into the TNLA.)

Eradicating Narcotics from Ta'ang Land

Since its inception, the PSLF/TNLA launched a war on drugs. The logic of a fight against narcotics was very much diffused throughout Ta'ang communities: "When the TNLA became 'on the ground' in 2012, everywhere the communities asked to work on the drug issue."[8] As analyzed in chapter 2, connected to a series of politico-military and economic conjunctures, the diffusion of narcotics was accompanied by a destabilization of Ta'ang families' social fabric. The issue of drugs circulated throughout Ta'ang societies, framed as an existential problem and as a genocidal technique deployed by the military government.[9] Production, trafficking, and consumption were approached as phenomena requiring a military response, and hence an ethnonational armed force, as a Ta'ang CSO member recollects: "some years ago, let's say at the beginning when it started out, the people supported them [TNLA] because of the issue of drugs and their willingness and policy to eliminate the drugs in the area . . . because the people wanted to be free from the oppression of other ethnic nationality groups and thought that now with our own army we could have been free."[10]

The eradication of narcotics was codified and ingrained in the disciplinary technologies of the armed force. Not just the uniform and by whom, and how, it would be worn or not worn, as we saw, but also codes of conduct and induction procedures. A soldier recruited in the initial years of TNLA clearly recalls how the war on drugs ranked first among the army's objectives: "Instructors and the code of conduct told us what our objectives were. It consisted mainly of three points: (1) the destruction of opium plantations; (2) the struggle against the Tatmadaw; (3) the struggle against the expansion of RCSS in northern Shan."[11]

At its third congress in October 2009, the Ta'ang political front based at the Thai-Burma border had identified its objectives in creating the TNLA: to achieve autonomy, ethnonational equality, and self-determination for Ta'ang nationals as part of a federal democratic union. The war on drugs was one of the political "objects" through which autonomy, equality, and self-determination would be embodied. But like any political object, in order to be performed "within" and "throughout" a Ta'ang Land, the fight against narcotics demanded a territory. In turn, controlling military means and forming the TNLA was more than a tool to achieve political objectives: it constituted a technical process to produce the territory throughout which Ta'ang Land should be freed from drugs.

Cleaning the Body

The antinarcotics logic and techniques unfolded and produced territory at different spatial and geographical scales. Drug addiction was seen as a question of both cleaning the body and cleaning the land. As an issue of individual deviation and lack of discipline, a familial plague disrupting the order of households and burdening Ta'ang women and men, addiction could not be disentangled from the delineation, identification, and targeting of the drug-addict body and community. Throughout 2011, the PSLF/TNLA drafted a policy that was eventually adopted at the 2012 midyear central committee meeting. An antinarcotics committee was formed under the basic principle of "no opium farms, no drug users, no drug sellers." Throughout the rest of 2012 the movement organized meetings with local Ta'ang authorities to collect their views and to "educate" communities about its antinarcotics policy.[12] Outreach initiatives and technical education sessions were held and a program of forced "rehabilitation" of drug users was planned. During the following poppy harvest, between January and March 2013, the PSLF/TNLA issued a series of orders criminalizing the figure of the drug addict, as well as cultivators and farmers under wage labor contracts in poppy plantations.[13]

While renegotiating and building authority through community dialogue, these activities contributed to reproduce drug-addict deviant population groups. Villages were fenced off and isolated for days in order to force out drug users, and then press them into military service for "rehabilitation." As drug addiction has been socialized, at least in part, as a genocidal threat and an "othering" technique deployed either by the Sit-Tat and affiliated militias or by rival EROs,[14] rehabilitation was and is mostly carried out through confinement and military disciplinary techniques. Governing drug users has become, on the one hand, a matter of "cleaning up drug addicts from the society, something the Myanmar authorities do not do"; and, on the other, one of preserving and regenerating the ethnonational social body.[15]

By no means has rehabilitation been a peculiarity of the TNLA, though. The RCSS/SSA-S and other EROs rival each other by instituting rehabilitation activities, at times as a form of forced conscription, at times as a social service to families willing to send a member of theirs.[16] Rehabilitation camps have been set up as part of the architecture of military camps and bases. For instance, up to 2019, Hu Sun, a village northwest to Kyaukme, hosted one of the largest RCSS/SSA-S bases in the area.[17] During a visit to Hu Sun, T., a mountain guide working in Kyaukme town, got stopped by some RCSS/SSA-S soldiers on leave while showing me around.[18] After sharing lunch with them, before moving back to the motorbike, T. reached back to the soldiers with a bottle of Shan rice wine. Recognizing one of them as the son of an influential business family from Kyaukme town who had sent him up to Hu Sun due to drug addiction, he later explained, T. had wanted to make sure to avoid any friction.[19] In other circumstances, rehabilitation camps have been set up as separate structures, sometimes even temporarily, for a month or two.[20]

Inside the camp, violence, military discipline, and hierarchy are not only structural elements ingrained in the creation and maintenance of the material and social articulations of the camp. They are diffused dimensions circulating in the techniques and practices governing the conduct of the camp members. Waging a war against narcotics via both disciplinary and governmental techniques entails a process of attempting to reproduce drug-free bodies and territories. "Attempting" here assumes an important meaning. For antinarcotics rationalities, techniques, and practices are by no means absolute. Many see them instead as problematic actions failing to cope with key problems, like the targeting of traffickers and substances in circulation:"[EROs] arrest and forcibly recruit the users, but in relation to the traffickers they only impose a pecuniary penalty and then the person can be back in business. . . . Another issue . . . is the fate of the seized drugs. What happens to the substances seized? The seized drugs always disappear and are never to be found again."[21]

Cleaning the Land

In the vision of the rebel movement, freeing Ta'ang Land from drugs could not be achieved without modifying broader economic and legal arrangements. While "in the past cultivation [had] occurred in discrete small plots of land in the forest," through land grabbing and labor exploitation the production of opium poppy had now morphed into large-scale plantations managed by "larger farms owned by militia groups under the Tatmadaw."[22] Forming and managing the TNLA in order to unhinge and transform opium poppy frontiers entailed the simultaneous calculation and production of a delimited, coherent, and contiguous territory.

The PSLF/TNLA operated systematic confiscations of poppy farms, plantations, and opium-smuggling hubs, thus reshaping the land and marking zones. Antinarcotics competences rested on three main branches of the PSLF/TNLA—(1) an antinarcotics special force, (2) the TNLA General Administrative Department, and (3) TNLA standard units)—all under the coordination of the antinarcotics committee but with different geographical areas of competence. The TNLA General Administrative Department was tasked with activities in the townships of more stable presence through TNLA standard units, while the antinarcotics special force operated in urban areas or zones bordering Tatmadaw and militia bases.[23]

Through awareness and education programs on the perils of addiction, farmers were incentivized and pushed to convert their fields to crops other than poppies. Vinyl posters schematizing narcotic-related threats, PSLF/TNLA's policies, and their scope of territorial jurisdiction were installed at the entrance of, or inside, villages. Drug seizure operations were combined with periodic public ceremonies, in which TNLA burned narcotics in front of local communities. TNLA's military manuals drafted by the antinarcotics committee codified not only penalties but also the ways in which seized narcotics would be disposed of.[24] For quantities smaller than ten grams of opium, one bottle of heroin, or two hundred methamphetamine Yaba tablets, the drugs had to be burned in front of the owner in a ceremony attended by the local community that would be recorded through videos or pictures archived by the antinarcotic committee.[25] When the seized substances would be of bigger quantities, the drugs had to be stored and later burned during larger public ceremonies organized each year for the celebration of the United Nations' International Day against Drug Abuse and Illicit Trafficking on June 26. In late 2019, a mountain guide working in Hsipaw described these shifts to me, noting how throughout the previous decade "the whole countryside of Hsipaw underwent a profound transformation. Before, this area was covered by poppy fields and opium was widespread here. But then, both the fact that armed groups have started to target farmers producing poppy, especially the TNLA, and in part also the development of tourism have resulted in the conversion of the fields to other crops. Especially corn."[26]

Focused mainly in the mountainous areas of Namkham, Manton, Namhsan, Namtu, Hsipaw, and Kyaukme, these territorialization practices led to conflict when overlapping with militia or other EROs' territorialization practices. In turn, different frontier areas of Ta'ang Land emerged. The areas of Pansay, southern Namkham, and Kutkai townships became liminal spaces of the PSLF/TNLA's antinarcotics processes and practices. To cope with these fluid territorial dynamics, the armed force was structured across Ta'ang Land in mobile headquarters, five regular battalions, one battalion for headquarters defense, and special forces (see also Keenan 2013, 124–125).[27] Battalions were organized by cartographically delimiting Ta'ang Land regions into contiguous districts, different from government ones, in which their mobile bases were located.[28] Ta'ang areas of southern Shan State were also included in these politico-geographical calculations of Ta'ang Land, but they were defined as frontier areas outside PSLF/TNLA's districts, given the predominance of Shan EROs' influence and control there. The relationship between margins and centers of antinarcotics processes and practices was a constitutive element of how the territory of the PSLF/TNLA would be calculated and produced. Talking about rural/urban distinctions concerning the relations between drug-free and drug-affected areas of the polity's territory, Secretary General Tar Parn La touches on this point by reflecting on the results of years of antinarcotics campaigns:

> In our land, where we have the full administration, no more drug users and sellers . . . we still take action every time. Last month we had about . . . ten rehabilitation camps . . . but we have some [drug addiction] in near the towns. . . . Even our people who stay in the village, if they want to use the drug, if they want to use the Yaba for instance, they just go to town . . . so we have that kind of problem, that is why we still have to work in this process.[29]

Managing Military Service

Techniques of recruitment, training, and ensuring the welfare of the armed movement's members constitute processes and practices of managing ethnic bodies and populations that generate territory effects. In a moment of social and political hardship for Ta'ang families and communities, with the constitution of its armed force, the PSLF started to practice recruitment on a household basis. Every Ta'ang household had to contribute to the war efforts with at least one member, male or female, between eighteen and forty-five years old.[30] Different caveats and exceptions to such policy were soon introduced, in an attempt to balance the goal of constituting an ethnonational force that would be tied to every Ta'ang household with the need to preserve families' workforce and livelihoods. For example, households without children and those with only one child and disabled, aging, or ill parents, could be exempted (although sometimes only temporarily, as their duty would be postponed[31]). Yet, exempted individuals could still be involved in service with the so-called Local Guerrilla Force (LGF), a TNLA militia of which we will see more in the next sub-sections.

Recruits have normally been assigned to battalions deployed away from their area of residence and/or recruitment, especially when the areas are under the presence of the Sit-Tat or other EROs. This practice is both inspired by a logic of amalgamation along ethnonational lines of different localized groups and the aim of controlling recruits by uprooting them from local contexts. This is done to counter desertion attempts, which may be incentivized and facilitated in areas geographically more familiar to the recruits.[32] As a village head in northern Kyaukme township notes: "Our grandparents were helping the Tatmadaw as porters, or cultivating land to feed them, or helping them with support of other kind. Therefore, today [the Tatmadaw] have maintained a relation with the village families and people. Often recruits [in TNLA] are sent far from their village—for example, people from Kyaukme area are sent to Namkham—so that they cannot easily flee at least for some time up until they start to know better the area where they are deployed."[33]

Besides creating relationships and bonds between place, troops, and local populations, managing military service and training techniques helps to generate and reshape ethnonational subjectivities and territories in other ways. For instance, by creating possibilities of population movement beyond the rural radiuses of the village and channeling livelihoods beyond the temporalities, routines, and rhythms of related agricultural forms of living. Military service molds a territorial scale beyond such lived experiences and configures a Ta'ang Land that places side by side recruits from different areas. In particular, recruits from "frontier" areas (such as southern Shan), being "rarer" than others, at times end up being treated with more consideration.[34] And "rarer" was exactly the word that a TNLA trainer used (in English) when pointing to a soldier coming from Kalaw township among a group of cadets and explaining to me how he, the rarer recruit from southern Shan, is often given preferential treatment during training. With few people coming from frontier areas, he elaborated, it was important to make sure their experience in the Ta'ang army would be a positive one.

Having a common training language represents an important territorial scale-building medium too. Practicing a single common Ta'ang language, which has been institutionalized as PSLF/TNLA's official and operational one, molds together different Ta'ang communities. For, while one single Ta'ang script has been codified (Kojima 2016), there remain different dialects and not all individuals are fluent in the same.

For some, military service provides a life-changing and personal development opportunity that is realized on territorial scales different from those available through Myanmar state institutions: "In the [TNLA military] camp you can acquire twenty-first-century knowledge and you can develop yourself. You do not have necessarily to graduate from university, but you can develop yourself and arrive to a point in which you are able to be critical."[35] This is not to idealize experiences of ERO recruitment and military service, which also often unfold as highly discriminatory and violent practices in ways that reproduce again ethnonational subjects and spaces. For example, in principle the PSLF/TNLA bars recruitment from non-Ta'ang ethnic people (unless they voluntarily request so). Also, women have been included in recruitment, but until 2019 they have been excluded from officer ranks and officer training courses. The point is rather that recruitment, military training, and service come to constitute techniques through which individual ethnic Ta'ang bodies, Ta'ang families, and a Ta'ang territorial scale named Ta'ang Land, in which a multiplicity of identities can be accommodated, are reproduced.

As touched upon earlier in relation to antinarcotics techniques, the body itself repeatedly comes to be constituted as an ethnic territory to be governed via military-related practices. For instance, in August 2018 a Shan CSO distributed birth control implant services and supplies to villages under the influence of SSPP/SSA-N and RCSS/SSA-S. The CSO's distribution campaign was part of larger health-related activities, such as trainings and household surveys on birth control and family planning practices. The Shan EROs' local commanders banned birth control implant services and supplies, as well as the CSO trainings from the village areas, in the belief that family planning measures might affect ethnic lineage and growth, thus impacting future recruitment efforts—regardless of the benefits that such interventions could actually have on population reproduction (Quadrini 2019). Women who had tried the implant services and CSO workers figured out alternative tactics to provide the devices and assistance, like arranging for women to travel to unsuspected locations to meet CSO staff.

"The Soldier Should Be Staying Freely in the Jungle": Maintaining Mobile Military Infrastructures

Before 2010–2011—that is, before the creation of the TNLA—Ta'ang areas rested under the influence and control of militia formations and the Sit-Tat. With the establishment of the TNLA, the Manton-Namhsan-Namtu triangle was designated a "black area" by the Myanmar state army. Given that mixed Ta'ang-Shan communities live here and, outside of the triangle, locals often supported the authority of Kachin and Shan EROs, the PSLF/TNLA has since refrained from building fixed infrastructures inside or in the vicinity of villages, even in its de facto strongholds of Namhsan and Manton. As Commander-in-Chief Tar Ho Plan elaborates:"[W]e do not want civilians to be suffering because of the army. If we set up landmines in the territory to defend our camp, then what about the people? If we use landmines that way, then the animals and civilians will suffer; that is why we set up the policy that soldiers should not be with civilians. The soldier should be staying freely in the jungle, not in the villagers' areas."[36]

This PSLF/TNLA's technique of not constructing territorial installments inside or in the surroundings of villages, and accepting to remain mobile, has been in part connected to the rationale of limiting violent repercussions on Ta'ang communities. This rationale is underpinned by the histories of violence experienced by Ta'ang populations with the unfolding of the four-cuts counter-insurgency strategies that led to the 1991 ceasefire. From Ta'ang perspectives, as seen in the previous chapter, the 2005 disarmament ceremony came to epitomize the apex of a process of deprivation of sociopolitical and territorial space, as well as the disappearance of security providers.[37] In addition, the militarization of the post-ceasefire decades highlighted the need to limit territorial displacement and preserve the integrity of ethnonational communities.

The Ta'ang territory has been constructed by the PSLF as a *dispositif* to face interethnic discrimination, military violence, and oppression, to which Ta'ang have been subjected. During the ceasefire years and after the PSLO/A disarmament, Kachin and Shan EROs are said to have envisioned a partition of Ta'ang Land.[38] In particular, the areas of southern Namkham, Kutkai, and southern portions of Hsipaw, Namlan, or southern Shan State townships like Mong Kung, are designated as frontiers of Ta'ang Land. Here Kachin and Shan EROs have practiced military recruitment of Ta'ang and have impeded Ta'ang communities' access to education or identification documents (Weng 2016). Southern areas have been reproduced as frontier spaces where different ethnonational political and territorial projects overlap. Spaces where civilizational orders at different stages meet, in which power relations are territorialized materially and symbolically, and that are characterized by disorder and violence (Korf, Hagmann, and Doevenspeck 2013, 31–33). As PSLF/TNLA's third secretary general frames these frontiers of Ta'ang Land and the people living there:

> [S]ince long years ago they have never had opportunities to study in school. Yeah, they are still . . . they look like wild, people from the jungle, like they are not up to date with the people in the northern parts. Between your country and the northern parts of Ta'ang Land there are maybe twenty years of difference in terms of development. We

are maybe twenty years late in development. But those [Ta'ang] in the southern parts are maybe forty to fifty years late from us. Very low education system. RCSS does not allow the schools. . . . They do not allow people to go to school, even the normal government Burmese school.[39]

Ta'ang Land is calculated through and on the political objectives of the TNLA. Such objectives have been "threefold—number one is to free all Ta'ang from oppression, second is to build self-determination [a Ta'ang State], and the third one is, together with the other ethnics, to build a federal democracy country to be in peaceful coexistence with other ethnics."[40] Constructing the TNLA as a vehicle to make space for, and freeing, Ta'ang ethnic populations from oppression allows the PSLF to generate an ideal and attempted space of autonomy and self-determination whose borders can remain fluid and unsettled: "Right now we do not have a fixed line of where is our border, or where Ta'ang Land is fixedly, specifically, but we have a territory and we have an army, and we will try to be able to defend and protect our people everywhere Ta'ang people live."[41] Such fluid and complex conceptions of a Ta'ang Land circulate among civilians, who see the Ta'ang polity as an archipelago whose "territories and places . . . are not close to each other. Namhsan, Manton, Namkham, Mai Kai, they are all separated."[42]

Maintaining mobile military infrastructure encompasses both social and technical-material dimensions. It requires armed movements to both govern the conduct of armed units and manage the material components of the camps or the landmines often used as part of such networked military infrastructures. It entails a certain political calculation of territory as a space to manage people and life: "If we had the permanent camp[s], then one would need to do like that [i.e., using landmines around camps], and if one did like that there would be problems for the villagers or their animals like cows or horses . . . that is why [we] are just mostly in mobile camps."[43]

But the fluidity of the Ta'ang polity's territory is as much spatially as temporally connoted. The diffusion of techniques that govern the relations among weapons, armed forces, and population on the basis of a logic of ethnonational security contributes to reproduce territory as a series of sociospatial relations that have to be constantly sustained throughout time, not just space. Given the mobile character of the armed force, among PSLF/TNLA's members there remains a diffused belief of Ta'ang Land as an achievement, a prefiguration that has to be worked through, and that will be realized in a future stage.[44] Governing and practicing the ethnonational armed force is understood as labor toward the constant reproduction of a territory. As an officer puts it, referring to the political vision of "the leadership":

> As this is the revolutionary time you do not have control on the territory so actually they [TNLA's leadership] are not focusing on

infrastructures, which is just the physical thing, you know . . . ? . . . The leaders are focusing on institutions because that can move, it can be mobile, and once you have strong mobile army institutions then you will be able to build infrastructures, once you have your own territory. . . . Building the armed force is like to get territory on our own, not given by others. Building the institutions is to get ready for when we get a territory. The army will carry out the revolution, the institutions are developing our people. But remember, PSLF and TNLA are the same. In Burma, since the governments build the authority or territory with the arms and they force the nationalities with their power coming from the army, from the weapon, the same did the Shan and we became even more minority of minority; all Burma has built authority from the gun, not from elections. Actually, authority in Burma comes from the army. . . . This is the mindset of Burma, it has become like this.[45]

PSLF/TNLA's rationality of ethnonationalism constitutes subjects in need of security and freedom from all forms of oppression. Security and freedom from oppression are pursued through both the spatial and temporal fluidity of TNLA's territory in opposition to other ethnonational territorial claims. The spatial mobility of TNLA's infrastructure is harnessed to construct the Ta'ang politico-territorial project as militarily more civilized and humane than those of, for example, the RCSS/SSA-S and the Sit-Tat[46]:

We do not build camps inside the village areas because there our people live. It is thinking about their security that we avoid to building camps in civilian areas and the camp has to be outside, 'cause otherwise this will affect the people. In this sense we differ from RCSS, since they have built camps inside villages: it is for the purpose of security that, they say, they built military camps to keep civilians safe, but at the same time it is their presence that triggers instability and clashes. If they are here [in Ta'ang Land], it means that sooner or later the Tatmadaw will come.[47]

The temporal fluidity of Ta'ang Land is harnessed to navigate the logic of *taingyintha* and "first comers"—that is, the idea that a certain population group can be identified as the original inhabitant of the land—in order to produce the polity's territory. The PSLF constructs its territory by tracing a genealogy of the ethnonational rebel polity and building on the political legacy of the PSLO/A under Captain Kham Thaung. Under the leadership of Kham Thaung, PSLO/A fought against the *sao hpa* administration as a postcolonial fallout of the colonial system of indirect rule—a system that had limited Ta'ang territory to Tawngpeng (today Namhsan). PSLO/A thus opposed the idea that the Palaung/Ta'ang polity should be limited to the colonial borders of Tawngpeng. It redrew the borders of the "Palaung State" by including parts of Mogok, Kyaukme, Momeik, Namhkan, and Kutkai townships, all areas that were argued to have been inhabited by Palaung/Ta'ang populations well before British invasion.[48] Such territorial claims, though, have remained highly disputed by other ethnonational territorial projects of Kachin and Shan politico-armed movements. Discussing rebel territory maps, an SSPP/SSA-N officer, for example, refers to an official demarcation agreement between SSPP, KIO, and PSLO finalized in 1978–80. The agreement established EROs' territories on the basis of a "concept of compound communities," meaning the borders of a rebel polity's territory should be demarcated according to where "the majority of the ethnics" live.[49] A criteria that, if applied, could potentially shrink Ta'ang Land back to the area of Tawngpeng. But one that, at the same time, if carefully reinterpreted, could allow the PSLF/TNLA to claim a different territorialization of the Ta'ang polity: a Ta'ang territory that includes more areas, and/or one that envisages the boundaries of the PSLF/TNLA's territory as being delimited by the presence of Ta'ang populations. If the polity's territory is where the Ta'ang populations are, older cartographic depictions such as PSLO/A maps that substantiate territorial claims become double-edged technologies. Similar to PSLF/TNLA's members, Tar Khun Yee, last vice chairman of PSLO/A, became particularly wary when, during a conversation, we ended up talking about old PSLO/A political maps:

> [W]e did have [a] map but we put it somewhere, I don't know. . . . The KIO/A and SSA did not agree, so this was not an agreement with other organizations, but it was what the Ta'ang leaders recognized. Because for example in Kutkai the Kachin said that it was their area . . . and maybe in Namkham it's all Shan now . . . that is why we are having problems with territory. . . . Like in Kyaukme . . . more Shan came and even in Namhsan . . .[50]

Grounding territorial claims onto older cartographies can both legitimize and delegitimize the technique of maintaining mobile military infrastructure. By redeploying the *taingyintha* rationality that underpinned PSLO/A cartographies, while avoiding fixing its borders, the TNLA's mobile infrastructure can be argued to build territory where Ta'ang communities live and are oppressed, even beyond PSLO/A's maps. The activities of civil society organizations at times can subvert, integrate, or reinforce the ethnoterritorial effects emerging from mobile military infrastructures. For instance, CSO activities in the ambit of human rights and international humanitarian law violations documentation entail a calculation of ethnic bodies and ethnic territories in order to frame and carry out reporting practices. The monitoring of indiscriminate shelling of civilians, landmine and explosive casualties, or porterage and human minesweeping by the Sit-Tat shapes ideas and practices of ethnonationality. In some cases, such as in the reports of the Ta'ang Women Organization (TWO), this can also provide a clearer geo-*graphic* picture of Ta'ang Land, one that is not necessarily in line with EROs' prac-tices but could nonetheless be exploited by different authorities, be it EROs for their territorial claims or military-state authorities willing to criminalize CSOs' work. For this reason, requests to share maps are usually met with reluctance, and restraint is exercised over circulating them: "I had a copy here in the office but we actually had to eliminate all the copies . . . because of the police increased security measures, especially after the fifteenth of August attacks this year [2019]."[51]

Similar to cartographic representations of the polity's territory, the logic of *taingyintha* and "first comers" is neither fully endorsed nor fully rejected, but rather played with. Military techniques, such as military staff manuals or anec-dotes used for training purposes, attempt to both oppose and reproduce the eth-nonational rationalities. One of the first entries of the PSLF/TNLA's staff manual, for example, opens by framing the Ta'ang as a "*lumyo*" in its own right. This is done in order to reformulate and to oppose *taingyintha* grids that construct the "Palaung" as a subminority within the Shan minority. Similarly, during training, anecdotes taking the Jewish people and the Israeli state's quest for a motherland as a role model are often used to motivate trainees. Normally nothing more than motivational rhetoric, behind which no support for Israeli policies and military action can be found, such anecdotes aim to explain how TNLA's activities are geared toward the creation of a motherland that can free Ta'ang populations from Sit-Tat oppression.[52] The ambivalence of this move of both rejection of ethnocen-tric oppression and reformulation of ethnonationalist stances forcefully emerges through the irony the anecdote is imbued with. On the one hand, the state of Israel has been one of the major exporters of the weapons, military knowledge, and technology through which the Sit-Tat has colonized the borderlands and pursued its ethnocentric political community. On the other hand, in its quest for the motherland, Israeli apparatuses have transformed Palestinian spaces into a battleground for the territorialization of its state authority (Weizman 2007).

The fluidity and mobility of the TNLA's military infrastructures fully embody this play and tension between the production of territory and the marooning of fixed borders and forms of territorializations, a tension that the TNLA's military administrative apparatuses reflect. In 2014 the Ta'ang rebel movement instituted its own General Administration Department (TGAD), the military body respon-sible for the administration of the areas progressively occupied by TNLA's troops. While the TNLA remained in charge of "entering" and "clearing the land," TGAD was made responsible "for civil society formation . . . for making development for the civilian," as a TGAD officer explained.[53] To do so—although TNLA's military infrastructures continued to remain highly mobile, avoiding installation in bases in or near villages, for example—the TGAD mapped Ta'ang Land into five districts, twenty townships, and fifty-nine zones. With boundaries that differed from those drawn by the Union government and resembling at least in part those established earlier by the PSLO/A,[54] districts, townships, and zones were named using Ta'ang toponyms. Owm Ta Mao district, for instance, was named after the Ta'ang name for the Shweli River, considered a landmark and treasure in Ta'ang cultures. Targeting lands that had to be made "free from oppression" regardless of their noncontiguity and distance from PSLF/TNLA's areas of presence, the activities of the TGAD relied heavily on the local political order of Ta'ang communities. Thus, in contested, "frontier" areas, where appointing and dispatching a TGAD officer would not be feasible, TGAD structures operated through the authority of local customary chiefs. In this sense, the TGAD is a territorial technique in itself, one that is ingrained into the TNLA and that is complementary to its mobile infrastructures, a technique that unfolds through a logic of ethnonational institutionalization of Ta'ang "ancestral" territories.

The Soldier Should Stay in the Village: Territory Effects of Fixed Military Infrastructures

Different techniques of governing military means linked to the logics of ethnonationality and their territorial effects can be seen in the ambit of the RCSS/SSA-S's expansion into northern Shan State. In late 2015, the RCSS's move north to the Mandalay-Muse artery significantly altered Ta'ang areas.[55] In an attempt to mark territorial authority and obtain indirect recognition from military-state apparatuses, the RCSS started to generate a buffer on the mountains abutting the strategic route to China.

Relying on Ta'ang communities and drug addicts' forced labor, the RCSS/SSA-S militarized Ta'ang areas on the hills through fixed infrastructures.[56] It constructed roads and enlarged existing ones, built fixed checkpoints, and disrupted paths in strategic ways so as to control and hamper the mobility of ethnic Ta'ang communities.[57] Permanent military landscape installments like camps for alleged drug treatment, fixed military camps and bases were built on the hilltops and inside villages under PSLF/TNLA or SSPP/SSA-N's influence networks. The RCSS/SSA-S introduced its own self-support mechanisms relying on the structures of local village headmen, who would now operate simultaneously within the landscape of three different EROs' authorities.

The RCSS/SSA-S adopted policies similar to those of the PSLF/TNLA—for instance, in terms of narcotics eradication—but its practices were informed by a more pronounced ethnonationalization of territory through discourses and practices of defense of Shan ethnicities. Discerning between ethnic Shan and Ta'ang areas and populations, the surroundings of its military camps on hilltops were mined, while strategic tea plantations or jungle areas were transformed into delimited no-go zones.[58] In Manton township, for example, in 2016 the Shan armed group established camps near an important local market (Mai Baw) and blocked the roads that connected it to hilltop villages in the area.[59] Movement was restricted and any rice purchase in excess of the quantities needed to feed a family was banned. Ta'ang households were specifically targeted on the basis of an assumed link with the TNLA. This severely altered the mobility and microeconomies of Ta'ang families given their reliance on the trade in charcoal, wood, and tea and the need to first collect these products in forested areas and later carry them to local markets. Similar to Sit-Tat's four-cuts counterinsurgency strategy, phone communication and internet access were also rendered more laborious given that people often had to reach specific hilltops in order to intercept signal and use their devices. This military strategy unfolded via the regulation of CSOs' access to rural populations: Ta'ang CSOs were prohibited from accessing and operating in RCSS/SSA-S areas so as to regulate population access to humanitarian and development services.

We can see the workings of such military infrastructures moving to the village of M.L., under RCSS/SSA-S' presence from 2016 to approximately 2021.[60] M.L. is a village of a hundred households, predominantly Ta'ang families, located a few kilometers northwest of Hsipaw, in an area not too far from the PSLF/TNLA strongholds. In 2016 the village came under the direct control of the RCSS/SSA-S. The Shan movement had built several camps and bases inside it: one on the mountaintop at the end of the village, next to the only road connecting M.L. to the next village to the north, and others on the surrounding forested hills. A small bamboo military post had been built right at the entrance of the village, blocked with a metallic white-and-red bar and a bamboo guardhouse next to it. Inside the guardhouse, uniformed RCSS/SSA-S troops alternated to check flows of people through the village: everybody moving in or out was registered on a notebook where soldiers kept track of the name, location of provenance, destination, and time of those passing by. To move out of M.L., people had to obtain a written permit from the chief of the village who had to fill in a form indicating the name of the person, traveling motivation, the direction of movement, itinerary, and duration. Depending on fluctuations in military hostilities, in addition to the chief's signature, the form may have to be stamped by an RCSS/SSA-S senior official in M.L. in order to be accepted as a pass at the gate. Again, depending on the security situation, at times the checkpoint may be left unmanned, or the form may not be required at all. After 5:30 p.m. normally no movement in or out of the village was allowed except those of the rebel group.

Inside M.L., the RCSS/SSA-S presence was intensely felt. Troops stayed in the camps on the surrounding hills and the base on the hilltop, but at the same time they were often stationed inside the village. From time to time, rotational shifts were arranged among RCSS/SSA-S camps/bases in different village areas. From time to time, soldiers may come to the village disarmed, leaving weapons in stashes inside the camp. A taxation system was imposed in the village, according to which truck owners, for example, were supposed to pay something like 200,000 kyat per year. Furthermore, some forested areas surrounding the village, especially those in the vicinity of the RCSS/SSA-S camps, had become no-go zones for dwellers. Yet many villagers maintained tea or other crop plantations there and thus lost access to the land. Often, phones were not allowed in public spaces. The RCSS/SSA-S commander had banned their use out of concerns that the Ta'ang rebel movement may maintain informants in M.L. Thus, it was mandated that Ta'ang dwellers' mobile phones be consigned to the head of the village, who would be responsible for storing them in his house. Yet, residents hid and used them inside their houses.

Hanging on the walls of some homes, one could see information posters provided by an international nongovernmental organization about landmine-contamination-related risks. On them, vignettes and explanations reminded people about the shape of different types of landmines, or the most common places they may be found. As a Ta'ang granny living in one of the last houses of the village recalled, pointing to the forest, the "RCSS laid landmines in the areas of their camps inside the village, out of fear that the enemies would come."[61] However, nobody was allowed beyond these houses, as the village base of the RCSS/SSA-S was located further toward the hilltop. Landmines' presence, or unknown absence, channeled population movements to more visible paths and primary roads, while justifying the imposition of no-go zones.

Composed of various discursive, technical, and practical elements, the RCSS/SSA-s' military infrastructures in M.L. profiled people and territories on an ethnonational basis. On a broader scale, networks of checkpoints, camps, bases, trenches, and mined perimeters materialized such ethnonational calculations of territory, both when they were manned by uniformed RCSS/SSA-S soldiers and when they were not, as constant reminders of a territorial politics of ethnic identity. Such networked infrastructure was not only an example of fixed defensive tactics and weaponized materials. It also represented armed assemblages made of a complex array of techniques and technologies that reproduced ethnic subjectivities and territories.

These techniques of governing military means have been underpinned by ethnonationalist discourses that frame Shan State as a territorial entity on the brink of fragmentation, due to interethnic disputes (see Ferguson 2021).[62] The RCSS/SSA-S has attempted to drop the indication of "South" from its official name, and at times referred to itself simply as the Shan State Army. Form here is definitely substance, for such a move expressed the broader ambition to make the ERO's territorial authority coincide with the boundaries of Shan State and argue for its role as the only guarantor of the territorial integrity of Shan State. Along these lines, Shan State is to be preserved by granting integration to other nationalities, while at the same time defending the Buddhist religion as a foundational trait of Shan society. Referring to the RCSS/SSA-S's aims, the head of a Shan CSO put these discourses squarely:

> The aim of RCSS is to get in control of the whole of Shan State—they will be protecting all the people living in Shan State, whoever you may be and whatever your nationality may be: Kachin, Shan, Palaung, Rakhine, etc. About the religion they say that they will not force the people to be Buddhists but they [RCSS] take care of the Buddhist religion and they will protect it. Whoever will attack the Buddhist religion they will stand in defense of it. . . . The Palaung people would like to build a Palaung state inside Shan State, but this is *Shan* State and, if you want to, at best you can have a Palaung region in which you can stay, not a state.[63]

Others instead look at EROs, the RCSS/SSA-S *in primis*, simply as an expression of opportunism and willingness to control business projects.[64] But even in this case, ethnonational discourses proliferate, and the Shan may still be seen as the older brother of other ethnic minorities in Shan State. Therefore, the mobilization of other identity elements, such as authoritarian political oppression or Buddhist religiosity and spirituality, complicate but do not necessarily supersede the workings of ethnonational rationalities. In the aftermath of violent events, such as shelling, civilian casualties, or landmine explosions, the RCSS/SSA-S units have carried out donations to communities handing in money and/or food to local Buddhist monasteries. While villagers and victims may be of varied ethnicities, donations are often directed to Shan monks and monasteries. At the same time, in prevalently Ta'ang villages, Buddhist monasteries have become a privileged place to hold public dissemination sessions concerning antidrug policies, to issue warnings about the use of landmines, or to proscribe access to certain zones.[65]

Stocking Weapons: The (Logistic) Life of Being a TNLA Weapon

The Ta'ang movement has combined the mobility of its battalions and headquarters with mobile and networked institutions under its armed branch, which attempted to make Ta'ang land a territorially coherent space from an arms control and military administration perspective.[66] Mirroring its structure—that is, a standing army supplemented by affiliate networks in villages able to quickly mobilize standing and reserve troops, or to collect taxes and secure new recruits—the TNLA's arms control practices have been heterogeneous and related to a certain calculation of territory.

In areas of more stable influence, such as the Manton-Namhsan-Namtu triangle, certain weapons are kept in stocks controlled by the group in mobile bases or encampments. Heavier weapons are stored in underground bunkers with trenches around, while service weapons are kept by soldiers in their barracks.[67] In addition, given their high degree of mobility, uniformed units often maintain their own munition supplies. In so-called mixed areas under the influence of other EROs or the Sit-Tat, firearms are also distributed to soldiers and the TNLA militias. TNLA weapons in these areas are mostly kept in caches hidden in the forest, in remote hiding places, or in small stashes located in the broader surroundings of villages. This two-pronged management system also meets the need to maintain the capacity for rapid mobilization as well as allowing troops to travel back to their home villages after duty.

These physical networked infrastructures are composed of different dimensions, of which the material location and structure is but one component. The standing-like nature of the TNLA is paralleled by the deployment of certain forms of record-keeping and formalized weapons management practices. Weapons are recorded according to their assignment to every different department and every different brigade and battalion, all through cloud databases to avoid information storage in one single hardware.[68] The TNLA keeps records of the weaponry held in stockpiles and the weaponry currently out with units or in localized stashes. Moreover, units on the move have to track their use of ammunition and the status of their arms, while at times specific types of weapons are assigned to dedicated personnel.

This concern with weapon stocks management and control has been formalized into training documents and training practices. While codes of conduct contain provisions concerning how to relate to civilians while carrying weaponry, a weapons manual has been produced to instruct members concerning weapons knowledge and identification, handling, use, and care.[69] Specific knowledge and training is passed on among the circles of explosive specialists, or dedicated units assigned to weapons safe storage. Besides codified rules and norms, care vis-à-vis the weapon emerges in the formalization of self-discipline measures about loss and misuse via training practices. Such measures are reproduced through the interiorization of the nonfungible, almost irreplaceable value of a weapon. Epitomizing something conveyed by many soldiers, a TNLA former combatant recalled what would happen in case of loss: "If you lost a bullet you had to pay five hundred kyat, while if you lost your rifle that would cost you around two thousand dollars."[70] Beyond the specific content of the rules, the clearly exorbitant sums—sums that in the case of a bullet would be exponentially higher to any form of (mostly nonpecuniary) daily compensation received by troops; while, in the case of a rifle, virtually no soldier would be able to repay—equate the loss of a weapon to an unpayable debt toward the armed force.

Training and knowledge distribution, though, should not be seen as smooth and even. In fact, the high mobility of troops in many areas has often created issues in relation to training.[71] Recruitment may be carried out while contingents are on the move, in areas that do not allow for extensive training, or in a situation of lack of arms to train new recruits with.[72] This, in turn, relates to how weapons management techniques are taught or not. In addition, weapons are not distributed lightly: new recruits are normally not trained with weapons until later stages; drug addicts pressed into service are not allowed to access guns until after the end of their "rehabilitation" process; and weapons are not provided to poppy-cultivating farmers.[73]

Localized hidden stashes are also managed through, and in a sense made of, techniques and practices, although they may appear as simple holes in the earth or improvized makeshift architectures. Stashes are normally assigned to trustworthy connections who hide and control guns via micropractices. This is not at all peculiar to Ta'ang areas, as a Mon informant notes, recalling his father's experiences with the New Mon State Party (NMSP):

> My father, for example, used to be that person for the NMSP army in our village. . . . He was given arms by NMSP soldiers and was entrusted to hide them and keep the weapons until the soldiers would have asked them back. We had a paddy field . . . and during the cultivation and harvesting times we . . . would live in the paddy for those months and my father would then move the arms he was hiding for NMSP. Sometimes he would hide the arms in the hay, sometimes inside piles

of rubbish. . . . I remember that in some occasions I went to the hay bale and rummaging into it with a stick I hit something metallic and hard, which turned up to be a rifle. Yet, my father would disassemble the rifles and the magazines, keeping the two divided.[74]

Arms control measures are in part combined, calculated, and practiced, together with modalities of arms acquisition and the weapons' technical potential. The processes of arms acquisition in fact should be seen as part of a continuum of techniques. The possibility to arrange new arms acquisitions rather than relying on craft-manufactured (i.e., artisanal weapons) or assembled firearms entails a different calculation of the spaces to be arranged in order to manage them and a different materialization of such spaces. This is especially true not only concerning firearms but also concerning ammunition, which must be more frequently resupplied and moved around. At the same time, techniques of arms acquisition and management combine with the technical potential of weapons, both symbolically and materially. Talking about the mountainous area of Namhsan, a TNLA soldier commented on recently acquired RPGs, noting that the "Tatmadaw . . . do not dare to come up to these areas 'cause they know they would be attacked otherwise. Recently we also just acquired new RPGs. That weapons is good also against helicopters of the Tatmadaw and their fighter jets. Actually, against those weapons we cannot do much . . . we do not know how the group acquired RPGs, this remains secret also for us."[75] Here the reference to the technical object and its potential highlights the production of a "secure" space—both in areal and vertical terms—not simply by virtue of an inherent quality of the object but rather connecting it to the (secret) capabilities to arrange their acquisition (and thus their storage too).

PSLF/TNLA's weapon stockpile management arrangements reproduce a fragmented composition of "core" areas—in which more stable, albeit movable, stocks are maintained—and "frontier" areas—in which stashes and other forms of diffused control are articulated. Even if weapon stockpiles and stashes are hidden, temporary, and mobile, their constitution and circulation nonetheless generates materializations of the territory effects distributed by the rebel movement.

This is also the case for weapons and armed forces of other EROs who reproduce similar techniques of governing the means of violence. The SSPP/SSA-N, for instance, has divided its self-identified ten-district territory into "areas fully administered by SSPP" south to the Union highway; two districts "only formally under SSPP, as RCSS invaded them after late 2015," which constitute the "frontlines between us and areas of overlapping administration"; and other districts in which "the SSPP overlaps with TNLA, KIO/A, the Kokang, and many government militias at the border with China."[76] Reproducing techniques of spatial calculation based on the identification of white-brown-black areas, territories invaded by RCSS/SSA-S or areas under the Sit-Tat and militias have been marked as "guerrilla warfare zones" by SSPP/SSA-N. Such territorialization unfolds together with arms control techniques, as an official of the Shan ERO illustrates:

> Stocks of arms are kept in the base camps on the top of mountains in the areas in the north, but in the south instead we hold warehouses and we maintain recording: which rifle is owned by whom, recording for each platoon, division, and regiment. Quarterly we also investigate on the status of the weapons. The main difference is that in the guerrilla zone there are very few fixed camps, as we need instead to go around and move, so there are no fixed stocks of arms there. We collect the weapons only in mobile camps and do not have stashes of arms in the forests or around the village. We maintain no stashes of weapons around the villages. After 2011 we started to set a system of village militias, though, after we broke up due to the Border Guard Force process.[77]

The Governmentalities of Village Community Militias

With the expansion of the RCSS/SSA-S troops north to the Mandalay-Muse highway in late 2015 and the eruption of hostilities in 2016, the PSLF/TNLA started to organize Ta'ang villagers into village militia formations.[78] Villagers were provided with basic weapon handling and use training, as well as landmine recognition—and in some cases deactivation—skills.[79]

Named Local Guerrilla Force, these militias were mandated to operate exclusively in their own village. Such territorial constraint, combined with, for example, reporting and intelligence-gathering duties that the LGFs would carry out under the TNLA's chain of command, functioned as a further demarcation of Ta'ang Land's village areas.[80] Moreover, these militias have been constituted especially in those areas that have been calculated as the borderlands and frontiers of the Ta'ang polity, as a Ta'ang village militiaman pointed out when describing the roles of the LGFs:

> I am inside the "BGFs." Not the "BGFs" of the Tatmadaw I mean . . . those of the TNLA. The group has its own militias under itself—we also have an administrative department and many more structures. There is also a foreign department, a supply department under the structure of

the defense department. The supply department also takes care of the issue of weapon supplies. The administrative department has to take care of the administration of the villages. If in the future, we will obtain our autonomy, then the BGF will be shifted under the purview of the borders and customs department. The BGFs of the TNLA are responsible for the supply chain to the troops especially and other tasks that may emerge. They are located in particular in those that are the "border" areas for us.[81]

In the following years the RCSS/SSA-S followed suit and started to develop its own militia formations, which were entrusted especially with investigation duties on drug abuse and trafficking. Likewise, the SSPP/SSA-N, which had long maintained community militias under local administrators of its own, intensified this practice.[82] In this sense, the PSLF/TNLA and the RCSS/SSA-S had been anything but pioneering. In fact, the constitution of different forms of militia related to ethnonational rebel movements had long represented a key feature of the armed conflicts in the borderlands.

Informed by the logic of mobilizing segments of civilian populations, and institutionalizing them as an appendix of armed and security forces, against a "common threat" to the safety and integrity of the ethnonational body of the polity, forming and managing village community militias has been a further technique through which the means of violence and the production of an ethnonational territory have been jointly controlled. Besides being a top-down military strategy, people's militias are sociotechnical infrastructures that manage the relations between weapons, humans, and space. In this sense, the governmental techniques of arms control and militia management that regulate these relations have been harnessed not only by rebel military apparatuses but by other actors, such as village communities themselves, to reproduce or contest rebel political communities.

Moving back a few years to before the conflict between the TNLA and the SSA-S, we can observe this through the T.H.K. village community and its militia in Kyaukme township. After the creation of the TNLA, and during its consolidation between 2010–11 and 2014, many Sit-Tat community militias in Ta'ang areas were dismantled. This was partly due to the TNLA's mobilization efforts and to militiamen defections to the TNLA. In the T.H.K. village militia, however, this was not the case. Here the Sit-Tat village militia dated back to the times of the current village head's grandfather—the chief being a man in his late fifties (in 2019). At the time of its disbandment in 2013, the militia counted 256 members from four different villages around T.H.K. The militia's territory in fact had been delineated by the Sit-Tat according to state administrative maps and organized into four units, one for each village, under the command of the T.H.K. village head. Members were divided into first and second lines, with the latter being unarmed reserve forces for emergency and/or security and community policing tasks, and the former being active members with firearms. The Sit-Tat had initially provided fifteen rifles to the militia group. The weapons, ten-shot magazine semiautomatic rifles, had been assigned to certain members, while others were also armed with craft-manufactured hunting rifles.

The Sit-Tat maintained control over weapons issued to civilians in mountainous communities as part of its multiple militia programs implemented since the 1950s. Sit-Tat-supplied firearms were controlled through record-keeping by the militia and the Tatmadaw Military Operation Command 1 in Kyaukme town. Both maintained forms indicating militia members and the serial numbers for each of their individually assigned weapons. The military would conduct a mandatory check of the arms supplied to the militia three times a year. Some militiamen of T.H.K. village were required to collect the Sit-Tat-supplied firearms across the militia's territory and carry them to the base in Kyaukme for inspection. As the current head of one of the villages involved recalls: "Everything was centralized in [T.H.K.], in terms of decisions, training, the firearms."[83] Village militia leaders had to report on the number of cartridges spent and the status of ammunition, specifying the circumstances of eventual supplies depletion. Thus, if individual members wanted to use firearms in their territory, to hunt or for other purposes, they had to obtain permission from T.H.K.'s village head. Sit-Tat concessions to use weapons for hunting, their mechanisms of record-keeping, weapons handling and management techniques, and weapons periodic inspections, all contributed to build Sit-Tat authority, to legitimize the armed militias in the mountains, and effectively reinforce the four village areas as part of military-state territory. Arms control here entailed a series of processes and practices constantly reproduced and linked to the idea of mobilizing local resources for local safety and security as part of the military-state Myanmar polity.

But since 2011, community militias in Ta'ang areas started to fear the possible escalation of conflict and their consequent involvement in hostilities due to the PSLF/TNLA's rearmament as well as the overlapping presence of multiple armed actors. Militiamen still had to travel to Kyaukme town three times a year to submit their weapons for regular Sit-Tat inspections, while the PSLF/TNLA progressively expanded its territorial influence and other EROs continued to maintain troops in the same areas. To avoid being involved in violent clashes in the area, in 2013 the militia decided to return its weapons and disband in agreement with the Sit-Tat. As the T.H.K. village head recalled: "The situation in the hills was not stable and it was not safe for us to hold firearms in the village. . . . So the army decided that we could give back the weapons."[84] Interestingly, not all of the four village community militia members disarmed and dismantled the militia. Militia members from the village closer to a Sit-Tat base decided to maintain the community militia.[85]

The T.H.K. community militia in this case used disarmament as a technique to reshape territorial arenas of control over the means of violence across multiple shifting and overlapping authorities. Consigning the weapons also entailed a halt to the techniques and practices that, technologically linked to and required by the firearms, contributed to compose both the militia and its territory as part of the military-state Myanmar polity. In the case of the T.H.K. community militias, one can see how civilian communities reappropriated and reinvented the techniques and practices connected to the formation of militias as a counterconduct, in a sense, to reshape local territorializations. While targeted by or supposedly benefiting from acts of governing the means of violence linked to the formation of community militias, at the same time they engaged with, problematized, and reappropriated logics of civilian safety (both discursively and practically) to navigate different governable orders and shift among governable territories in the (un)making.

Humanitarian Arms Control and the Politics of Explosive Devices

The rearmament process of the PSLF/TNLA added a further layer to an already intricate conflict landscape, as the governmentalities of village community militias suggested. Rebel formations, Sit-Tat troops, different militia formations: a kaleidoscope of armed actors overlapping and contesting each other turned Ta'ang areas into one of the borderland regions most affected by remnants of war. In particular, with the conflict scenario growing in complexity, landmines and explosive devices became more and more tangible markers in the land. They are objects that may demarcate territory, through which claims of territorial authority could be substantiated and opposed. In this context, the technical implications of weapon technologies and armed collectives opened spaces for producing territory.

Weapons' Active Mediation

Patterns of use of landmines and explosive devices generate a symbolic, diffused presence of the weapons. They project the arms beyond their actual physical location. Landmine casualties and the presence or absence of landmines not only deter in a physical manner, they also spread fear as a diffused control method.

In this sense, landmines have been materializing "nonremoval" zones in which armed actors attempt to limit access. This has occurred especially in relation to the militarization of areas targeted by land-grabbing operations for extraction and business development opportunities. The actual (or at times assumed) presence of landmines provides a reason to prevent the return of communities to their home villages and can be harnessed to manage the territory of the rebel polity.[86]

But beyond purely tactical purposes, the material presence of explosive devices and their explosion operates as a marker of, and a claim over, territory. The landmine's technical properties as a weapon require and activate techniques that aim to manage the relations between weapons, armed forces, and populations. EROs' explosive specialists are normally trained to avoid civilian casualties as much as possible, in order not to waste precious munitions. In fact, explosive items constitute a valuable asset to be preserved. Thus, in many cases, military units recuperate devices that have not exploded and can still be functional. Yet sometimes troops forget to recuperate the weapons, they miscalculate offensives, or it becomes too onerous to take the device back, and they are left in the ground. In case unplanned explosions involve civilians or their cattle, for example, EROs have both provided compensation to and requested compensation from victims, their families, or the owners of the cattle triggering the mine.[87] The armed actors usually *provide* compensation when victims or affected households are of their same ethnicity, while instead more often *request* compensation when they are of a different ethnicity.

When compensation is *provided* to the victims' household, the landmine and the explosion contribute to reproduce ethnonational territories by institutionalizing the responsibility of the rebel movement to take care of the polity's population and space. At the same time, *providing* compensation reproduces an "othering" effect that distinguishes not only between ethnically identified populations but also between areas where compensation is to be provided, and areas where it is not. Instead, compensation *requests* are issued to affected households on the basis of an institutional logic of hardware and assets management through which landmine destruction is constructed, first and foremost, as an event jeopardizing EROs' ethnonational cause.[88] When explosive items explode without hitting any military target, often unit commanders may be deemed responsible and may have to cover up the event, since equipment and supplies are recorded and have to be accounted for. In this sense, landmines, explosions, and compensation requests contribute to reproduce territory as a space calculated on the basis of a logic of ethnonational preservation. Civilians, for their part, often disguise their wounds, try to hide their cattle, or deny involvement to pretend they had nothing to do with a mine explosion, so as to avoid any compensation request or retaliation.

At the same time, landmines and their unplanned explosions may work to put into question the territorial authority of armed actors by exposing a lack of proper institutionalized control over the means of violence. A mountain guide in Hsipaw, for example, questioned EROs' conduct:

> They have said that, when the conflict ends, they will take care of the landmines and of clearing the areas from the landmines. But I mean, I do not think this is true because sometimes they do not even know where they are or even where they put them. In fact, sometimes the soldiers of one group injure themselves with their own mines. I think this is in part due also to the fact that they have replacements of troops and when new ones come, they do not really know.[89]

The question of "delimitation" of EROs' territorial presence is one that is strictly intertwined with ethnonational claims over territory:[90] be it present or absent, the landmine provides an opportunity for reactivating such claims through its social and technical composition. Explosions trigger issues of authority over demining—and over the governing of landmines, explosives, and the remnants of war more broadly—that delineate a political field for the "delimitation" of the polity's territory. While, in discourses about demining contaminated areas, civilian perceptions of who should be clearing the land once conflicts end are usually articulated around responsibility criteria (meaning, those who planted landmines should decontaminate), EROs turn that very logic of responsibility on its head. If civilian perspectives highlight the use of landmines as a burden imposed by armed authorities whose legitimacy remains debatable, EROs' governmentality revolves around a humanitarian responsibility to ensuring the creation of a "humane," safe space for life. In turn, if civilians react to the question of "who should demine" with comments that underline how "whenever an army commits a crime they do not take responsibility,"[91] EROs harness the same question to argue that EROs must demine "their own" ethnonational territory.[92] The head of a Shan CSO based in Kyaukme, for instance, talking about demining, recalled one of a series of periodic meetings to which CSOs and village heads had been summoned by the RCSS/SSA-S to carry out advocacy sessions on narcotics and mine-related threats: "They said in a meeting . . . that it is their responsibility to demine where the area is under their control. They will take this responsibility in the areas where they control, to keep people safe."[93] ERO-related perspectives frame demining as a field of discourses and practices to calculate and govern ethnonational territories according to humanitarian logics of keeping people and land safe, as PSLF/TNLA joint secretary general Tar Parn La remarked: "That is our land right . . . ? . . . our land, so we will do that. If we do not do it, our people will get a problem on that. But we know that we may not have enough knowledge or wisdom or education that much to solve all of the struggles, but we will ask [those] who know and who are qualified in those fields."[94]

Decontamination as Territorialization

The construction of demining as a technical matter not only allowed EROs to produce and claim authority over ethnonational territory, but it also opened up room for other forms of counterterritorialization. As noted by a CSO's head:

> The responsibility of clearing conflict areas from landmines and explosives must remain on the government and the armies of the EROs. One day, when Myanmar eventually will be at peace, the armed actors will be leaving the landmines on the ground. Local people will be completely without knowledge about the dangers related to landmines and how to deal with this issue, so I think it will be the task of CSOs to be leading the process of putting pressure on the government and the Tatmadaw and EROs so that they take care of this problem and demine contaminated areas.[95]

At times landmine-contaminated village areas have been demarcated by local community members. Although this activity is usually mostly informal, localized, and volatile, in some cases local dwellers have marked areas suspected to be populated by landmines or explosive items, while in others they have come back to their villages carrying explosive remnants and weapon parts to be recycled for other purposes. People have marked sites, sometimes areas, using disparate improvized methods: flags, bamboo sticks tied together with a rope to form an X, tree markings.[96] Nonetheless the unstandardized nature of such signposts in the land often leaves them void of meaning. At the same time, the landmine, as a technical object reproducing its own ecology, makes it so that marking a single site often is not enough: triggers and wires may remain undetected, and at least a polygon area, rather than a site, should be marked.[97] All of these issues are summarized by an SSPP/SSA-N high-ranking officer who also runs a CBO working on landmine contamination in northern Shan:

> When the item found is on the side of a village we demarcate it also besides informing [villagers] . . . we demarcate it with bamboo sticks and go to explain to the villagers not to go close to there. We use also internationally recognized signs when we have them there or when we can bring them there, but otherwise, if it is too far, Halo Trust [i.e. a humanitarian mine action INGO] goes and brings signs. But this is funny, so to speak: villagers do not write and read, so often they are not used to international signs. The international signs have the problem that they [villagers] cannot read in Burmese or in English often, so at times [it] can be useless unless they are given exact knowledge about it. At the same time, even localized signs can be misinterpreted. For the Karenni, for example, an X on a tree means that that

tree carries honey bee and has already an owner thus people are not allowed to take honey from that tree. At the same time the X made of two bamboo sticks means "danger." A horizontal stick crossing a cross made of two bamboo sticks or a sort of a gate with two vertical poles and horizontal sticks means ownership of an area. So, it very much depends on local wisdom.[98]

Landmine contamination, as it appears clear in the words of the SSPP/SSA-N interlocutor, has been at the center of a myriad of human and nonhuman agencies and collectives that attempt to govern the weapons and related processes by deploying, refusing, and/or reappropriating techniques and practices linked to humanitarian logics. CSOs in particular have played a key role in spreading techniques to shape people's conduct in relation to landmine contamination via trainings-of-trainers. These educational activities disseminate techniques that entail adjusting one's own behavior in terms of where to go, how to behave in case casualties are found, or what to see as a weapon and what not. Reasoning about differences between the pre-NCA period—when humanitarian mine action had been in its infancy in the borderlands (especially in northern Shan)—and the last years, a Ta'ang CSO member pointed to how "people have started to mark dangerous areas with marking signs or to inform other villagers and be aware of the dangers; they do not go to suspected areas or have stopped picking up from the grounds things they do not know and find there."[99]

The territorial effects of such techniques become all the more apparent when considering that, on the one hand, CSOs' role as implementers of INGO humanitarian mine action programs has been performed often on a northern Shan regional scale and that, on the other hand, activities of mine risk education have been accompanied by surveying and mapping village areas through forms, self-drawn community mapping exercises, and danger signs were needed. In addition, with the assistance of these same INGOs, nontechnical surveys have been carried out. These have occurred in accessible areas territorially designated by national government authorities and/or in contested areas. Activities have been implemented via networks of regional ethnic CSOs negotiating access with EROs and village heads mediation. Here, again, the words of the SSPP/SSA-N officer-cum-CBO's head illuminate the complex territorial effects reproduced: "Whenever we [CBO] do an MRE [Mine Risk Education] session, we carry out a village mapping. If they [villagers] saw items we do a map and then inform Halo Trust. The map in this sense means only the data that Halo Trust will transform into GIS maps. SSA already knows about it when the village is under their administration."[100] Far from being a result of the agency of actors alone, techniques of humanitarian arms control reproduce in a diffused manner, and reinforce or problematize, existing power relations. In teaching villagers how to behave in case an explosive item is found, for example, CSOs teach them "how to contact authorities nearby: heads of villages, monks, people who have power to contact ERO authorities; some areas sometimes are covered by three different armed groups at a time. So the head of village or monk has to call only one ERO to come and take the landmine away. The decision is on the authority that you trust."[101]

Humanitarian techniques of governing explosive devices have reinforcing and/or contesting territorial effects. From EROs' perspectives civilian knowledge about contamination is often tantamount to explosives removal,[102] as well as consequent deterritorialization processes in turn. Independent demarcation signs are removed by ERO troops in contested areas. Civilian access to, and the diffusion of, CSOs' humanitarian mine action educational and victim assistance activities are often hindered, controlled, managed. This occurs via the designation of accessible and inaccessible village areas and an informal system for regulating access that runs through the practice of identifying customary village heads as the basic territorial unit and authority. At the same time, CSO access and mine action delivery regulation becomes not only a process to redraw territory but also one of producing it through the management of weapons and armed forces in their relations with civilians according to logics of humanitarian control. The regulation of CSOs' activities becomes a channel for mediation between the means of violence and civilian communities. It constitutes a source of information as CSOs operate as a "spokesperson" of bottom-up instances to EROs as well as to the media. It is a governmental channel to reproduce and remold rationalities not only among civilians but also in the complex relations between leadership and rank-and-file divides: "I told you, the top leader is just in the office right . . . [he] cannot see all of the troops. That is why the civil society is important."[103]

Information concerning landmine deployment and contamination has often been circulated by ERO units under the logic of preserving communities of civilians. Such information takes the form of not clearly specified physical zoning or diffused proscription to access areas on the grounds of a lack of safety.[104] This technique of governing access does not only work to regulate mobility but also to construct territory as a peaceful and safe space while dismissing EROs' responsibilities should anything go wrong for civilians due to contamination. The practice of informing civilians about landmine use is part of EROs' trainings and discipline techniques, as both a way to reproduce control over armed forces and weapons, and a way to justify population control through humanitarian logics. Information to communities circulates through village heads and their subheads, which are normally in charge of a certain number of households in larger agglomerates.[105] During these information sessions, civilians are instructed on topics concerning not only contaminated areas but also how to report to EROs' channels. The different dimensions of the territorial effects generated by the diffusion of techniques of informing communities are recalled by the head of a Shan CBO operating in Kyaukme:

> RCSS also reaches out to villagers to tell them that they are not allowed to go to certain areas in the places where they live because it would be dangerous for them. They also once told us that, after battles and armed clashes or operations by them or other armed groups, the soldiers of their group take care of removing the landmines and explosives that may result from the clashes. They even have a dedicated crew that does so, that is tasked with clearing the areas after battles and operations. . . . People that live in RCSS control areas they can live peacefully and can work in the tea plantations and pick up tea leaves—but just a little bit far away the life is not peaceful at all because RCSS cannot provide security to those areas and their villagers.[106]

Nonetheless, people affected often ignore or resist such territorialization practices through counterconducts. They reject ERO humanitarian protective claims, arguing that "if there is one armed group it means also that another one may come at any moment. So actually, it would be much more secure if nobody was here!"[107] They access areas that represent their life places or place informal marking signs to prevent ERO units access to those areas.[108] Although in reverse, traces of this reinvention of humanitarian logics and techniques can be glimpsed in misunderstandings among soldiers and civilians telling them not to access contaminated areas:

> Accidents very often occur because in some villages EROs do not trust communities and, vice versa, communities do not trust EROs. . . . People might disregard the orders or warnings because of lack of trust or livelihood needs: the same is true also for EROs, who do not trust information about landmines passed on by the villagers, thinking that the community just does not want military forces to access certain places."[109]

Managing the Effects of Military Means

As the politics of managing explosive devices aptly shows, military means generate social effects to be managed. By illuminating how women's gendered everyday free labor in both the household and the armed force underpins militarization processes, for example, recent academic and nonacademic research has underlined that armed movements, and related conflict processes, are sustained by welfare practices (PWO 2006, 2010, 2011; Hedström 2020). But the diffusion of armed assemblages also *demands* social welfare techniques to cope with the effects of military means, not just to sustain their reproduction. In turn, welfare techniques to manage the effects of military means reproduce ethnonational territorialities.

This is particularly apparent in the management of the wounded and dead. The PSLF/TNLA's General Staff Office, the body managing PSLF/TNLA's staff, was endowed with a military care department dedicated to the welfare and assistance of Ta'ang soldiers. This department has been in charge of providing medical care and some form of socio-psychological assistance to people physically and/or mentally injured via a network of disguised, privately run guesthouses in areas of Lashio or other towns, as well as by arranging forms of livelihood assistance, or youth education for the wounded and fallen's families.[110] While such clinics and guesthouses are accessible to members of the PSLF/TNLA in general, the logic of ethnonationality imposes that a line be drawn between private health matters and ethnonational communities' health matters. For example, when he suffered some prostatic issues, Second Lieutenant D. was permitted to take residence in one such guesthouse in Lashio, but he had to arrange and sustain his cures privately. His brother—working as a mine worker in the Wa Self-Administered Division at the time—had to send him money in Lashio.[111]

Similarly, other forms of caring for the well-being of ethnic populations and territories via the armed collective can be seen in the constitution of the TNLA's mobile health units dispatched to ethnic communities lacking medical access, or the setting up of a so-called "land convention department".[112] Sometimes referred to as "geographic"—others as "land"—convention department, this PSLF/TNLA division was set up with a view to surveying and mapping Ta'ang Land so as to "defend the land of our people" in the aftermath of armed conflict entailing land grabbing through militarization and landmine contamination.[113] Some TNLA members were sent to Yangon University and to Thailand to study cartography and land-surveying techniques. The department was tasked with the mapping of Ta'ang customary land tenure systems' geographical extent, including the surveying of communal forests, free hold land, and forested areas not covered by customary tenure or other regimes that could have been turned into conservation areas.[114] Another key role of this PSLF/TNLA branch has been that of assisting Ta'ang people to officially register community land so as to avoid the application of the Vacant, Fallow and Virgin Land Management Union law that declared all vacant, fallow, and virgin land state property unless such land has been officially registered with the state (Gelbort 2018). The PSLF/TNLA's department devised a system to digitally record and issue documents certifying different land statuses in order to avoid the erasure of land held under Ta'ang ethnic customary tenure statuses or its transformation into privately held land.

To sustain these welfare techniques to cope with war-related effects, and in an attempt to expand beyond and overstep the armed authorities surrounding Ta'ang areas, formal taxation has been implemented by the PSFL/TNLA. The Ta'ang movement established standardized fee tables that categorize people, things, and space—such as types of commercial vehicles—to be taxed in territorial districts. Behind the articulation of a taxation system stood the ethnonational logic of ownership of the Ta'ang motherland and the rearticulation of revenue redistribution as a measure to reclaim that ownership from the invading forces of the Bamar-led state institutions. As PSLF/TNLA officials argue: "It [is] since long long years ago, right . . . ? . . . this is our mother land, right? It is just the central government that takes all of the taxes. We are the only indigenous owner and we do not get anything. We collect the taxes because this is our land, they use our land. . . . who uses our land, has to pay to us. And we look after our people."[115]

Taxation remained intimately related to the governing of the means of violence under those that have been called "war funds." The fiscal idea and practice of the war funds can be traced back to British colonialism in Burma whereby, throughout World War II, colonial authorities extracted a tax to fund war efforts of the British Empire. Underpinning the war funds is the logic that a statelike institution, throughout its territory, would extract public resources to be allocated to the maintenance and reproduction of the armed force. This rationality of stateness, in a sense, can be seen as countering the so-called business for peace approach promoted in the last decades by the Sit-Tat, according to which EROs would exchange business concessions of a different nature in ethnic lands for a halt to hostilities. The business for peace approach has been in part directed to the corrosion of the link between EROs' legitimacy and taxation and the elimination of forms of double taxation in the borderlands. As ERO taxation over people and resources has been based on the legitimacy that the EROs retain in light of their ethnonational struggle, the Sit-Tat tried to delegitimize the role of EROs in the eyes of civilian populations through the offer of business opportunities. Conversely, the war funds taxation arrangement in part reminds contributors that resources emerging from an ethnonational territory have to be allocated for the maintenance of weapons and armed forces fighting for the defense of that polity's population and territory.

The PSLF/TNLA's war funds system was imposed through censuses of Ta'ang Land. Via village heads, a census was periodically conducted in order to draft tables of households, people, and properties of each village and calculate war funds along with recruitment duties.[116] Official receipts were issued to households as well as at mobile and temporary checkpoints, while official payment-request letters were systematically issued to businesses in major towns of Ta'ang Land. Yet, this taxation system imposed by the PSLF/TNLA has not been taken at face value. While some people keep tax receipts in order to oppose future monetary claims at checkpoints, many others have refused to pay. Shan people in particular, but not only. In August 2018 a Shan woman in Namkham refused to pay taxes the TNLA deemed owing in accordance with its own rules and was arrested in a raid for reporting the case to the Myanmar police. A Ta'ang head of village likewise remains hostile to such war funds, noting that "armed actors only take money from the population but never provide anything to us. In our village the security of the people is provided by the Tatmadaw, in case we need we refer to them."[117]

As common for other EROs, individuals arrested for tax evasion, as well as other crimes, like drug use and trafficking, are held in bamboo prisons inside PSLF/TNLA camps. The Ta'ang politico-armed movement's judicial policies and regulations have been based on an adaptation of colonial penal and civil acts (at least until the 2021 coup). Emphasis, like for other EROs, has been placed on its internal rules and judicial mechanisms that civilians can resort to in case of members' misconduct. In this sense, a network of district and village administrators has been set up that has to report violations occurring in their territories to the public relations department of the TNLA. The department contacts village heads and administrators daily, but villagers and CSOs dispute the accessibility and accountability of such mechanism and often omit claims of misconduct. While these techniques and practices of the ERO aim to reproduce a certain sense of institutionalization, their reach is highly contested on grounds of a discrepancy between different social and geographical scales of the movement: "If some human rights violation occurs at the upper level it is condemned, but at the village or local level . . . not really."[118]

Blunt Rebel Rule in Frontier Assemblages

Part I of the book illuminated how, throughout the two decades of ceasefire and disarmament (ca. 1991–2011), Ta'ang land witnessed frontierization and territorialization processes in continuation with longer colonial and postcolonial histories (reconstructed in chapter 2). In such processes, the governing of military means was key to making frontiers and territorializing borderlands. Sit-Tat governmental assemblages of military means naturalized areas of these borderlands as spaces outside, or on the fringes of the Myanmar military-state polity. The four-cuts and related military infrastructures, in particular, turned Ta'ang areas into spaces to be reterritorialized by erasing recalcitrant rebel polities present "over there," at the frontier, while "incorporating" civilian ethnic minority populations. Death and military violence continued to be primary modes to territorialize the state polity. All this in an overall politico-economic context characterized by moments of agrarian dispossession, extractive economic projects, and brokerage networks underpinning the trade in narcotics.

The PSLF mobilized political rationalities of ethnonationality, narcotics eradication, and humanitarian security to manage the acquisition of weaponry, and to form the TNLA. Diffused throughout society, these logics emerged from historically and spatially nuanced relations among the different components of the leadership of the resistance movement, civil society actors, and larger frontier assemblages. Amidst a landscape of annihilating military violence intertwined with capitalist arrangements that would expand without the need to care for workforce and populations (Callahan 2003; Sarma, Faxon, and Roberts 2022; Prasse-Freeman 2023a), and in the face of the crumbling of the PSLO/A, forming the TNLA became the major political arena for the PSLF to reconstitute a Ta'ang polity. The PSLF lacked the governmental apparatuses needed to define and enhance "Ta'ang" populations at a large scale across the borderlands. Therefore, it harnessed the militarization of those very "Palaung" populations—populations that were being bluntly massified, divided, and made "surplus" by the military state, militias, and other rebel entities—to shape the bio-geo-political identity of the "Ta'ang" polity.

Governing the encounter between weapons and human bodies produced a political community detaching itself from the political-geographical projects of competing state and rebel polities. Yet, the PSLF/TNLA carefully avoided the definition of any clear-cut border. It was the reproduction of its margins, of its frontiers, via the management of military means, that allowed it to constantly reproduce the Ta'ang polity's territory. Ta'ang land overlapped with, and coexisted with, competing territorialization practices of other rebel movements. At the same time, nonmilitarized forms of contestation and resistance by unarmed actors enacted reterritorialization practices. While these chapters foregrounded the political geographies of territorialization and counter-territorialization processes, the second part of the book will delve more into forms of space other than territory that emerge through the complex politics of governing weapons and military means at the edges of state and nonstate authority.

Part II
WEAPONSCAPES

FIGURE 7. RCSS/SSA-S soldier taking a selfie during the celebrations for the inauguration of a new temple, Kyaukme township, 2019

NETWORKS (WEAPON BIOGRAPHIES)

Midafternoon. In the open-wall living room, sharp beams of dry-season sun-rays pierced through the cracks of the west-facing sheet-metal door, lighting up Yar's skirt in different spots. Sitting across a low mahogany coffee table, she lightly touched her htamein,[1] reaching out to feel the sunlit leg areas where the craft mine shrapnel had jabbed into the flesh underneath the embroidered cloth. Drawn shutters on the three remaining sides of the room enclosed a silence into which the deep barking of the Persian mastiff, down at the gate of the safe house where Yar stayed, penetrated like the echo of distant bursts from the past. And as the other hand, palm wide open, reached up to her right temple to mimic with a light slap the dizziness and unbearable pain that had caused everything that had happened afterward to seem remote, Yar's story reached the moment of the explosion. One of the various encounters with weapons her body had experienced throughout her life (Buscemi 2024).

It was late April 2022 and she had reached Chiang Mai province just a few months before, more or less a year away from the February 1, 2021, coup d'état. After a few months spent trying to settle the terms of the research fellowship that had granted her a chance to eschew incarceration in Myanmar, she had been forced to go underground, once and again, into the safehouse at the outskirts of town where we were now speaking. This time it had not been persecution from the Special Branch or the Sit-Tat to make her flee, though. Being the most influential Ta'ang pro-federal democracy female activist, as well as a close relative of several of the PSLF/TNLA's leaders who had put me in touch with her, Yar's presence in town had soon reached the ears of the RCSS/SSA-S Foreign Affairs Office in Chiang Mai. After harassing her on multiple occasions, one night they had tried to kidnap her. Now in her fifties and with many years spent in different leadership positions across the Ta'ang rebel movement, she remained a key focal point coordinating the activities of the Ta'ang political party (TNP) and CSOs with those of the PSLF/TNLA. Her last endeavor in this direction had culminated in the creation of the Ta'ang Landmine Action Network: a platform that emerged after the coup to coordinate humanitarian mine action efforts across Ta'ang land. It was an initiative that she felt strongly about, because of her first encounters with the destructive realities of weapons: "I myself have suffered from a landmine explosion. And also my younger brothers died because of a landmine."[2]

She was born in Homein, the village at the edge of Namhsan township where the weapons seizure with which the book opened took place. Her parents had been among the founders of the PSLO/A. After a counterinsurgency offensive by the Sit-Tat, during which both of them were killed, she and her brothers went to live with their uncle. Some years down the line, Yar's uncle started teaching them how to craft and use explosives: being in their early teens, the three of them would have better chances to blend in, pass undetected through checkpoints, and carry out hit-and-run attacks on regime camps and bases. And it was during one of these guerrilla operations that she was injured by the deflagration of one of her own explosive devices, whose pieces and debris she could still clearly feel, as if they had remained stuck in the legs' flesh.

"When approaching work meetings"—she now noted, sliding backward on the sofa in a reflex—"for a long time I had to calculate an appropriate distance between me and other people, so as to accommodate for twitches and tics in my legs." After medical treatment, and nowadays sometimes too, Yar would find it difficult to walk and sleep: sudden tremors and muscle stiffness took hold of her legs, while sleeping would have to be reduced to four-hour intervals to let the whole body reckon with the presence of the shrapnel and avoid falling into the stiffness that longer sleep intervals would have brought along.

In the aftermath of the accident, afraid the Sit-Tat would come after her family, she decided to leave northern Shan State. A close friend of her mother, a woman working with the Karen National Union (KNU) at the Myanmar-Thai border, adopted her. Although active mainly as a political officer in coordination efforts between the PSLO/A and other rebel movements, in the Karen border-lands Yar continued to receive military training. Like others in her batch coming from northern Myanmar, she was struck by how many good and new guns were available here in the early 1990s: her parents and uncle had long worked as relay to transport weapons from here up to the north, but all the Ta'ang movement could afford was a handful of older decommissioned pieces from Royal Thai

Army surpluses that never really reached the hands of trainees but would go straight to frontline fighting troops. So now, unacquainted with this degree of exposure to the gun, she felt her mind and body starting to be affected by the newly manufactured M-16 she would be trained with.

The instructors would first teach how to disassemble and reassemble the rifle, while showing trainees how each piece combined to work with the others. She and her comrades, Yar remembers, stood in some kind of awe. Everybody would gather closer and closer around the trainers, their eyes wide open, marveling at the different pieces and asking for more information: how long the barrel was, how far the rifle could shoot, how fast it was, how many different calibers existed. How could such a small thing, the projectile, and the gun too, kill a person? How could it give power? A kind of technical excitement lingered throughout the training sessions, one marked by a sense of hope, as if that small thing could finally grant a chance to bring down the Sit-Tat; as if it could grant a chance, and a space, for a better political future too. Yet, the following training phases—when the body was to be confronted with the actual weight of the rifle, with its ergonomics, or its recoil, throughout marches, drills, active shooting exercises, long-range hikes across mud and rivers—wiped off all the patina of enthusiasm, reminding them of how, in order to manage and use it, you first needed to become one with the gun. But becoming one with it, Yar noted, also meant that the power the rifle granted you with had to be constantly balanced, not to end up abusing it like military regimes had long been doing in Myanmar.

For her, the rifle—with all the adrenaline, excitement, fun, and joy that hitting the target would bring—catalyzed so much emotional and bodily energies that it would obscure, almost erase, other possible ways of doing and being. Ways that she continued to feel alive in herself: "Until today my problem is that there are two parts of me. In my brain and in my heart too. The left side does not want to be in armed conflict, the other side instead is very interested in conflict." The explosion that had injured her in that first encounter with weapons as a child still stood there, to signal how weapons and armed violence remained part of processes that could not only build but also destroy a political community. Although she had embraced the weapon because of her love for her "Ta'ang land and family," bodily familiarity with the gun resurrected the death of the political community too. At least in part. Every time she felt her body becoming one with the M-16, she could not help but remembering how her parents and their political project "had been killed by the soldiers because they held the guns." And so had both her brothers died, one after the other, in landmine explosions while fighting for the PSLF/TNLA.

By recounting her encounters with weapons, Yar's words provide a self-ethnography of armed assemblages. She tells a personal account of both the networked spaces through which weapons flow, and of the networked spaces that managing the encounters and entanglements between guns and humans generates, with their peculiar scales and places. The weapons part of these stories are more than mere technical objects of violence, such as a landmine or an assault rifle. In fact, as her story progresses, it becomes clear how the explosives and rifles at the center of her memories constitute complex networks made of technical matter, forms of knowledge, techniques of control. Not only do they flow through networks. They also produce network spaces with their scales and places across these borderworlds. Through her words, one can see how the practices of acquiring arms and governing an armed force are constantly related to calculations and materializations of space well beyond the reproduction of territory that we explored in the first part of the book.

While part I focused on how the governmental assemblages of blunt rebel rule produce political geographies of territory, part II delves into other dimensions of space: networks, scales, places. This chapter opens with an exploration of the specific contributions of armed assemblages in shaping the borderworlds as network spaces. It provides an analysis of how the spatialities and modalities of violence materially codified in weapons require different conflict actors to govern their presence, and their encounters with humans, in manners that are highly political. To do so, the chapter takes as entry points the biographies of some of the most widespread weapons in the Myanmar borderlands. Similar to Yar's story, these biographies illustrate how armed assemblages are constantly produced by and productive of networked spaces.

Weapon Biography One: "Dead" and "Alive" Landmines

What is this? Some years before my conversations with Yar, during an interview with PSLF/TNLA officials, I asked about the heavy burden landmines pose for civilians in Ta'ang land. One of the general secretaries of the PSLF/TNLA replied by showing this picture (Figure 8) (Buscemi 2021b). Empirical evidence, he claimed, of what he defined as a "bomb-mining" (in English, while speaking in Burmese) that the RCSS/SSA-S had set up on a road leading to a Ta'ang village in Namkham township.[3]

What is this? I asked him, when I first looked at the picture. The weapon appeared to me as a craft-manufactured landmine. A cylindric item, with what looked like a green lever attached to a metal cable. Red and blue plastic wires fastening them to a wooden stick planted into the ground to the side of a road leading to a village on a mountaintop. But could it perhaps be what specialists call an improvised explosive device (IED) instead? In doubt, with three different

FIGURE 8. "Persistent" explosive device manufactured through factory-grade tools and machines, Namkham, November 2019, author's fieldwork.

possible answers on the table—(1) a craft-manufactured landmine; (2) an IED; (3) a "bomb-mining" that RCSS/SSA-S set up on a road—some months later I turned the same question to a weapons expert.

What is this? This is something very interesting and strange, the expert replied. It is what we would call a "persistent" mine that does not use battery power but instead relies on a strike detonator only. It has most probably been manufactured through a factory-grade process rather than craft assemblage, or mere handcraft production, and shows "an increase in sophistication of mine manufacture which I am now seeing for the first time among nonstate armed forces in Myanmar."[4] Another definition. Four by then. All different, but all about the same thing apparently. A thing that started to emerge as something that is not reducible to anything else (Bourne 2012). It is more than one but less than many (Law 2002).

So again then, what was that? One could say, as Jairus Grove has replied in his article "An Insurgency of Things," that that was "the weaponisation of the throbbing refuse, commerce, surplus, violence, rage, instant communication, population density, and accelerating innovation of contemporary global life" drawing on "[s]urplus weapons, postcolonial injustice, e-waste, nationalist identification, or just rage . . . all drawn into amplification to create the IED and unleash its explosive potentiality" (Grove 2016, 342, 348). It is not *just* a landmine, but it is also a landmine in a sense. It is not just an IED, because its various components stick together in a different manner than an IED's elements would, but it is also not an unicum. Staying with Grove, the key point was, and is, exactly the impossibility to clearly draw its contours. It is an explosive fractal coherence: in the sense of an entity that cannot be reduced to a single determined identity or unit, but at the same time maintains a certain coherence constantly reproduced by a myriad of associations spanning across time and space.

The borderworlds of Myanmar are full of such explosive fractal coherences. Myanmar is in fact one of the few areas on earth where landmines are not only actively used but also actively produced by both state and nonstate armed forces (Landmine Monitor 2019). A wide range of landmines populate the borderworlds. Those most commonly encountered are industrially manufactured landmines and craft-manufactured ones assembled through a cacophony of commercially available and/or military surplus hi-(and less hi-) tech things. A third genre, though, to which the one in figure 8 belongs, can also be found: mines manufactured through factory-grade processes, that is, through machining and tools rather than by assembling commercially available pieces together, but not on an industrial scale.

Industrial Landmines

Industrially manufactured landmines have been produced partly in Myanmar and partly in other countries (Moser-Puangsuwan 2000). Until 1988–89, the Tatmadaw relied mostly on imports and military assistance programs in partnership with the armed forces of other states in order to acquire landmines. In particular the military regime was able to access USSR-designed landmines and copies of the latter provided by China (Selth 2000a).[5] In addition, further military assistance programs and deals aimed to restructure the armed forces allowed to obtain landmines from the UK, US, India, Italy, and Yugoslavia (Landmine Monitor 2004; 2019, 3; see also Callahan 2003, 176).[6] Transfers of a different nature run parallel with what constituted the forerunner of an import substitution program aiming to lead the Sit-Tat toward eventual autonomy in small-arms production, including landmines manufacture (Selth 2000a; Picard et al. 2019; Vining 2019). Since 1957, West Germany provided technical assistance to the Ka-Pa-Sa in order to build several factories mostly manufacturing small arms and ammunition. These factories, agglomerated in an industrial complex in the area of Prome and Magwe, included a site for the production of military and civilian explosives and a TNT-manufacturing plant, built in the late 1960s and early 1980s, respectively (Selth 2000a). While technical assistance was sourced from West German manufacturers, landmines produced in these facilities have appeared to be copies of types and models with which the Sit-Tat had familiarized through acquisitions via military assistance and transfer deals.

As the European Community and the US decided to impose an arms embargo on Myanmar in the aftermath of the 8888 uprisings,[7] the Sit-Tat sought to diversify its weapons acquisition processes and accelerate the wide-ranging substitution program that throughout the 1990s and 2000s led the armed forces to attain a certain degree of self-sufficiency (Picard et al. 2019, 47). This process included transfer agreements with the Russian Federation, Pakistan, Israel, China, Italy, and Singapore (Selth 2000a; Vining 2019).[8] After the collapse of the Communist Party of Burma (CPB) in 1989, China in particular became a major supplier of landmines for the Sit-Tat. In addition, Israeli and Singaporean technical assistance was key for maintaining and building additional manufacturing capacity, while agreements with China led to the construction of a new landmine factory in Meiktila (Selth 2000a; Vining 2019). To date the production of landmines has reportedly been consolidated in factories located at Ngyaung Chay Dauk, western Pegu (Bago), and in the area of Magwe (Landmine Monitor 2019).[9] Eventually these processes reproduced a "heartland" of landmine production and borderlands without any consistent industrial hub (and where landmines instead would mostly be deployed).

While the processes of industrial production tied to the Sit-Tat already provide a sense of the difficulties in clearly defining an entity to be called a "landmine" (its lines being blurred by technology transfers, design adjustments, or copies of models already in circulation), by moving to the acquisition practices of rebel movements and militias these difficulties become even more starkly outlined. For nongovernment forces the diversion of landmines produced in third countries, or the leakage of surpluses linked to the armed conflicts in Vietnam, Laos, and Cambodia, has represented a major channel of landmine acquisition. In particular, besides mines of US, Soviet, and Chinese manufacture, copies made in Vietnam and Thailand also circulated at the Thai-Burma border before the 1990s. Although only the CPB could rely on direct provision of landmines from China, seizures from Sit-Tat supplies represented the other main way to obtain a weapon technology that was otherwise unusual for some rebel movements— particularly those that lacked territorial influence, connections, and funds, and were not enmeshed in the borderworld flows of consumer goods and narcotics trading.[10]

Factory-grade Landmines

Yet, armed movements and militias have also manufactured landmines by them-selves. In this sense, the borderworlds have been characterized by both factory-grade mine production through machining and mines craft-manufactured through assemblage. Two major rebel movements, Khun Sa's Mong Tai Army (MTA) and the CPB, have been deemed to hold some form of industrial manufac-turing capacity. In particular, the MTA maintained furnaces and lathes, machining tools, and explosive stockpiles (like TNT) at its base in Ho Mong, where cop-ies of industrially manufactured landmines where produced (in particular the POMZ-2 model; Selth 2000a, 17). With the demise of MTA in 1996, part of the production assets and stockpiles were seized by the Sit-Tat while another part, including weapons specialists and manufacture technicians, further circulated to a faction that rejected the Sit-Tat-forced surrender of the MTA (the Shan State Army-South, later RCSS/SSA-S).[11]

Other rebel movements have not been able to produce factory-grade landmines but, throughout the decades-long armed conflict, still managed to consolidate craft-manufacturing expertise. Workshops for weapons repair and repurpose, as touched on earlier, have been established alongside arms stockpiles in different areas under the influence of EROs. The major armed movements have acquired such capacity through a mixture of long-standing experience in gunsmithing and explosives management; knowledge diffusion through exchanges, assistance, and cooperation with other armed actors; and trainings by foreign former or active military personnel, mercenaries, or volunteers.

For example, military-related knowledge sharing and training occurred at the Thai-Burma border throughout the late 1990s and early 2000s where the Palaung State Liberation Front force, before actually constituting a full-fledged armed wing, would train with the Wa National Organization/Army (WNO/A) based in the border areas of Mae Hong Son.[12] Before the fall of Manerplaw in 1995, the main Karen rebel movement (KNU/KNLA) maintained workshops for weapons craft-manufacture and stockpiles in the compounds of the headquarters (Selth 2000a). Later transferred to Mae Sot and other locations along the Thai border when Manerplaw was swept away by Sit-Tat operations, such bases served as hubs for the ethnic armed movements part of the interethnic umbrella organization National Democratic Front.[13] ERO contingents were based there and could share knowledge and practices of landmine craft manufacturing among revolutionary groups familiar with different kinds of military and civil explosives.[14] Before the February 2021 coup, this occurred particularly in Laiza—the headquarters of KIO/A on the Myanmar-China border, where different rebel movements have been hosted and trained in the last decades—or in specific areas of the Wa Self-Administered Division (SAD).[15]

Reliance on craft-manufactured explosive devices by politico-armed movements increased especially throughout the second half of the 1990s. As we have seen, the accessibility of weapon flows throughout the borderlands had started to change. Concerning industrially manufactured landmines, this was tied to some developments in particular. In 1987 the Thai Royal Army began demining operations at the Thai borders. With the conflicts characterizing the Laotian, Cambodian, and Malaysian borderlands fading away, Thailand also initiated the procedures to sign and ratify the Ottawa Convention in 1997 (McCracken 2001). As part of this process, stockpile destruction and security measures accompanied by legislative reforms concerning landmines were undertaken. Similarly, in Cambodia, the Cambodian Mine Action Center (CMAC) was launched in 1993. CMAC later became a primary vehicle of landmine decontamination and destruction, in part overlapping with the EU SALW control assistance program to Cambodia,[16] as part of larger security sector reform processes (Tholens 2012). More broadly the production of landmines decreased worldwide in the same years, and the stigmatization of such arms as pariah weapons was gaining momentum through the campaign to ban landmines and the consolidation of the mine ban treaty. For example, US landmine manufacturing came to a halt, stopped by the Clinton administration (Grove 2016), and it has also been argued that the international ban on landmines was eventually backed by state authorities partly because of the fact that industrial mines were believed to be an obsolete technology no longer appropriate to modern warfare (Beier 2011, 171). In addition to a considerable reduction in stocks and sources, industrial landmine acquisition became more difficult because of state territorialization processes through counterinsurgency strategies and practices, adopted by the Sit-Tat, which altered the geographies of the borderworlds. Besides military operations, these included the distribution of formal and informal business concessions to leaders of ethnic rebel movements, or people connected to them, based along main arteries or nodal places in border areas in order to generate indirect territorial authority and control via partnerships between the Sit-Tat elites, business companies, and rebels turned paramilitaries (Brenner 2019; Meehan 2016a; Woods 2016).

Craft-manufactured Landmines

Politico-armed movements without industrial landmines or factory-grade production capacity have focused on the craft manufacture of explosive devices. The devices crafted by EROs through assemblage are usually varied and unstandard-ized, although they present certain shared components.

They combine a triggering mechanism/device with a sensitive explosive chemical (priming compound) that provides the flame to ignite an explosive propellant; shrapnel made of different waste material; and a container to water-proof the assemblage and manage the encounter of these "elements" with the elements. A vibrant kaleidoscope of heterogeneous material objects combines in a way that each makes a difference in relation to the other. They engage in mutually constitutive relations that illuminate how the resulting explosive entity is at one time more than one but less than many (Law 2002, 3). Craft landmines are usually either victim-activated or radio-controlled. Most craft mines manu-factured by EROs require the use of a battery, since they deploy commercially available electric detonators.[17] Electromagnetic tributaries such as cell phones, car/motorcycle locks, flashlights, or other remote controllers transmitting and monitoring signals are used as triggering devices. Through the electromagnetic spectrum and the battery power, these provide current to the detonator by elec-tronically closing a switch. Electric detonators and batteries may be sourced from the construction industry or the metals and mining sector; bought on the civilian market; provided by movements' supporters; or extracted from different kinds of vehicles. The same goes for the required assortment of copper and metal wires. Similarly, the explosive propellant is self-manufactured through commercially available components such as agricultural chemicals, used in agribusiness, or TNT and other explosives finding use in large-scale construction and miner-als extraction. EROs buy them from legally operating businesses, directly or at times setting up dummy companies that can import such goods from abroad or on the Myanmar market. Otherwise, explosives may also be acquired through theft from industrial construction sites or mine complexes, or through informal transactions with third parties willing to steal and resell them. At times smug-glers have bargained commercial explosives or detonators to be delivered to rebel movements in exchange for informal agreements, allowing the former to freely operate throughout certain ERO influence areas. The shrapnel exploded by the device is usually composed of different materials like nails or bolts, shotgun pel-lets, or fragments of metal waste. A great heterogeneity of things has been used to provide to such conglomerates of matter some sort of fractional coherence (Law 2002, 2), from metal food carriers typical in Myanmar and Thailand to plas-tic pipes, bamboo pieces, glass or plastic drinking bottles, plastic wraps/casings, mortar rounds and other kinds of unexploded ordnance, or remnants.

This "insurgency of things" (Grove 2016)—in particular, electric detonators, batteries, primer compounds, and propellants—has been closely connected to Sit-Tat's counterinsurgency strategies. On the one hand, "buying" armed actors' collaboration through offers of preferential access to the drug economy, the Sit-Tat generated what Patrick Meehan has called a "limited access order." This in turn regulated the system of rents tied to narcotic flows and hampered the financial means of some rebel movements or parts of them, thus regulating access to cross-border flows of goods (Meehan 2011, 389). On the other hand, the granting of business concessions to rebels turned into paramilitaries/militias, and the development of economic partnerships with large agribusiness and construction companies, allowed the proliferation of sites and complexes to source the things needed to manufacture explosive devices. This also consolidated access networks that would have the possibility to channel imported goods of a different nature from the extended Chinese and Thai borderworlds.[18]

"Dead" and "Alive" Landmines

In the borderlands of Myanmar craft mines are seldom referred to as IEDs, even by humanitarian mine workers. The ontological recognition of explosive fractional coherences like craft mines as "improvised explosive devices" has been primarily consolidated through categorization efforts begun by the US Army in 2006, in the midst of the wars in Afghanistan and Iraq, for the purpose of identifying, tracing, and recording the use of craft-manufactured bombs (Beier 2011; Grove 2016). Especially starting from the 2000s, military expertise has circulated among rebel movements in Myanmar via the presence of different heterogeneous organizations set up by former military personnel—especially from the US Army, but not only—or militant political movements.[19]

These movements or associations—of which the Free Burma Rangers is just the most famous and conspicuous example—have often included former US Army personnel with previous active duty in the US wars in Afghanistan, Iraq, and northern Syria. Based mostly in northern Thailand but operating throughout the borderlands, these volunteer organizations understand themselves as humanitarian services movements aiming to provide mostly emergency medical care and humanitarian assistance in the form of shelter, food, clothing provision, and human rights documentation. While operating in different manners, as we will see more in depth in the next chapters, many of them have delivered different kinds of trainings to rebel movements besides direct involvement in humanitarian relief operations. The humanitarian character of such services has been blurred by an instrumental understanding of weapons and armed violence that underpins the inclusion of weapons handling and use trainings to members of ethnic armed organizations that would act as humanitarian relief agents and/or trainer-of-trainers. Weapons recognition trainings and manuals have been part of these humanitarian assistance activities, in a view to provide ERO members with skills to document human rights violations. In turn, education activities and materials part of these trainings have tended to consolidate terminology defining craft-manufactured bombs as IEDs.[20]

Nonetheless, normally they continue to be termed mines,[21] most often "battery" mines, by EROs. Using the term "battery mines," and practicing their manufacture, EROs try to parcel off the issue of battery-powered craft-manufactured mines from that of industrial landmines and victim-activated bombs more generally, in order to legitimize their use. The case of the PSLF is particularly informative in this sense.

In 2007, with the movement based at the Thai-Myanmar border and gradually initiating a rearmament process,[22] the PSLF's leadership decided to voluntarily adhere to a ban on landmines by signing a deed of commitment with the Swiss NGO Geneva Call. In the same years, other EROs were doing the same, in part in an attempt to boost international recognition amidst the harbingers of the "democratization" process initiated by the military regime with the road map to "discipline-flourishing democracy" (Geneva Call 2007). Clauses of the landmine-ban deed of commitment made direct reference to a prohibition of any victim-activated explosive device, drawing from the definition of landmines in article 2.1 of the Ottawa treaty (Geneva Call 2019). When conflict reignited, after the creation and consolidation of the TNLA in 2009–2011, in Ta'ang areas of northern Shan a problem arose. As the TNLA's general secretaries clearly recall:

> At that time we told them [Geneva Call] that for the landmines we could not avoid [using] them, right . . . ? . . . because we are still fighting . . . to attack the army, so we use landmines, we cannot avoid. However, we had already committed ourselves that we would not use landmines, so that is why we . . . use just the "active landmines," not the "dead landmine." We just use the remote landmines—if you put the battery, it is alive . . . if you do not put the battery then it is not alive; even [if] you put the battery, if you do not chip the remote, nothing happens. But sometimes, at first [there] was something that got a little bit problematic . . . like, if near the landmine you called with a cell phone, when the cell phone is ringing it could become a problem; or maybe if one went with the car or motorbike and stood nearby the landmine, when you start the engine sometimes it can go wrong.[23]

In this sense, Marshall Beier has underlined how the stigmatization of landmines as barbarous weapons has been crafted through a set of technological deterministic and essentialist logics that worked to parcel off an entity called landmine from broader problems of nonviolence and disarmament (Beier 2011). This occurred by defining mines as technologically backward because of their technical incapacity to discriminate humans (combatants/noncombatants) and the difficulties in separating the weapons from their environments. Such views depoliticized landmine use. They extracted such a backward, pariah weapon from a larger pool of "technologically advanced" arms constructed as legitimate instead (Beier 2011).

In Myanmar, throughout the years, the Sit-Tat has refused to access the mine ban treaty, adopting a different instrumentalist approach that repoliticizes landmines use—by arguing that it is their indiscriminate use, rather than their use per se, that constitutes the real problem when it comes to landmines. EROs instead operate to separate the battery mine from the category of landmines. Through rationalities and practices of explosives craft manufacturing, they appropriate the deterministic logic of humanitarian arms control vis-à-vis victim-activated arms. That is to say, they condemn the use of landmines but mold such stance to argue for the discriminatory capacity and nonpersistent character of battery mines. The insertion of batteries and techniques of remote control, as dimensions of the explosive device, are argued to unfold the slippery residual ontological space at the interface between, on the one hand, industrial landmines and victim-activated IEDs and, on the other, legitimate weapons that allow for discrimination. In other words, battery mines are argued to be discriminatory, unlike victim-activated landmines or IEDs.

And yet, exactly because of the fluid composition of battery mines, it is not so simple to remove batteries, conserve craft mines after they were laid, or even to remotely control them. Like any other weapon technology, they combine into fluid entities that do more than the EROs expected. In this sense they are also indiscriminate and environmentally integrated. Due to the very process of composition and how the elements come to constitute each other, they become inherently integrated—more than one but less than many, so to speak. Removing and reinserting batteries at ease, once craft mines have been laid, would intervene on the ecological niche of the weapon. Likewise, radio or other kinds of interferences can alter the environment and trigger them. They cannot actually be hibernated and then exhumed at will, as the EROs claim. Battery mines' persistence is contingent on different compositions. Depending on how terrain, elements, or the tree canopy combine, the battery can last shorter or longer. Equally, if the batteries sit in water, are cheap, or are poorly assembled, they may stop functioning. Charges could be eaten by ants. Persistent or so-called dead mines, like the one in figure 8, instead integrate differently. Manufactured through machining and factory-grade techniques, persistent factory-grade mines do not need battery power and employ a strike detonator. However, whatever the type, heavy rainfalls and landslides can move mines around, while the wooden sticks used to set them up can deteriorate and fall.

Such mixed vibrant combinations of commercially available hi-tech stuff and military surpluses are both generated by and generative of ecological niches. They are and become dimensions of the environment. In this sense, landmines in Myanmar are the quintessence bomb-mine. Because not only does one not know with precision what they are, but also how many and where they are: a largely unknown and dispersed contamination of largely unknown and dispersed technical objects.

Weapons—be it booby traps, battery mines, persistent mines, or industrial landmines—are inherently geographical. As another interlocutor commenting on the picture of the "persistent" mine noted, while reacting to the question that opened this section ("what is this?") and providing their own analysis:

> The first thing is not the device itself, but it's placement. There are a number of issues here. This is not where/how you'd expect a device to be placed. It's clearly visible on an open dirt track, and even at night, the stake would stick out like a sore thumb; even civilians would clearly spot this as something that doesn't belong, much less hardened infantry troops who would be looking out for booby traps, mines, etc. . . . [An] option would be it is not meant to maim/kill anyone at all, but rather it is just a warning; it was placed in such an obvious manner as to warn people to not come anywhere near this area, as there will be other mines/IEDs (or not; just the threat is usually good enough).[24]

Weapons appear to be productive of networked spaces, since they can alter the configurations of a familiar geography and reformulate it in mutual relation with their own internal technical coherence (Grove 2016, 8); while, at the same time, they are produced by networked spaces too, since their consistency or inconsistency emerges out of surrounding assemblages, of which they are formed and with which they enter into association. A certain resonance exists between the explosive artifacts and their networked spaces. On the one hand, mines are built and distributed throughout the very architecture of the milieus of Myanmar's borderworlds. Discursively they are made to resonate with the environments of landmines' unavailability, humanitarian mine action, and landmine bans; while materially they resonate with those of "ceasefire capitalism" and socioeconomic distributions of relations of arms production and control. On the other hand, compositions of mines as technical objects and human agency reshape spatial relations in nonstandardized ways. They are used not only defensively but also offensively. And in any case, landmines lead to a reformulation of the spaces emerging with them. The ways in which these weapons are set and the ways they act are not always the same and combine in unpredictable manners.

Weapon Biography Two: AK-type Rifles

FIGURE 9.   AK-type assault rifles of an SSPP/SSA-N unit, July 2020, author's fieldwork.

The most common weapons in the borderworlds of Myanmar are Kalashnikov-type assault rifles (AK-type) (figure 9). Throughout the Cold War period and well into the 1990s and mid-2000s, weapon flows in the borderworlds were characterized by the circulation of surplus weapons from regional armed conflicts in Vietnam, Laos, and Cambodia. In the 1960s and 1970s, considerable flows were channeled into the region by the US and USSR, which injected their own production weapons (Tholens 2012, 101; Capie 2013, 92). Moreover, in the 1960s the Chinese Communist Party (CCP) supported communist organizations in Laos and Vietnam, as well as Thailand and Malaysia. In Myanmar, in 1967 the CCP agreed with the CPB to a ten-year aid program consisting of full-scale support in terms of training, logistics, medical services, and weaponry (Smith 1999, 248).[25] It has been estimated that in the aftermath of the Vietnam War in 1975, Vietnam and Cambodia alone were home to two million weapons and 150,000 tons of ammunition (Capie 2013). At the beginning of the 2000s, before the European Union's Assistance on Curbing Small Arms and Light Weapons (EU-ASAC) arms control program was carried out, figures referred to the presence of 500,000 to one million SALW in Cambodia only (Tholens 2012, 71).

Surplus weapons were trafficked to Myanmar's borderlands especially via Thailand's northwestern areas, in particular Mae Sot, Mae Hong Son, Chiang Mai, and Chiang Rai. The circulation of such weapons rested on negotiations among different links and nodes of smuggling networks. For example, when in the second half of the 2000s weapons availability and access became increasingly scarce and difficult (especially in the southeastern borderlands),[26] this downward trend resulted from a combination of technical measures of physical security and stockpile management tackling surpluses in Cambodia, stricter state counterterrorism and counterinsurgency strategies in Thailand and Myanmar, and military-private business partnerships that were expanding state authorities' influence in the borderlands (Meehan 2011; Brenner 2017b; Buscemi 2019).

In this context a "new" channel of weapons acquisition and circulation emerged. At least two rebel organizations—the UWSP/A and the KIO/A—developed the capacity to produce their own rifles (and other types of weaponry) to adapt to the changing landscape. Traces of these processes and assemblages can be seen in figure 9, which depicts AK-type assault rifles held by an SSPP/SSA-N unit lined up on the veranda of a house in northern Shan.[27] They all seem to be AK-type rifles, but they are not. The second, third, fourth, and fifth rifle from the left are AK-type rifles that, without in-depth examination, could be of German, Russian, or Chinese production. The first, sixth, seventh, and eighth rifles from the left are so-called Type 81 clone rifles produced by the UWSA.

In the late 1990s China's People's Liberation Army (PLA) substituted the Type 81—a rifle combining elements of other AK-style firearms—with a newer assault rifle (the QBZ-95). At this point the PLA engaged in collection efforts to withdraw the Type 81 from army stockpiles. Yet, Yunnan-based military personnel transferred surplus Type 81 weapons to the UWSA (Capie 2013, 99; Lintner 2015, 133). The UWSA could now substitute older rifles—especially the Type 56, which had been inherited in large quantities with the crumbling of the CPB in 1989—with the Type 81. At the same time, in the second half of the 1990s, and especially throughout the 2000s, China-Myanmar relations improved and Chinese-manufactured weapons being transferred to rebel groups in Myanmar became a sensitive topic.

To adapt to the situation, the UWSA set up a small production and assembly factory through machinery, components, and expertise sourced across the border in China. It began assembling two different models of UWSA Type 81 indigenously produced.[28] Two hybrid rifles not in circulation before were created. One model is almost an exact copy of the Type 81 (first and eighth from the left in figure 9). Another one combines components of other rifles for reasons of ease of manufacture and the possibility to use spare components or components accessible on the black market (sixth and seventh from the left). This second variant incorporates the action (operating components) of a Type 81 rifle and the gas block mechanism of a Kalashnikov-style rifle. Being a hybrid and a completely different rifle, the UWSA Type 81 became noninterchangeable with other rifles, meaning that, for example, single components could not be so easily exchanged in case of wear, or magazines other than its own could not be used with it reliably.

Similarly, in 2009 the KIO/A set up local manufacture through machinery and engineering assistance sourced across the border. Up to then weapons acquired through UWSA's channels had represented a key source for the KIO/A. Nonetheless, it remained logistically difficult to travel to UWSA areas and then move weapons back to Kachin.[29] With the heightening of tensions in connection with the military regime demands to transform EROs into paramilitary forces under the control of the army after the promulgation of the 2008 constitution, the KIO/A hired engineers and technicians to construct a weapons factory located in an upland area in the outskirts of Laiza (KIO/A's headquarter). Apparently, the production site began manufacture in 2010.[30] After completion, the factory was accessible by civilians, but a year later access to the area was restricted by the KIO/A. Besides assembling a variant of the Chinese Type 81 (called M-23), another weapon patterned on the AKM rifle was also produced. This rifle has since then been widely circulated to EROs in northern Shan.[31] In addition, the factory in Kachin managed to produce and assemble other weaponry such as rifle-mounted grenade launchers, marksman rifles, machine guns, and mortars.

Nonetheless, the weapons populating the borderworlds of northern Shan State are neither artifacts to be encountered there, nor the independent products of technological development and diffusion (Bourne 2012). In the picture, there is another interesting detail that shows further transformations of the weapons throughout their processes of circulation: the ribbon tied to the muzzles of the second and eighth rifle from the left.

In northern Shan, rifles are often customized with red and/or white cotton ribbons tied to different components. Attached to cars, motorbikes, and other technical objects, these ribbons are amulets connected to Buddhist and animist practices of spirit or ghost worshipping that are especially widespread among Shan populations (Tannenbaum 1987; de la Perriere 2009; Ferguson 2021). Rebel movements' soldiers often visit spirit shrines when they are expected to go to war, and newly acquired weapons are brought to them to undergo ceremonies performed by mediums. As the Nats—as spirits are commonly referred to—can act against or upon humans, as well as upon events, by negatively affecting or altering them, the ribbons tied on the guns' muzzles are intended to provide a sphere of protection to the carrier of the weapon and the weapon itself after propitiating the spirits—for example, by allowing bullets to be ejected and projected correctly when one shoots, or by being spared from malevolent beings' actions.[32] Similarly, combatants visit mediums to ask spirits' permission to use their weapons or to donate wood or bamboo replica weapons to the spirits in order to gain their favor.[33] Or at times shrines are built in armed actors' forward bases in frontier areas, where periodic donations are carried out, including the submission of replica firearms to the local Nats.

The red-and-white amulet is one among many other techniques that are used to obtain protection and invulnerability for the people that bear them, such as alchemy and tattoos (Tannenbaum 1987; Ferguson 2021). Nicola Tannenbaum has shown how these techniques reproduce the values, moralities, worldview, and power relations that are expressed through animist and Buddhist integrated religious systems (1987). Systems of political authority in Shan State, like those of rebel rule, often legitimize themselves by embracing such values, moralities, worldviews, and power relations.

The power embodied by the amulet—which becomes an integral element of the weapon—is more the effect of specific discourses and practices than a quality of the person providing or carrying the amulet. Similar to "protective" tattoos, the power of the amulet derives from three main bundles of (micro-)practices (Tannenbaum 1987, 701). First, it derives from the status of the medium performing the process and providing the amulet. The medium's condition itself, though, is the result of a combination of the practice of discipline and self-restraint according to Buddhist precepts with a dialectic of generosity and offering (whereby those with more bestow upon those with less, while those bestowed upon perform offerings to seek refuge/protection and display gratitude). Second, power derives from the constant awareness/acknowledgment of the authority of the medium, the medium's lineage of teachers, the spirits, and the Buddhas with their teachings and precepts. This awareness-cum-acknowledgment is practiced by both the medium and the person seeking protection and refuge—the former actually channeling the offering made by the latter to the superior beings. And, lastly, power is an effect of the respect paid to the Buddhas and their teachings, as well as of the Pali scriptures recording the Buddhas' words, which are recited during the ritual and at times inscribed in the amulet. In any case, amulets, similarly to tattoos, create a barrier and (spatial) sphere of invulnerability that protects the bearer, fends off dangers, and impedes elements of a different sort to enter. Such sphere of protection and invulnerability is essentially made of respect for the authority of the beings involved in the ritual. Such respect and the sphere of protection that it produces circulate together with the amulet. The amulet-weapon in a sense contributes to "certify" the political authority and power exercised through/in it.

Such spiritual practices are to be understood as entangled dimensions of the technological artifacts and associated techniques of fighting and violence (see also

Ferguson 2018). On the one hand they represent a personal and collective mechanism to cope with the insecurity and violence linked to the gun while shaping the technological possibilities of weapons (both those that one carries and those that one confronts). On the other hand, they are the expression of local agencies and forces (Ferguson 2018). For while patronage and protection requests may seem the most immediate purpose of worship practices, they also entail prestige mechanisms of arms control and channel sociocultural and spiritual norms of (non)violence (Ashkenazi 2012).

Power relations are reproduced as an effect of the micropractices of the ritual via which conducting the conduct of the gun becomes protected and authorized. These practices reproduce authority and power. The guns and the violence inherent in them have to be inserted into and subordinated to the values, morality, and worldview of the cult of the local Nat spirts integrated with Buddhist religious systems. Weapons, like other technical objects, have to be managed accordingly, because they could be affected by the spirits once they enter into their domain, or could be affected by other beings as they step out of the spirits' domain of protection. Presenting the technologies of violence to the spirit and/or attaching ribbons through the figures of mediums authorizes a certain use of force in the domains of the Nats, and reproduces relations of authority out of respect for them outside their domains. These social and spiritual agencies and forces are not immanent, autonomous powers that can prevent or alter events but dimensions that engage in mutual relations with technologies of violence. In this sense the weapon is neither understood as a tool purely subservient to the human, nor as a completely autonomous technology, but rather as a technical object embedded into a system of norms, values, and practices that actively intervenes to reproduce a sphere of protection and authority. Likewise, violent practices unfolding are also understood as more than a function of combatants or enemies' agencies alone.

Moreover, amulets and practices of spirituality contribute to reproduce and reshape the geographies of arms control and authority in synergy with guns. This spatial dimension has to do with the mutual interrelation between territories and places—of which spirituality is a key defining (geographical) component—and the technology. The two redefine each other. In fact, in Shan State, villages are understood as the domain of the Nat's spiritual sovereignty, the worship of which constitutes broader territorial ties and boundaries while feeding into a sense of place for people (Ferguson 2018). The territorial boundaries of rebel movements' sovereignty are renegotiated through the use and control of arms and armed collectives. At the same time, the spirits and spiritual spheres extend to the domains of human-technology relationships and dynamics of violence when it comes to control over weapons and armed units. The guns become the object of the moralities and sociocultural norms connected to both religious (Buddhist) and spiritual (Nat) practices, as well as a channel for extending these moralities and sociocultural norms into what are contested frontier areas for rebel movements. The gun can become a token of spiritual sovereignty pushed to the front lines when spirts have been honored back to the shrine in the village one comes from. Or it can become a token of compliance with the local spirits and of the legitimacy that derives from it.

Inflecting the violent functions of the gun through spiritual and religious practices speaks not only of societal norms and practices of arms control but also of how the gun then contributes to reproduce a certain geographical understanding of spiritual power and political authority attached to it. Neither the presence of weapons by themselves, nor the relations among various actors alone, produce social and spatial relations. Rather, weapon technologies and techniques of control combine actively through various practices to mold the borderworlds' networked geographies of (dis)order.

Weapon Biography Three: Type 11

FIGURE 10.   A Type 11 assault rifle of an SSPP/SSA-N combatant, Wan Hai, November 2019, author's fieldwork.

The Type 11 is a derivative of the German Heckler and Koch HK33 assault rifle (figure 10). It has been produced in Thailand through licensed production agreements and for several years has been the service rifle of the Thai police and Thai army. While definitely less widespread than AK-type rifles, the Type 11 has been sourced in Thailand and eventually circulated into Myanmar via gray weapons transfers and loopholes in legislation.

Since the 1950s Thai state authorities started to engage in a series of territorialization processes aiming to reorganize people and resources along territorial lines and make the Thai borderlands more legible (Peluso and Vandergeest 1995). The Thai military and police forces held a key part in these strategies and were reinforced through weapon supplies and training provision offered by the US (Peluso and Vandergeest 1995, 413). A sort of internal borderline was generated via militarization of the Thai borderlands for counterinsurgency and forestry management purposes.[34] Checkpoints and military bases delimited this buffer area where the army could retain de facto authority over security and in part commercial matters (especially in relation to trade and forestry) (Peluso and Vandergeest 1995). Heads of security forces were periodically rotated in and out of the area, but this did not prevent them from becoming enmeshed in the cross-border economies of the Thai-Myanmar borderlands, including weapons smuggling. Flows of different kinds of weapons—in particular, M16-style and Type 11 assault rifles—occurred especially via this sort of inner border area on the Thai side of the borderlands. Until the late 1990s the most prominent sources of weapons in circulation were linked to state security forces in one way or another. For example, weapons that were part of covert supplies destined for conflict actors in Cambodia were withheld and smuggled when transiting through the Thai borderlands. Batches of official transfers supplied by third countries (mainly the US) to the Thai police and army, or segments of official imports operated by Thai-registered trade companies as part of procurement deals, were diverted from stocks. Decommissioned arms waiting for substitution, due to declared wear or malfunctioning, and weapons seized during security forces' operations also leaked (Phongpaichit et al. 1998, 137–142).

Moreover, the Cobra Gold multinational joint military exercise first held in 1982 by the US and Thailand, and later involving other state armies, represented a further relevant mechanism of weapons flow since the early 1980s.[35] The exercise generated surpluses that would then be smuggled to rebel organizations in Myanmar. It would trigger the movement of considerable quantities of weapons throughout Thailand and allow for different forms of diversion. At the same time, Cobra Gold would also represent an occasion for arms procurement and entailed at times the accumulation of weapons left in Thailand by the participants in the exercise. The primary beneficiaries of these flows would be the KNU/KNLA, KNPP/KA, RCSS/SSA-S, and previously Khun Sa's MTA.

Such weapon flows have been strictly intertwined with cross-border economies in narcotics and consumer goods. Throughout the 1960s–1980s in particular, consumer goods legally acquired were collected in large warehouses located in the buffer area before being inserted into the Myanmar black market. Once across the border, the goods would have to proceed through the areas of rebel movements, militias, and the Sit-Tat. In exchange, these armed actors would request businesses' middlemen to facilitate the acquisition of weapons in Thailand.[36] Similar mechanisms and exchanges characterized the flow of narcotics as well as the issuing of business concessions by rebel movements and paramilitaries in their areas of influence. After 1989, for example, when logging authorizations in Thailand were withdrawn and the Myanmar regime instead granted permits triggering an influx of Thai logging companies, the latter were found to be involved in arms flows to nonstate actors in exchange for informal agreements securing operations in southern Shan State or for partnerships in drug trafficking (Phongpaichit et al. 1998, 141).

Since the late 1980s, political and economic relations between the Myanmar military regime entourages and the Thai army and state authorities improved (Brenner 2017b). Given the close intertwinement between Thai-Myanmar military-business partnerships and nonstate armed actors in these border-worlds, for rebel movements it became more important to attempt to conceal the origins of their weapons. Masking markings on weapon components has been practiced in various ways, such as pouring rubber, metal, or other substances on arms' serial numbers, in particular those originating from Royal Thai Army stockpiles (like the Type 11 in figure 10).[37]

Concealment practices of identification markings by nonstate actors have been tied in part to the fact that many weapons were acquired in the first place via corruption and the exploitation of loopholes in Thai firearms legislation. The amount and type of firearms that can be imported and possessed at a time by Thai-registered dealers has been limited by firearms laws.[38] At the same time, government agencies, state enterprises' personnel, and designated individuals supporting the activities of state authorities have been legally allowed to obtain a permit to hold and carry a weapon.[39] In addition, arms acquired by state-affiliated officials have often been exempted from the overall norms applying to civilian firearms possession that ban automatic and semiautomatic weapons.[40] This generated a large pool of state-affiliated officials with a potential right to initiate procedures to obtain a permit. These permits in turn granted the possibility to import a firearm through authorized dealers. Corrupting eligible state-affiliated figures allowed to accumulate signed requests for permits to import a weapon, while corrupting local registrars allowed gun dealers to lawfully purchase the weapons through the granted permits. These became consistent practices for circumventing the import and possession limits legally imposed on gun dealers.[41] Once imported via legal permits, however, no limits constrained the retrade of firearms by gun dealers. This in turn granted them with the possibility to either legally resell the arms, provided these complied with the type restrictions imposed on civilian firearms acquisition, or to smuggle them underground (Bohwongprasert 2020).

Weapon Biography Four: "Yat-Thai" Rifles

FIGURE 11.   RCSS soldier with a "Yat-Thai" rifle in Loi Tai Leng. Source: Vining (2020).

The considerable munition shortages that many rebel movements experienced since the second half of the 2000s was attested by the fluctuation in prices at the Thai-Burma border. Here, around the end of the 1990s an AK-type or M-16 rifle would be sold for around 4,000 baht (ca. 108 USD), while one round of ammunition cost 3 baht each (ca. 0.10 USD). In 2008 the cost of an assault rifle had reached 10,000 to 15,000 baht (ca. 316–474 USD), and bullets came at 10–15 baht (ca. 0.30–0.50 USD) each (Weng 2008).

While only the UWSA and KIA, as explained earlier, had managed to achieve a certain degree of production capability, the majority of the rebel groups struggled to obviate this paucity, particularly concerning small-arms ammunition. Thus, seizures during guerrilla operations and corrupt sales by Sit-Tat personnel with access to military stocks, or stationed in forward bases, became two relevant sources of ammunition acquisition for many EROs.[42] Yet, and this holds true nowadays too, when a seizure would be carried out, or a sale by Sit-Tat officials arranged, a series of issues concerning rifle-ammunition compatibility could emerge.

Sit-Tat-produced standard infantry assault rifle series are chambered for 5.56x45 mm ammunition. In theory, this is the same caliber that M16-style rifles use.[43] M16-style rifles represent a considerable portion of the rifles held by rebel movements throughout the borderworlds—in particular, those who emerged in southern Shan State, Karenni, and the southeast.[44] One would thus expect that Sit-Tat ammunition could be used on M16-type rifles, so as to cope with the ammunition scarcity. Nonetheless, Sit-Tat-produced rounds of ammunition have technical specifications that cause malfunctions when used in the actions of non-Sit-Tat series rifles, even though equally chambered for the same caliber, like M16-style ones. At the same time, these 5.56x45 ammunition cannot be used even with the other most widespread type of rifles held by EROs, meaning AK-type firearms that are fed by ammunition of a different caliber (7.62x39 mostly). Rebel movements thus would have ammunition available but would lack 5.56x45-caliber rifles in which these can be used: essentially, if Sit-Tat-produced ammunition was used with M16-style rifles, the actions of the latter would break.

Two additional problems have overlapped with this. First, even if supplies of non-Sit-Tat-produced 5.56x45 ammunition could be sourced—such as bullets produced in Thailand, or in the stocks of the Royal Thai Army that at times are accessed through deals involving Thai officials—there would still be a scarcity of reliable well-functioning rifles. This is because the M16-style rifles held by EROs are mostly quite old surplus weapons, many of which use worn-out components prone to breakage or broken parts. Second, Sit-Tat-produced, 5.56x45-caliber rifles could actually be used with non-Burmese-produced ammunition. Yet, these rifles are not easily available in large quantities, as their acquisition is limited mostly to seizures from the army. They do not allow for easy interchange of components—among them and with other weapons—due to poor manufacturing quality and lack of sufficient spare parts. And last, ERO combatants are slightly reluctant to use Sit-Tat-produced 5.56x45-caliber rifles due to technical and tactical preferences.

Therefore, at this point, craft-manufacturing and assemblage practices come into play. This occurred in particular in Loi Tai Leng, the headquarters of the RCSS/SSA-S built on several mountain ridges located literally on the Thai-Myanmar border. Here the RCSS/SSA-S has long maintained stockpiles of old components and surplus guns as well as gunsmithing workshops.[45] Local gunsmiths are employed in these workshops where surplus firearms are repaired, adapted, and modified.

A rifle called "Yat-Thai" is manufactured here using AK-type rifles as "baseline" weapons (figure 11). The main structural components of surplus AK-type rifles are maintained, fixed if needed, and adjusted in order to be assembled with likewise repurposed parts of other firearms that will allow the Yat-Thai to fire 5.56x45 mm Sit-Tat-produced ammunition. Magazines and barrels of either Sit-Tat-produced rifles or old M16-type guns are modified and adjusted so as to be assembled with functioning AK-type actions and gas operation mechanisms, while wood furniture pieces are crafted locally. This way, the 5.56x45 mm Sit-Tat-produced ammunition—which does not cause malfunctions in repurposed AK-style baseline components—can be used.[46] When combined with stashes of old M16-style rifles, few pieces of seized Sit-Tat-produced guns, and AK-type rifles, local gunsmiths' technical knowledge, and practices of components modification, assembly, and craft manufacture provide a potential solution for using otherwise unusable Sit-Tat-produced ammunition.

Gunsmithing practices, techniques, and knowledge have developed differently in different parts of the borderworlds of Myanmar. In Chin, Sagaing, Kachin, Karenni, and northern Shan State, though, gunsmiths have operated mostly in relation to the craft manufacture of muskets for hunting (until the 2021 coup). For example, as a hunter in the area of He Pu village (Kyaukme township) noted when talking about his rifle: "[I]n the village there are people who hand craft the firing mechanism, while most of them go buying the barrel in Hsipaw."[47] Metal barrels are crafted in Hsipaw, where they can be bought in a specific marketplace. In these areas, individuals who are unaffiliated with rebel movements have had virtually no access to semiautomatic or automatic firearms: "[V]illagers only keep hunting rifles and 'shot-guns,' but not semiautomatic or automatic arms."[48] Located just outside the town, He Pu is under the influence of the Sit-Tat and, even though many different EROs maintain links in the village, hunters "have to report to the army about how many firearms" they have.[49] Villagers have to report the possession of weapons to the village head, who has to request a permit to the Sit-Tat local base in order for the person to hold and use the gun. Subsequently, every year the head has to report about the number of firearms held in the village.

Similar procedures are also in place concerning the craft manufacture of firing mechanisms or other components. Gunsmithing practices have evolved differently here. In part because of the different modalities in which the authority of rebel movements and state apparatuses has been territorialized; in part due to the relative inaccessibility of semiautomatic and automatic weapons, and the noncirculation of techniques, machines, and tools of components craft manufacture.[50]

In southern Shan State and Karen areas, conversely, gunsmithing practices have been fertilized by the possibility of accessing and participating in events occurring in the Thai part of these borderworlds, such as military trainings with state armed forces and agents, including weapons handling, management, and maintenance; local factory-grade manufacture of weapon components through computer numerical control (CNC) machining in Mae Sot, and other border districts; or the accessibility of replacement weapon parts sourced from the airsoft gun market in Thailand.[51] This cross-fertilization occurred in particular throughout the 2000s. In the aftermath of the terrorist attacks of 9/11, the US registered a considerable increase in the production and commercialization of AR rifles—a civilian version of M16 types—whose production had been previously falling due to a federal ban on assault weapons (Watkins, Ismay and Gibbons-Neff 2018). This in turn made possible the development of a considerably larger market of replica firearms, especially AR-15 airsoft rifles, not only in the US but also across Southeast Asia. Here production came to be concentrated especially in Taiwan. Airsoft rifles produced in Taiwan have been of relatively higher quality as compared to others. Therefore, when Taiwanese-produced airsoft weapons were injected into the Thai civilian gun market, this offered a potential good source of components for substitution in automatic/semiautomatic weapons, in particular M16-style rifles. In addition, CNC machining capability was developing in parallel in Thailand thanks to the spread of internet coverage and auto-CAD programs that would be used in CNC machine setup and use. In Mae Sot in particular, a bustling border town key for weapon flows, this allowed rebel groups to forge copies of components of M16-style rifles, at times using the legally accessible airsoft pieces as models.[52]

Although perfectly functioning, weapons like the Yat-Thai or other repurposed ones are not considered to be the same as industrially manufactured pieces due to recurrent glitches. While there are differences among the status of the arms issued to different rebel force units, assembled refurbished firearms are normally relegated to rear positions or bases and camps outside frontier areas, where different armed actors overlap and fighting is more intense. In this sense one can see how there is a sort of deterrence and institutionalization function that the weapon operates beyond its actual technical status. The geographic management and deployment of weapons becomes a symbolic and material mediator of territorialization practices. The constitution of fixed military installments, headquarters, and security provision at/of such places and surrounding areas has to be calculated in combination with the networks of firearms and ammunition availability and craft manufacture possibilities.

Weapon Biography Five: Pyi Thu Sit Weapons

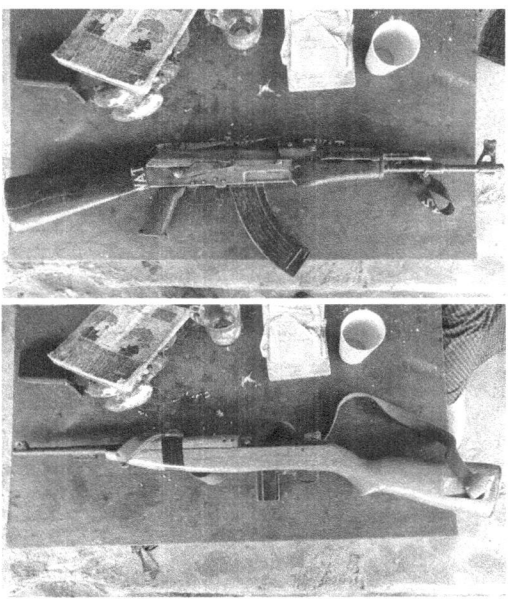

FIGURE 12.   Assault rifle (above) and M-2 carbine (below) in Pang Kun village, Kyaukme. Author's fieldwork.

Pang Kun is a village located forty kilometers northwest of Kyaukme, on the mountains straddling the road leading to Monglon and eventually to Mogok. The M-22 assault rifle and the M-2 carbine (figure 12) are part of a larger number of weapons that were distributed to the members of the local Pyi Thu Sit (village community militia) by the Sit-Tat in 2017. A first Pang Kun community-based militia was created by villagers and the Sit-Tat in 1976. Even in those years, the antecedents of militia formations in northern Shan were already numerous and stretched back to the late colonial and pre-independence period. In the 1930s the *sao hpa* of Hsenwi,[53] for example, had authorized local leaders under his authority to create armed formations in the areas bordering China (Buchanan 2016, 6). Throughout the 1940s and especially in the aftermath of World War II and Burma's independence, these and other militias created by political leaders or important customary figures stood somewhere in between personal security forces and political-movements-cum-militias used as mobilization tools in domestic political rivalries (Buchanan 2016).

However, more immediate precursors of the Pang Kun village militia should probably be traced to the following decade, when a combination of events resulted in the deterioration of the security landscape of Shan State. Political insurgencies along ideological and ethnic political lines, mutinies in the Sit-Tat, Kuomintang (KMT) invasions, and dacoity (armed banditry) pushed the *sao hpa* to train men into the so-called Shan levies and organize volunteer defense forces into villages (Buchanan 2016). In 1956 the Union of Burma government devised an official program—the Pyusawhti program—as a security sector strategy to coordinate local police, administration, and the army with residential or mobile militias in village areas and to support the Sit-Tat in counterinsurgency operations. With the unfolding of Shan and Kachin ethnonationalist rebellions, after the 1962 coup d'état the Sit-Tat experimented with different arrangements of militia strategies, among which was the so-called Pyi Thu Sit (People's Militias), that is, community-based formations linked to the army.

Yet, it was only with the advance of the CPB into eastern Shan State in 1968 that Pyi Thu Sit plans for training, procedures, and weapons distribution were institutionalized. These techniques had emerged especially in connection with the progressive adoption of concepts linked to People's War doctrines developed via Sit-Tat officers' study visits and weapons procurement negotiation meetings in Yugoslavia, Germany, and Israel (Maung Aung Myoe 2009; Buchanan 2016). According to John Buchanan, the Sit-Tat claimed that—between the military coup of 1962 and the 1974 transfer of power from the regime's revolutionary council back to the parliament—village militias were formed in 1,831 villages with the distribution of 15,227 firearms (2016, 11). Notwithstanding the different trajectories of subsequent militia programs after their codification into the 1974 constitution, the name Pyi Thu Sit has become a sort of catch-all term in local parlance, throughout Myanmar's borderworlds especially.

In 1976, like others in similar villages, the Pang Kun militia was created mainly as it found itself along a main artery—the road leading to Mogok and the "valley of rubies"—strategically located vis-à-vis several rebel movements' territories (such as the Ta'ang PSLO/A, the SSA, and the KIO/A). "Here at that time there was a lot of fighting and therefore we needed the militia to protect the village," a current militiamen recalled, with an old rifle in his hands and a photograph portraying the original Pyi Thu Sit's members together with Sit-Tat officers hanging over his head on the wall of the village hall.[54] The militia remained active until 2010, when the roaming presence of EROs, a progressive escalation of hostilities, and the obsolescence of both militiamen training and militia weaponry had brought into question the very need to maintain a militia in place: "Our weapons were old, we would not feel safe and we decided to disarm."[55]

Since March 1995 the militia had come under the control of the First Military Operation Command (MOC-1) in Kyaukme, created by the Sit-Tat as part of broader restructuring and territorialization processes (Maung Aung Myoe 2009, 78–80). Known locally as Sa-Ka-Kha,[56] the MOC-1 Kyaukme has been responsible for the consolidation of base areas, the establishment and maintenance of communication lines, and the training and management of militias—including arms control. The MOC-1 has also retained financial and administrative (but not judicial) authority throughout these processes. Thus in 2010 the militia head traveled to the Sa-Ka-Kha base to tell them the arms they had were too old to cope with a situation that was growing more unstable by the day. The army agreed to disband the militia. Weapons were collected from all the members and consigned to the military base some thirty kilometers away. Some members remained in the village, others left, while still others went serving in other village militias.

Yet, in the following years, and especially after the signing of the so-called Nationwide Ceasefire Agreement in late 2015, the situation further deteriorated. As we have seen in the first part of the book, the RCSS/SSA-S encroached on the mountainous areas where the PSLF/TNLA had been expanding its presence since 2010–2011. Heavy clashes between the two rebel movements fighting each other to maintain influence in the area soon followed. Now villagers were cyclically on the flight, at times abandoning Pang Kun en masse to then come back when hostilities would subside. Forced recruitment practices further aggravated the situation—for example, in December 2016 twenty-four villagers were detained by the TNLA as the village had failed to provide the recruits deemed due by the rebel movement.[57] Thus in 2017 the village head traveled back again to the Sa-Ka-Kha base in order to reconstitute the community militia. Thirty men from the village were sent to the Sit-Tat base for a monthlong military training, at the end of which firearms and ammunition were distributed. Every member was consigned a weapon together with 120 rounds of ammunition (four magazines in some cases).[58]

The weapons distributed, like the M-22 rifle and M-2 carbine, were "not the same as those that they [Sit-Tat] have in service." They were weapons recycled by the army after being seized in police or counterinsurgency activities and stocked away to feed paramilitary and police programs.[59] The M-2 carbine in figure 10, for example, presents a craft-manufactured stock and an old selector mechanism for automatic firing, which are modifications commonly made in gunsmith workshops of EROs in the southeastern borderlands. These modifications could indicate that the gun was seized from rebel movements and then recirculated.[60]

Militia weaponry was held by its members, who were responsible for properly storing and managing them in their houses. A record-keeping form was issued to the head of the militia upon consignment of the weaponry, a copy of which was maintained by the MOC-1 in Kyaukme, and from time to time army units would come and check the arms and ammunition. Occasionally, shooting trainings were allowed in the area but had to be authorized by the Sit-Tat in order to resupply ammunition afterward.[61] Militiamen's service was organized into duty shifts of normally two days per week and some compensation mechanisms were put in place, so as to allow them to keep conducting their work activities.

The militia was structured in four units, one for each of the villages in the surrounding areas included in the program. Nonetheless, these remained under Pang Kun militia's chief, normally the headman of Pang Kun village, who was in direct contact with the Sit-Tat units and to whom the other villages had to report concerning security matters.[62] For example, although drastically diminished throughout the last years, the area was visited by tourists from time to time, and when this happened the authorization for overnight stays and accommodation had to be issued by the militia head in Pang Kun. At times the militia moved around, but in any case it operated throughout these four village areas and was not authorized to venture elsewhere. After 2016, these areas came under the influence of the RCSS/SSA-S that, counting on the Sit-Tat's more lenient attitude, positioned itself on the mountains along the Kyaukme-Mogok road, at times even coordinating or collaborating with the army. Since its creation, though, the militia had no military encounter with EROs and, as its chief remarked, "we were free in a sense but at the same time we were controlled by the Tatmadaw."[63]

Throughout the trajectories of Pang Kun's community-based militia, weapon technologies moved, entering and exiting the space-place of the village. In these networked processes, weapons have been combined with other techniques and technologies of control: the record-keeping paper forms, training techniques, shooting ranges, duty shifts, or individual in-house storage and management. Weapons changed in mutual relations with such techniques of control and were not the same at every point of these trajectories. They were old but still operating until the Ta'ang rebel movement started to grow and be active in areas surrounding the village. They became older, useless, potential sources of insecurity and peril afterward. As mediators through which Sit-Tat's authority territorialized the space-place of the village, the weapon technologies became actually old only when the TNLA started to undermine state territorialization practices. At that point the old weapons could act in a different, unexpected way. They could become a strategy to secure the village via militia disarmament and disbandment. Later on, in 2017, when the RCSS/SSA-S—which maintained better relations with the Sit-Tat at that point—consolidated its presence in the area, the reconstitution of the militia and the reacquisition of weapons became a mechanism to link again the village to prevalent authorities and insert the four village areas into state, rather than rebel, networks of territorialization. This way, weapons and arms control techniques on the one hand allowed the militia to obtain Sit-Tat recognition of the four village areas as disconnected from EROs' territories and, on the other, reinstated Sit-Tat authority and legitimacy vis-à-vis them. Neither the presence of weapons by themselves nor the relations among various actors produced these spatial relations. Rather weapon technologies and techniques of control combined actively through various processes and practices to shape the networked geographies of Pang Kun's security landscapes across time. We now turn to the geographies of scale that the armed assemblages foregrounded in this chapter generate.

FIGURE 13. RCSS soldiers taking a picture with local residents during a village festival. Kyaukme Township, Shan State, Myanmar, June 2019, author's fieldwork.

SCALES

The armed assemblages we encountered in the previous chapter are performed at different gradients of territorial resolution, at different territorial scales. They generate a scale politics of spatiality concerning where, meaning at what geographical resolution, control or uncontrol over weapons and armed collectives unfolds. In this sense, I understand scale in political-geographic terms as a territorial arena, both discursively and materially produced, where social and spatial relations of authority and power are negotiated, contested, and regulated (Swyngedouw 1997, 140–141). Processes and practices of managing armed assemblages that are performed by humans and weapons constantly rescale authority over military means and violence in the borderlands. On the one hand, such processes and practices produce, and operate at, multiple encompassed (but hierarchically not determined) territorial spaces at which authority over armed assemblages is contested, negotiated, regulated. On the other hand, processes and practices of control reproduce struggles over scaled modes of governing weapons.

One can get a sense of such scalar granularity thinking back to the swirling spirals of ceasefires, disarmament, and rearmament sweeping Ta'ang areas that were analyzed in part I. With the wave of ceasefire agreements, the splits among EROs, and, in some cases, the transformation of EROs' factions into pro-regime militias that characterized the 1990s and 2000s, the Sit-Tat performed a double bind, a double move. On the one hand, it worked to centralize. On the other hand, it worked to delegate. It centralized the control over community militias, their weapons, and training, mapping arms control onto the territorial extents of village tracts and villages, while working through the territorialized networks of its bases at the scale of the borderlands to enhance command-and-control structures spanning Myanmar. Meanwhile, via ceasefire capitalism agreements that functioned as a bridge between its political and economic centers, like Yangon, and the cross-border economies of Kachin and Shan (Woods 2011; Prasse-Freeman 2021b), it worked to legitimize and indirectly control militias and ERO splinter factions, thus instituting an informal limited access order also concerning weapon flows and acquisitions at the borderlands.

Instead, EROs witnessed a process of fragmentation at different scales. Erosions, ruptures, disconnections characterized inter-ERO alliances across the borderlands, as well as single rebel movements and their territories. Inserted in the circuits of ceasefire capitalism, some EROs, for example, grew more reluctant, or simply less able, to facilitate weapon acquisitions and rearmament processes in their immediate sphere of influence (such as in the case of KIO/A, SSPP/SSA-N, and KNU/KNLA in relation to the PSLF). At the same time, fractures occurred throughout rebel movements' social and territorial arenas, emerging especially from (and along the lines of) youth sections. Contestations and frictions concerning questions related to ceasefire accords, disarmament, and rearmament unfolding at ERO's ethnoterritorial scales provided—directly or indirectly—a means and field of struggle around which ethnonational youth and civil society organizations, or even reshaped EROs' components such as in the case of the Ta'ang movement, would take the initiative and forge new cross-border scales of activism and resistance. These dynamics intersected and operated at different scales, as in the case of part of the former leadership of the PSLO/A that, settled in the township of Manton, accepted to partake in the Sit-Tat's nationwide practice of militia institutionalization. Becoming a militia in Manton township, they aimed to cover up a territorial void at the heart of Ta'ang Land that could have otherwise been taken up by non-Ta'ang militia formations thus hampering new Ta'ang resistance efforts had they refused the regime's scheme. And again, in connection to the Shweli hydropower projects and the Shwe Gas and Oil Pipeline that contributed to rescale territories from Rakhine to Yunnan, throughout the 2000s, transnational companies negotiated operational access with militias and EROs like the PSLF/TNLA. These negotiations further entrenched the territorial authority of EROs, militias, and the Sit-Tat that had been directly or indirectly involved in their constitution or institutionalization.

Against this backdrop, the rearmament of the PSLF, as we have seen in chapters 3 and 4, unfolded in an inherently inter- and transscalar manner. A movement made of new young political figures and segments of the old PSLO/A entourages, both

embedded at the inter-ERO NDF's headquarters and in Shan State, first contested
and rejected the ceasefire and disarmament occurring in Ta'ang areas of northern
Shan, and then merged with disgruntled elements of the former PSLO/A com-
mandership and armed force still back in Ta'ang land. In so doing, they operated
at and forged cross-borderland and intergenerational political scales. Rearmament was envisaged and orchestrated across the scales of cross-border activism at the Myanmar-Thai borderland and those of inter-ERO armed politics in
Karen and southern Shan. Eventually, the emergence of the TNLA came to add a
further tessera in the mosaic of ethnonational rebel movements. This generated a
further territorial scale along those of the military government and militias. The
rearmament was, at least in part, an exercise of navigating scales of struggle in a
struggle over scaled modes of controlling military means. This could be seen, for
example, in the way that the PSLF at the Thai-Myanmar border acted to link up
with former PSLO/A members that facilitated the recirculation of weapons and
soldiers deserting militias to join the TNLA in Ta'ang areas of northern Shan. The
TNLA's first territorialization attempts here produced frictions with local militias
in their areas. At the same time, for the movement it was equally important on
the one hand to act at the scales of village communities and households while
seeking to solidify a broader ethnoterritorial identity to mobilize recruits and,
on the other, to work for an alignment with "regional" EROs like the UWSP/A.
The movement had to navigate amidst a complex scenario involving the KIO/A,
SSPP/SSA-N, and especially the UWSA in the north, where these EROs reasserted
their role in the production, channeling, and provision of weapons, equipment,
and training. Part of the political rationale behind its recent wars with the RCSS/
SSA-S (2015–2022), for example, had to do exactly with this struggle over scales:
a struggle to rescale Ta'ang land as an autonomous territory rather than a zone of
autonomy under Shan State, or rebel, rule.

Such scalar politics of controlling military means, though, is never a pure matter of institutions, actors, human agencies, or authority. Diffused technologies
and techniques of conducting the conduct of weapons and armed forces constitute practices that contribute to the undoing and sedimentation of existing
(and future) political scales (Hong 2017, 7). They unfold struggles that result
in changing arrangements of territorial authority on violence (Smith 1992, 74).
Such scales are not fixed but made and remade through the very rationalities,
techniques, and practices of controlling military means that at one time forge
territorial scales of struggle and produce struggles over scale.

To analyze this scalar politics of spatiality, I understand rationalities and
techniques of controlling military means as fields of struggle as well as a connecting tissue across scales. As the first part of the book showed, the rearmament process of the PSLF and the wars that followed were closely entangled

with frontierization and territorialization processes enacted by the military-state apparatuses in the borderlands for several decades (see also MacLean 2022). Redeploying militarized techniques of land-settlement programs it had drawn from the Israeli Defense Force's Nahal Brigade as well as from the British colonial machine, the Sit-Tat has been zoning the borderlands into "white," "brown," and "black" areas since the 1950s (Maung Aung Myoe 2009; Kozłowska and Lubina 2021). This zoning technique was part of a broader frontierization-cum-territorialization of the political communities living in the borderlands. It naturalized these zones of the borderlands as outside or inside the Myanmar military-state polity. Spatial zoning in white-brown-and-black worked to produced areas of the borderlands as frontiers of the Bamar polity, that is, areas where a sociocultural process of "Burmanization" has already been initiated or still has to be initiated. The concrete terraforming of these divisions between the insurgent/rebel polities and what have been thought of as the communities belonging to the military-state polity was performed via the four-cuts strategy and the key deployment of free-fire and scorched-earth techniques. In this context, the chapter looks at how techniques of governing military means that have characterized war in Ta'ang Land after the PSLF rearmament produce a politics of scale.

The first three sections revolve around rebel counter-frontierization techniques by delving into the politics of recruitment, landmines use and control, and the production of militarized and mobile humanitarian spaces. The last three sections shift the focus toward techniques of managing armed assemblages through which local communities navigate these weaponscapes by looking at the creation and dismantling of community militias, the politics of landmine contamination referrals, and civilian attempts to shape rebel forces' conduct.

The Scale Politics of Recruitment

Armed mobilization and recruitment can be understood as a technique of managing an armed collective that counters the frontierization and (re)territorialization operated by the Sit-Tat. Recruitment is often carried out along ethnonational logics. In this sense, recruitment is rationalized as the transformation of ethnic bodies into soldiers, and ultimately armies, as a means to protect the very same recruits and ethnic populations living in EROs' territorial entities—that is, both the populations allegedly represented by a given ERO and "other" ethnic people living in EROs' territorial scales. For this reason, from Ta'ang perspectives the acronym DDR—disarmament, demobilization, and reintegration—has become a taboo, and with it recruitment of Ta'ang by other ethnic armed forces as well:

> The problem with DDR is the fact that, if we are going to give up our own army, then the people [Ta'ang] will be recruited by others. If there is no army [of the Ta'ang] then we will get recruited by the Kachin or the Shan armies. In RCSS, for example, they used to have many Ta'ang in their ranks—those are especially Ta'ang from the south of Shan State. So actually, it is better to have your own army than no army at all, because otherwise this is going to mean that other armed groups will recruit our own Ta'ang people. This has already occurred in the past when the PSLO/A had disarmed in 2005 and before.[1]

Practicing recruitment, different scales of ethnic populations and territories are reproduced. This occurs linking back and forward, into past and future ethnonational histories. Ta'ang communities and areas of southern Shan are seen here as frontiers of Ta'ang Land via the prism of recruitment as both a practice that can dismember and one that can delineate ethnonational populations and territories. This is not to say that EROs' forces only number same-ethnicity individuals among their ranks. In fact, as we have seen, Ta'ang people have often been subject to recruitment by non-Ta'ang EROs. Rather, it is to say that ethnonational logics are activated to recruit at local scales in order to configure village areas, as much as households, as local scales inserted into broader ethnonational territories.

Such logics and practices of recruitment can be used simultaneously to unify populations living in ethnoterritorial entities and to divide them. In fact, the logics of ethnicity, and recruitment along ethnic lines, contribute to engender distinctions across class scales as well as to divide people at the same class scale (Campbell and Prasse-Freeman 2021a). This has occurred, for example, with the recruitment by the RCSS/SSA-S of Ta'ang peasants living in southern Shan State, who are usually considered backward populations and belong to the lower socioeconomic strata of society. Ta'ang individuals in the RCSS/SSA-S normally are recruited for porterage or remain in the lowest ranks. While recruited from villages of southern Shan as part of the ethnoterritorial scale of RCSS/SSA-S, they are at the same time divided along the intertwined lines of class and ethnicity. Informing recruitment and the formation of a Ta'ang army with the logics of ethnicity instead helps including the "Ta'ang from the south" and the frontier territories they live in—both often seen as socially and economically backward among the PSLF/TNLA rebel polity—into the ethnonational and territorial scale of the PSLF/TNLA.

Individuals and communities across scales, while targeted by or supposedly benefiting from acts of governing military means, refuse practices of recruitment and rework them. An example here arrives from the experience of the Ta'ang village head of K.L., a mixed Ta'ang-Shan village north to Kyaukme, to the east of the main road to Mogok.[2] In 2016 the village came under the RCSS/SSA-S, with the Shan ERO building a camp on top of the hill overlooking the village area. The village head had to start arranging so-called war funds in the form of rice and money to be provided to the RCSS/SSA-S. Yet, the PSLF/TNLA continued to remain in contact with him, either via phone cables or visiting the village undercover, to bring forward recruitment drives as usual. The PSLF/TNLA, as we have seen already, has maintained a recruitment practice according to which each Ta'ang household has to provide recruits. K.L. village was no exception, but the village head tried to navigate these recruitment practices:

> We have to pay the RCSS in rice and money; but we, as Ta'ang, have also to pay tributes to the TNLA. TNLA sometimes comes here to the village, wearing civilian clothes, sometimes they send us letters. Generally, we are obliged to do what they ask us to do, we cannot say no. For example, once some time ago they came and asked for recruits and told me I should have sent to them a number of people. They would have come to pick these people up and I had to arrange who would go within the village. But after talking with the people of the village I decided I did not want to send anybody to them to join the army of the TNLA, so I had to find a way to convince other people from other villages to go. Ultimately, I paid some families in other villages to send their people.[3]

While the logic informing the practice of recruitment is one of ethnonational ties, and K.L.'s village head does not openly contest it, he attempts to counter and circumvent recruitment by shifting village scale locally. He reworks the ethnic logics that require Ta'ang village communities to provide recruits to the TNLA to bypass the coercive element through which the ERO imposes recruitment.

Discriminating Explosive Devices from Landmines

To counter Sit-Tat's zoning military techniques, rebel movements have long relied on explosive devices. As touched on earlier, discursive practices have circulated among EROs about their armed forces not using landmines in their territories. This argument is directly connected to the materiality of handcrafted explosive devices' design and manufacture. These are argued to be the only devices available and accessible by resistance forces (as opposed to industrial mines available to the Sit-Tat mostly), to be less durable, and to be potentially capable of discrimination given that they are battery-activated and not victim-activated devices.

Drawing an explicit parallel with the Claymore-type mines—that is, land-mines that can also be used in command-detonation mode, and that thus far remain virtually permissible in that modality under the Ottawa treaty—an SSPP/SSA-N officer points out how, since he joined in late 2015, he "noticed that SSPP/SSA-N uses more bomb-mines, like Claymore landmines remotely activated, rather than victim-activated landmines. Claymore-like landmines are used to defend the camps and to ambush, but not for demarcation. I mean, home-made Claymores."[4] EROs in fact refuse the labeling of their "battery mines" as victim-activated devices that are indiscriminatory and environmentally embedded, like IEDs or industrial landmines.

These distinctions have been used by ERO leadership to navigate the international scales of humanitarian mine action and arms control. Various rebel movements have condemned the use of victim-activated landmines, either directly or indirectly, subscribing to international standards and participating in activities of INGOs operating countrywide in Myanmar. Some (like the PSLF/TNLA) have banned the use of landmines in their territories and subjected themselves to the international monitoring activities of the Swiss INGO Geneva Call by signing unilateral deeds of commitment. Undersigned by Geneva Call's head in Geneva as witness, and the government of the Republic and Canton of Geneva as custodian, the deeds of commitment draw from the Ottawa treaty: they demand the ban of any victim-activated explosive device. At a local food stall along the bumpy road that leads from Namtu to Namhsan, during a break from the suspension of the motorbike we were traveling with, I once asked a TNLA officer and a soldier accompanying us to the ERO's temporary mobile headquarters about the Ta'ang army's stance vis-à-vis landmines. After a moment of silence filled with indecision, the officer wished to clarify:

> Actually, according to the Geneva Conventions, the use of landmines is prohibited and we have subscribed to the Geneva Conventions so we are not allowed to use landmines.[5] We do not even have landmines actually, so we cannot use them, 'cause we do not even have landmines. The problem is that they last for too long, they can last for up to one hundred years. So, what we use instead are those landmines that we can do, and those do not last for long. Those are the landmines that . . . you can decide when they are going to explode. We do not use the other landmines. . . . Also because this would create problems to the population as those mines last for too long and create many problems.[6]

Through processes and practices of "making" and controlling the mines, ERO leaders try to legitimize the use of battery-powered craft-manufactured mines, their users, and their users' territorial authority. They attempt to produce rebel scalar arrangements of territorial authority on weapons and violence. On the one hand, they shift "up" to international arenas of humanitarian arms control in order to legitimize their organizations by banning victim-activated mines throughout their ethnoterritorial entities at the borderlands. On the other hand, by banning mines in their territories, they shift "down" to legitimize themselves and consolidate some territorial autonomy vis-à-vis the Sit-Tat's military state, which notoriously rejects the application of international mine action standards at a state scale. At the same time, EROs continue to use battery mines in their territorial entities but retain some space to justify their use vis-à-vis their polities.

Constructing landmines as discriminatory and nonembedded is not a purely discursive remark, for in fact it connects directly with the technical object's nature and its potentialities. Exactly because of this, though, things often proceed differently at the scale of landmines' ecologies of violence. The TNLA rank-and-file soldier listening to our conversation did not answer back to the officer, but he would be in the best position to remind his comrade about how, at the scale of local areas, where battle-hardened ERO members operate, practices of assembling and using mines, the deployment of explosive specialists, and the ways in which its various elements interact with one another make it complex for the explosive item to be as straightforwardly "command-detonated" and not environmentally embedded as it is framed, legitimized and legalized. At the scale of the ecologies of violence of the mine, the battery mine often defies rebel scalar arrangements of territorial authority over violence projected onto it from other territorial arenas.

Adjusting the logic of discrimination occurs across other scaling processes as well. In practicing advocacy, documentation, and reporting on a range of sociopolitical issues, ethnonational CSOs reproduce territorial arenas of activity. Vocalizing discontent for the negative impacts on human beings, livelihoods, and ecologies in their areas of operation, CSOs condemn landmines in toto while advocating and reporting at different scales: throughout international media, in interethnic CSOs and cross-border fora, at subnational and countrywide INGO coordination mechanisms, or within local communities (McCabe 2019). They avoid drawing any distinction between victim-activated and non-victim-activated explosive devices, neither referring to rebel movements of the same ethnicity nor to other EROs.

Instead, in understanding, operating at, and (discursively) reproducing other territorial arenas—namely, the scale of the ethnoterritorial entities at which EROs wage war on the Sit-Tat—some do draw a line. While Sit-Tat mines used in the borderland's ethnic territories are argued to be victim-activated, underground, difficult to see, and left there polluting the future, given their industrial nature as well as the modalities of use and control, those deployed by EROs of the same ethnonationality are said not to be the same. Elaborating on the TNLA's techniques of landmine use and control in village areas across Ta'ang Land, a member of TWO is keen to redress what she sees as an inaccurate use of the word *landmines*:

> Landmines . . . for me they are not really landmines, we should rectify this, as in reality those they [PSLF/TNLA] use are craft-manufactured explosives, mostly with remote control. The TNLA used to say that they use them only for offensives against the Tatmadaw. So the TNLA, when they use the remote control landmines they inform the villagers in advance—they reach out to the head of the village and tell him to let the villagers know that they do not have to go in this or that area. If villagers or animals get hurt due to the landmines the TNLA provides compensation, something like . . . 200 USD to each victim and sometimes they help the victims go to the hospital. So, before they use the mines they inform the chief of village.[7]

Mobilizing the logic of discrimination between civilians and combatants serves to delegitimize the territorial authority of the military state and the Sit-Tat by underlining their lack of respect for international standards throughout the territory of Myanmar. Under the same logic, the use of command-detonated devices and techniques of control aiming to discern civilian targets and spaces legitimizes the territorial arena of the ERO in relation to the village areas where the TNLA operates. Simultaneously, this allows one to condemn landmines—and contribute to building subnational scales of humanitarian mine action and civil society activism—while not positioning oneself in open contestation with EROs.

The weapon—the landmine and/or explosive device—here becomes a field for a broader scalar politics of controlling military means, a current and prefigurative politics that entails an ordering of conflict actors and their territorial arenas of authority over violence on the base of a civilizational trajectory whose parameters are given by the discriminatory potential of the technology. Here the violent, institutionalized history of the Sit-Tat, with its racialized conception of both nation and statebuilding (conflated), runs together with industrial landmine production and their barbarous deployment in the processes of colonization of Myanmar's borderlands; while the seemingly less advanced histories of ethnonational politico-armed struggle are shifted to a more advanced stage via a (seemingly technically less advanced but) more advanced technology. Processes and practices of making and controlling the battery-powered mine in a way that distances it from other categories of pariah weapons aim to both harness international mine action processes to legitimize ethnonational territorial scales and continue to legitimize the use of explosive devices. Such politics of governing military means reshapes the scalar arrangements of territorial authority at the borderlands. It rearranges the relative importance of different arenas of territorial authority over violence.

"Humanitarian" Spaces as Counterfrontierization

The deployment of four-cuts and zoning techniques has triggered, and triggers, massive population displacements, pushing entire communities to flee their homelands or live under enormous pressure in nearby areas. Against the logic and praxis of enemy/target massification that such techniques perform, rebel movements have long worked to produce humanitarian spaces of rescue and refuge. To this end, key have been the support and activities of the Free Burma Rangers (FBR), a humanitarian-missionary organization that—not without controversy[8]—defines itself as a "multiethnic humanitarian movement" operating in Myanmar, Iraq, Syria, and South Sudan (Buscemi 2024). Many rebel movements, though not all, have been partnering with the FBR to train and constitute relief teams through which they generate humanitarian spaces in combat zones to counter the frontierization and territorialization of the borderlands. For many years, the PSLF was among these, sending some of its troops to join FBR trainings. With the consolidation of the TNLA in northern Shan State and the formalization of a politico-military alliance with the UWSP/A around 2014–2016, though, in recent years the Ta'ang movement has stopped taking part in FBR activities.

Started by Dave Eubanks, a former US Army special forces officer and evangelical Christian pastor raised in a Thailand-based missionary family (Horstmann 2019), the FBR emerged in the midst of the Sit-Tat offensives in the Karen borderlands that led to the crumbling of Manerplaw in 1995. With the Sit-Tat's strategy and techniques of territorializing borderland areas unfolding,[9] it was practically impossible to attend to population care. The "realization of a so-called 'humanitarian space' to which people could have fled and in which they could have been given respite and refuge from conflict was simply impossible, nonexistent," as an FBR member put it.[10] The very territorialization practices of the military state made it impossible to identify a front line behind which to create a "humanitarian space inside Burma."[11] As a practice to cope, relief teams would thus have acted as units embedded in ERO military operations so as to bring medical and humanitarian assistance to populations caught in conflict areas.

The main logic behind the training and formation of relief units thus was, and still is, one of generating a mobile and temporary "humanitarian space" of relief and rescue. Such logic reflected the militaristic practice of operating roving special forces units capable of proceeding toward conflict/fighting in order to relieve troops, rather than the latter having to move away toward a "no-war" zone. At the same time, this technique was also informed by a humanitarian rationality. The creation of relief teams has been underpinned by the need to manage people and space in conflict zones: that is, the need to preserve a body of civilian population and alleviate suffering so as to allow life to carry on in a landscape of full-on war. Besides an overall biopolitical paradigm of rebel rule taking life as the primary object of politics and governing, EROs relied heavily on civilian communities to sustain themselves. Thus, and even more so in relatively sparsely populated geographies like those of the borderlands, depopulation would seriously affect their war effort. Therefore, as a ranger explained recalling the origins of the FBR activities, the model of teams was "based on the idea of military special forces teams that have to provide relief to civilians and troops and documentation in conflict areas. The idea is to go toward the areas of conflict, toward fighting, rather than creating a space where people have to come. Essentially the main tasks are to provide medical assistance and evacuation and to provide documentation in otherwise inaccessible zones."[12]

Managing not to have people displaced by Sit-Tat strategies and preserving civilians has required adopting techniques to manage and conduct armed units. KNU/KNLA units were the first to be trained, but throughout the 2000s other EROs were allowed to send their members to training areas at the Thai-Myanmar border. In addition, the "multiethnic humanitarian movement" scaled up the organization of trainings in different Myanmar borderlands, from Chin to Kachin and northern Shan. The creation of such "humanitarian spaces" has proceeded hand in hand with the circulation of practices, techniques, and discourses. A lieutenant of the SSPP/SSA-N offers a window into such link when talking about the trainings he was commanded to take part to:

> "[T]hey provided us with training on first aid and on first response medical assistance, rescue techniques, swimming and survival techniques let's say; land navigation and map reading—very important especially; human rights and humanitarian law, in order to be able to report about violations by the Tatmadaw. Because besides rescuing, the other main task is reporting about abuses and violations by the army. During the trainings they also provided us with a bit of weapons management and use skills and training because we have to be armed in the operations. . . ."[13]
>
> "Armed in relief operations?" I interrupted.
>
> "Our main task is to provide humanitarian support and medical assistance, first of all, and secondly, also to collect information and

report about the Tatmadaw presence and human rights violations . . . or similar behaviors of the army. Of course, we are armed when we go on mission. It could not be possible to go without weapons. You have to understand that 'without the weapon in your hands you are nothing in Burma.' Because there are so many 'weapons groups' that you can't do anything without it. . . . They also provided us with landmines and explosives training. They taught us how to deactivate and clear a landmine. So we know how to deactivate a landmine if we find one. They also provided us with weapons training to recognize the arms we see on the ground. Self-defense courses were also provided with this Hollywood actor that came in . . . Viktor something . . . he showed us how to quickly disarm somebody pointing a gun at you [he says while reaching for a YoutTube video of the actor performing the move in a short movie]. The landmines training actually is also provided by SSPP to soldiers but the one that [they] gave was of higher quality I think."[14]

While temporally provisional, these "humanitarian spaces" are produced at local battleground scales through the circulation of military techniques that operate first and foremost at the scale of single bodies. Such techniques essentially manage the way in which armed men and women conduct themselves, and the weapons they act together with, in practicing relief tasks that target civilian populations. Medical assistance; mapping; documentation of human rights and humanitarian law violations; weapons identification, use, and handling; as well as minor landmine and explosive items management are all techniques of governing the entanglements among bodies, individuals, and weapons in a concerted manner.

The practices of training and conducting the conduct of relief teams to generate humanitarian spaces unfold at and shape different scales of controlling the means of violence. The FBR carries out trainings in its camps maintained at the Myanmar-Thai borderlands. Such camps—as well as the circulation of weapons and military training related to them—are made possible thanks to linkages the movement has built at different territorial arenas of control, namely those of Karenni and Karen EROs, and Thai military and state authorities in border provinces like Chiang Mai and Mae Hong Son. Performing the trainings then revolves around the mobilization of people, resources, and knowledge at "international" scales. Courses are taught by former US military personnel, medics, or other professionals with previous missionary or military experience in places such as Afghanistan, Iraq, and Syria, who spend part of the year at the Thai-Myanmar border. Financially speaking, the FBR sustains its activities by linking up with international networks of Christian charities/foundations that mobilize funds, people, and connections, and could be seen as part of transnational communities of charitable Christian-led activities (Horstman 2018). During their first deployments, the relief teams being inducted are accompanied by FBR trainers, while occasionally the latter act embedded with EROs too. Further suggesting the creation of a transborderland scale of operation, the movement performs its advocacy and documentation network throughout different Myanmar warscapes both independently and drawing from relief teams.

From another angle, ethnonational rebel movements' leadership and commands send rank-and-file soldiers and officers to be trained at the Thai-Myanmar border areas, or arrange joint trainings, all in order to generate humanitarian spaces at the scales of their territories. Furthermore, members of different EROs at times have been formed into interethnic relief teams in northern Shan—for example, among KIO/A, PSLF/TNLA, and SSPP/SSA-N. The performances of such joint teams contribute to create regional scales across the territories of the different EROs involved and in their theaters of operation where "humanitarian spaces" are generated.

Different multiscalar arenas and arrangements of territorial control over the means of violence are reproduced in the process. The FBR as a multiethnic humanitarian movement connects with members of different EROs and CSOs at its borderworlds and international scales in the material organization and implementation of trainings that are nonetheless nested in ERO territories and the Thai border provinces. The relief team members then operate at different territorial arenas carrying out humanitarian spaces. The creation of humanitarian spaces at the scale of the battlegrounds occurs by jumping up to transborderlands and international scales to reaffirm EROs' territorial arenas.

The trainings and creation of "humanitarian spaces" are informed by different rationalities that become a connecting tissue, and at the same time are variously reworked, across scales. In particular the formation of relief teams is underpinned by a spiritual component linked primarily to the spiritual/religious (Christian) dimension of the FBR. As V., a longtime member, stresses:

> Ours is definitely a movement underpinned by Christian values and definitely the way we act, and what we train ethnic minority groups on, is based upon Christian teachings, but I would say that the training we provide has also a spiritual component. This spiritual component basically rests on the idea that you have to change your heart in order to redress a situation of oppression and lack of freedom. This country has actually experienced openings to democracy or even the spaces of ethnic minorities have experienced democratic situations, but often we have seen political leaders and elites turning against these values. And

essentially this was because their heart was corrupted. If you change your heart you can be able to change the political situation, as heart is the most powerful of the motivations. If you want freedom, the start-ing point is to achieve spiritual freedom, and that can be achieved only [by] changing your heart. So, we kind of try to change hearts during the trainings. In any case, the majority of the teams that we train are Buddhists, animists, or in any case not Christian, but still the spiritual component of the training has great appeal to them as well.[15]

The Christian and spiritual bases of the trainings contain complex material and discursive histories of projects of governing the conduct of individuals and com-munities that see their roots stretching back at least to the work of missionaries in the Thai-Myanmar borderlands. Together with a militaristic understanding of freedom that undoubtedly underpins the very idea of managing guns in a man-ner that allows people to carve out spaces of refuge, rescue, and ultimately politi-cal autonomy, such religious dimension represents the other main controversial issue raised by the FBR praxis (Horstman 2019). This logic is characterized by Christian values of the transnational charitable communities the movement is linked to as well as of its missionary connections. Yet, it is also inflected by a spiritual component concerning inner individual change that is promoted at the transborderland scales generated by FBR activities. Participants, who often are not Christian, relate to spiritual inner change at the scale of individual hearts and minds through their own religious paradigms, and the relief teams are instilled with the idea that change at the level of one's "heart" will bring change to "spaces of ethnic minorities" or the "country."

Similarly, layered militarized histories unfolding at distant temporal and spa-tial scales—such as those of US imperial wars or ERO rebellions—characterize the military techniques that are diffused through trainings and the constitution and deployment of the relief teams. In this project of governing military means that is infused with humanitarian aspirations, the militarized rationality of preserving civilians so as not to jeopardize military operations in battle theaters has been remolded by EROs in order to possibly contain population displace-ment in their territories and counter the Sit-Tat's processes of frontierization and reterritorialization of the borderlands.

The (Un)Making of a Village Community Militia

The four-cuts strategy and the zoning techniques of the Sit-Tat at the heart of these processes of frontierization and reterritorialization have relied heavily also on networks of village community militias. Moving back to Pang Kun village, which we encountered in chapter 4, we can look at the politics of scale produced by the ways in which local communities navigate techniques of constituting and managing militias and their weapons.

As we saw, the village community militia in Pang Kun, some forty kilometers northwest of Kyaukme, was dismantled by its members in 2010 amidst the building up of the TNLA and the intensification of hostilities with the Sit-Tat, only to then be reconstituted in 2017 against the backdrop of the fights among the PSLF/TNLA, SSPP/SSA-N, and RCSS/SSA-S.[16] With the re-creation of the village militia, certain logics and practices of arms control were also rekindled in the territorial arena of the four villages that had been placed under the Pang Kun community militia by the Sit-Tat's Sa-Ka-Kha in Kyaukme:[17]

> The arms are controlled by the militia and by the army, in the sense that from time to time they come and check the weapons and ammunition. We keep the weapons always with us, every member keeps them at his house and has roughly 120 bullets, four magazines each. We keep a list of the arms and the Tatmadaw does the same. . . . They keep a list in which arms and ammunition have been registered.[18]

With the militia back in the village, governmental measures were reintroduced to manage weapons and militiamen as well as the relations between them and village populations. The record-keeping form maintained by the community militia and weapons management techniques, for example, turned the scale of every armed militiaman's household into the repository of the weapons, as every militiaman normally has to keep their firearm and ammunition at their house.

At the same time, the formalities of arms management would encompass the militia and its territorial scale as a whole, with the record-keeping form and associated duties placed on the militia's head. The criterion for selecting the militia's head was neither an ethnonational one nor a military one; rather, the village institution would have borne that duty, and Pang Kun's village head (as per state territorial administrative scales) was to become automatically the head of the community militia as well. Yet, maintaining a formalized weapons and militiamen management system was not a mere family or village matter, for the Sit-Tat's Sa-Ka-Kha based in Kyaukme could have occasionally shown up for an inspection.

From time to time, the militia self-organized shooting trainings by setting up a local shooting range: practicing shooting, though, was carried out in a formalized way, requiring here again some record-keeping and bureaucracy given that ammunition depletion would need to be tracked at the scale of the village and justified at the scale of the Sit-Tat's Military Operation Command base in Kyaukme. Militias' firearms have not been used for purely military purposes.

The arms, in fact, could be exceptionally good for hunting as compared to the handcrafted muskets that rural hunters normally can acquire. However, the conduct of the weapons and the gunmen has been conducted with the logic of formalization and institutionalization in mind. As the weapons and ammunition are institutional "militia firearms," their use for other purposes has to be recorded. The bullets, for example, are resupplied by the Tatmadaw from time to time and have to be counted in order to justify missing ammunition, or otherwise one has to find other ways to substitute them.

Being a militiaman entails also a certain politics of identity that unfolds together with the reproduction of different scales via discourses and practices of formalization and institutionalization, as a Pang Kun community militia member noted:

> They [Sit-Tat] do not pay us; what we do here is completely voluntary and we have to organize to serve in shifts so that every member can also keep conducting his normal jobs and life. Normally we have shifts of two days per week. Nonetheless, even if they do not pay us, we are under their control. We perform only security tasks and do not have other prerogatives. We do not decide anything, it is the chief of village and the Tatmadaw that decide what we have to do. In any case we do not have any justice-related task, we do not function as judges. The chief of village [Pang Kun] is also the chief of the militia.[19]

The technique of militia personnel management through duty shifts suggests how being a militia member is a matter of diffused practices and of navigating and reproducing various scales. For example, villagers that are members of the community militia are often compensated via a system of community contributions—in kind or in the form of agricultural labor assistance—at the scale of the four villages included in the militia's territory. Contributions are meant to obviate the lack of financial support by the Sit-Tat hierarchies at the broader state scales. The "burden" of holding guns here is governed by formalizing shifts and instituting mechanisms of coping in a context where the Sit-Tat's control is not structured via the official military/administrative integration of the community militia into the Sit-Tat.

The remobilization of the community militia in 2017, with its recirculation of techniques and practices of controlling the means of violence, has also reshaped scalar arrangements of territorial authority over the means of violence. A Kyaukme-based mountain guide's experience aptly illustrates this dimension. Often working with tourism companies, S. has long had to drive his clients to a village adjoining Pang Kun. Before the reconstitution of the community militia with its epicenter in Pang Kun in 2017, S. would contact the head of the village where he planned to spend the night during trekking tours or would simply travel there and seek the head's permission to be hosted by a village household. In April 2019, two years after the militia was reinstated, S. walked to the same village with a tourist and, thinking of spending the night there, visited the village head to request the authorization. Confronted with S.'s request, the chief replied that he had to first inform and seek permission from the head of Pang Kun's community militia in order to be authorized to "host the foreigner in the village, since they [the militia] were those in charge of security" throughout the four villages now part of the scheme.[20]

The techniques and practices connected to the insertion of firearms as technical objects into the four villages contributed to territorializing the scale of the militia's territory. As we saw in chapter 4, the disarmament and rearmament of the village community militia—with the halt or restart of the techniques related to the weaponry distributed and the composition of the militia—were practiced by different actors. Disarmament and rearmament were practiced to reshape the scale of the four villages, the militia's territory, and to insert such scale into one scalar arrangement of territorial control over the means of violence or another.

Landmine Contamination Referral: Where To?

Landmine and explosive remnants contamination has been another peculiar trait of the Sit-Tat frontierization-cum-territorialization processes. Landmines have been deployed especially around the (often very extensive) perimeters of its bases, camps, and key infrastructures. Practices of landmine contamination referral—meaning the practices of reporting the presence of an explosive device, be it a landmine, explosive remnant, or other—provide a snapshot of the shifting and overlapping scales reproduced in Ta'ang areas of northern Shan where multiple rebel movements and the military state have been battling and living with each other.

People encountering explosive items on/in the ground might face different referral paths.[21] INGOs carry out nationwide humanitarian mine action programs by outsourcing activities to ethnic and local CSOs that implement them in their respective areas of operation. Through mine risk education, they have usually instructed civilians to reach out to the local village head at first. Village heads often request assistance from the CSOs that operate in their village. Normally it is left up to the head of the village, however, to decide to whom to refer the presence of the item and request its removal. INGOs and CSOs leave the decision to the village heads in order not to jeopardize their position vis-à-vis different actors.

In fact, multiple armed actors reproduce their territorial scales of authority relying on village-level local leaders. Thus, in areas of overlapping presence, the latter become focal points for the rebel movement aspiring to represent their own ethnonationality while also being the main reference figure for other EROs and/or the Sit-Tat's apparatuses present in the broader area. Village heads play a key mediatory role among communities, INGOs and CSOs, and armed and state actors. They may refer the presence of the explosive item to one ERO, to the Tatmadaw, or to the state General Administrative Department (GAD).[22]

In areas with only one armed actor, heads of villages usually refer to one among the ERO present, the Sit-Tat or the GAD. Even in such seemingly straightforward cases, though, referral paths can be circuitous. Informing EROs or Sit-Tat units might require arranging an appointment with the commander responsible for the broader territory, either in a camp in the village area or at a base elsewhere. Reporting to local government administrative structures, where present, means referring to the village administrator component of the GAD. Based at the same village scale of the head of the village, the GAD village administrator is nonetheless inserted into a different scalar arrangement and would then follow the state administration referral chain until reaching the Sit-Tat military command territorially responsible for the removal. In other cases, the CSOs active in the areas somehow directly or indirectly constrain referral paths. Mine action and related activities—such as victims assistance, mine risk education, first aid, and some (minor) nontechnical surveying—have been implemented via networks of regional ethnic CSOs. These CSOs, though, stand and operate at the intersection of different scales. An extremely illustrative case can be heard in the explanation that a high-rank SSPP/SSA-N officer provides when talking about the Shan CSO he founded in Lashio:

> [W]ith the CSO that I run in parallel . . . we have a mine risk education project in cooperation with Halo Trust, UNICEF [United Nations International Children's Emergency Fund], OCHA [United Nations Office for the Coordination of Humanitarian Affairs]. We provide mine awareness training since 2016 in a four-year project that will end next year. . . . We could say that SSPP implemented through my CSO—we work sometimes as brokers for SSPP and at times we facilitate in relations between the SSPP and the government and Tatmadaw, or SSPP and INGOs. When we discover the presence of an IED [improvised explosive device] or ERW [explosive remnant of war] we mostly inform the SSPP authority, the township administrative structure of SSPP. Then they come to see the item. If they can solve the problem and eliminate the bomb, they do it. If they cannot, they will inform to the SSA. If the item is near the township areas controlled by the government or Tatmadaw we inform Halo Trust.

If the item is to be found inside or near jungle areas we inform SSA. If the item is in town areas we inform Halo Trust. . . . [M]y CBO provides technical assistance to the SSPP administrative structures also. Training on MRE [mine risk education], training on reporting system and also on mines mapping.[23]

Village heads find themselves in an even more spinous position in areas in which one or multiple EROs and the Sit-Tat overlap and attempt to exercise authority. Here, the presence of landmines and the dynamics of referral by village heads contribute more conspicuously to construct the mine, its site, and the village as a territorial arena grafted onto different scalar arrangements. Referring the presence of an explosive item found on the ground to an ERO contributes to scalarly construct the item's site, and the village area, as within that ERO's territorial arena of authority. By referring the mine to the ERO, though, the village heads risk criminal sanctions under section 17 of the infamous colonial-period-drafted Unlawful Associations Act (1908), signaling that the village remains simultaneously nested into state territorial scales of authority.[24] At the same time, such referral dynamic reshapes the landmine, the site, and the village area in relation to yet other scalar arrangements: those of the rebel movement of one's own ethnic identity, or those of other EROs, depending on to whom the explosive item has been referred. Reporting to the Sit-Tat or GAD instead deconstructs the village as part of rebel territorial scales. Such referral path avoids legal retribution, but, on a personal and local scale, it is equally fraught with the risks of delegitimizing the precarious and complex connections that village heads maintain with EROs.[25] Different nested and intersected territorial resolutions of controlling the means of violence are activated by the presence of landmines and the dynamics of referral.

The position of chiefs has thus come under increased pressure, especially since 2015 with the arrival of the RCSS/SSA-S in Ta'ang areas and due to the overlapping presence of multiple EROs and militias. This is linked to the fact that a whole other series of practices and techniques are reproduced through the institutionalization of the village head's role. So much so that "no one wants to be head of [the] village anymore."[26] The reproduction of governmental practices leaning and layering on the institutionalized position of the head of the village has eventually brought to a reconfiguration of such scale of struggle. Struggles over the scale of the village as a geographical resolution for channeling different practices of governing the means of violence that operate at and reproduce larger territorial arenas—recruitment, landmine referrals, mine action implementation, etc.—have been met with a counterconduct. In highly contested and fluid areas at the intersections of multiple territorialization projects, communities have progressively assigned the role of village head to youth.[27] In particular this has occurred among Ta'ang mountainous communities, in part due to the fact that many of them, especially older generations, speak only Ta'ang languages and might have problems with Burmese or Shan.[28] Youth, on the other hand, do normally command Burmese and can relate with different armed actors. However, many often refuse or flee to avoid becoming village head.

The rationalities and techniques of conduct that inform, and are tied to, different referral practices are differentiated across scales. These rationalities represent a capital to reproduce territorial scales of authority for different governmental actors and their governable orders. Framing the referral of explosive items through different rationalities of control, rebel movements and the military state construct the village as a territorial arena that is part of different scalar arrangements of authority. At the same time, rationalities are deployed with a view to navigate across such scales.

Referring the presence of explosive items in the village to ethnonational rebel movements is linked to an ethnonational rationality that envisages ethnic territories' security and safety as being ensured by ethnic politico-armed organizations. EROs construct such rationality both when explosive items are argued to be a matter of military strategies and tactics, part of wars to secure ethnoterritorial entities and communities of different ethnicity living there, or when landmines are argued to be a threat to civilian populations living in an ERO's territory. This clashes, of course, with the realities of the borderlands in which ethnoterritorial scales and related rationalities are not fixed in time and space. From the perspectives of village heads and ethnic CSOs, the referral of explosive items to an ERO and the understanding of the village as inserted into one ethnoterritorial scale or another may be connected to familial, amical, or other social ties that the single person maintains with rebel movements purportedly representing their same or another ethnicity.[29]

At another scale, that of national and subnational mine action programs, referral is guided by a humanitarian logic of removing and disposing an item that does not belong to civilian areas in order to preserve human beings. This is channeled through chieftainship at the scale of village areas where people finding an item report it to the village head. Here INGOs and CSOs, while assisting the chiefs, deploy bottom-up and "do no harm" logics and delegate the decision on what referral path to undertake to village heads. They act this way in order to avoid indirectly framing the village into one scalar arrangement of territorial authority or another, thus risking negatively affecting the village head's positionality.

The scales of struggle are further complicated as the Sit-Tat and state apparatuses frame the conduct of practicing referral as an institutional prerogative at a state scale and, in case of unlawful contact with EROs, as a criminal activity according to state legal scales. Village heads (by referring via one path or another), and the other actors and entities involved, reaffirm the scale of chieftain-ship and the village, and contribute to reproduce different scalar arrangements of territorial authority over violence in which the village is inserted. Civilians on their part often refrain from reporting at all, in order to avoid being enmeshed in troublesome dynamics. The dynamics of practicing referral become exercises of producing and/or jumping across political scales, while also navigating a scalar politics of controlling explosive items.

Complexly (nonhierarchically) nested and overlapping ethnonational scales are constantly reproduced via local practices of victim assistance. In humani-tarian mine action, ethnonational CSOs have mediated between INGO/NGO activities on the one hand and local ERO orders on the other, to access areas and implement victim assistance activities, for example.[30] Such mediation unfolds mainly through the reproduction of ethnonational links between the ethnic CSO and head of villages, key communal or religious figures living in village areas throughout the CSO's territorial scales of operation. These figures allow the CSO to bridge the regional and countrywide projects of INGOs into local areas by configuring an interface with EROs. Nonetheless, this does not unfold linearly.

In Kyaukme, for example, there have been three main CSOs/CBOs: K., N., and O.[31] K. is closely interconnected with the Shan Nationalities Democratic Party; N. is closely related to the Shan Nationalities League for Democracy (SNLD). And both are Shan community-based organizations. O., a Ta'ang CBO, maintains links with Ta'ang organizations and TNP, the Ta'ang political party. Landmine victim assistance has been shaped by the circulation of ethnonational logics and linkages reproduction, as a member of Ziwita Social Assistance Association, another CSO—this time nonethnically profiled—explains:

> [I]f somebody steps on a landmine in remote areas under the control of TNLA, the two Shan organizations have great difficulties in reaching the zone and operating, in part also because they are afraid of meeting the TNLA. The same is true for [O.] which is often concerned with avoiding as much as possible RCSS areas. Needless to say that this creates holes and overlaps in the provision of civilian humanitarian support to those who have been affected by landmines. That is also why [Ziwita Social Assistance Association] expanded the activities of [the] organization so as to include also humanitarian action for those who were affected by landmines.[32]

Governing landmine victim assistance unfolds through and shapes different overlapping and nonhierarchically nested territorial scales as it proceeds along ethnonational rationalities and linkages. To provide assistance in one area requires mobilizing linkages at broader ERO territories; or, alternatively, to pass on the provision of assistance to a different actor at the same township scale, so that they can mobilize the linkages needed at other rebel territorial scales.

Negotiating Rebel Military Conduct

Caught in the midst of Sit-Tat massifying violent practices as well as armed conflict among resistance actors, civilian communities have often attempted to influence the conduct of rebel armed forces. In this sense Ta'ang land can be illustrative due to the multiple rebel movements that in recent decades have been overlapping at these latitudes.

For example, throughout 2018 armed clashes intensified between the RCSS/SSA-S and the SSPP/SSA-N. The latter, since the expansion of the RCSS/SSA-S in northern Shan in late 2015, had progressively strengthened ties with the PSLF/TNLA to contain the other Shan ERO. Toward the end of the year, with the yearly figure of people temporarily displaced by conflict reaching 21,900 and the numbers of clashes between rebel movements escalating,[33] requests for a ceasefire between the two Shan EROs started to circulate. Although the PSLF/TNLA has played a key role in the context of the armed conflicts and the Shan EROs have been enmeshed in different borderland alliances and arrangements,[34] only the RCSS-SSA/S and SSPP/SSA-N were involved in prenegotiations.

Channeling previous consultations with Shan civil society and religious figures, the SNLD political party issued official letters to the two EROs calling for a halt to hostilities and offering mediation. The SNLD secretary general commented on the letters noting that "[b]oth sides [SSPP and RCSS] announced that they would carry out the interests of their nationalities and started long lasting revolutions. At present, the clashes are between Shan nationalities. The first to suffer is the public. Therefore we have requested, on behalf of the public as well as monks, to both sides to cease fighting."[35] Negotiations for this intraethnic (albeit inter-ERO) bilateral ceasefire unfolded via the Committee for Shan State Unity (CSSU), a broad political coordination body formed in 2013 and including Shan EROs, political parties, youth organizations, CSOs, and other civil society networks.[36] This forum constituted an arena, on a Myanmar national scale, in which different Shan ethnonational communities with their territorial scales of struggle converged during the negotiations.

An important role was played by the Tai Youth Network (TYN), a CSO whose creation is worth spending some words on, as it illustrates the transversal and cross-border character of ethnonational scales. TYN was created at the end of

2013 on the occasion of a Tai Youth Conference organized to bring together various representatives of Shan youth throughout both Thailand's and Myanmar's borderlands. The network was conceived as a coordination body and CSO that was to represent Shan youth spanning the borderlands into CSSU. TYN members' backgrounds were the most disparate. Among them, for example, were Shan former students and migrant workers that, coming back from the Thai borderlands, had mobilized to support the SSPP/SSA-N during the offensives launched by the Sit-Tat in 2011–2012.[37]

Through CSSU "civilian organizations were able to have a forum inside which they could exercise some form of pressure on the politico-armed organizations so as to push them toward the [signing] of a ceasefire."[38] Struggles over controlling the two ethnic armed movements—which at one time also entailed struggles over the two EROs' territories and how to calculate them, since a key point for reaching a ceasefire concerned the demarcation of troop presence and boundary lines of their respective territories[39]—not only occurred at different territorial scales but also contributed to their reproduction.

Ethnic CSOs' roles in attempting to influence the conduct of ethnonational rebel movements' armed forces have emerged in other fields as well. Talking about its relationship with the TNLA, for example, the Ta'ang TSYU underlines the logic of preserving Ta'ang civilians from the side effects of armed struggle as one underpinning its attempts to approach the TNLA when seeking to influence the ERO's war efforts:

> When it comes to fighting and the problems created by the war, we do not try to reach out to TNLA . . . to ask them to stop fighting . . . because that would simply be useless. What they have told us has been that these are none of our issues and responsibilities and that we should not deal with these things. What we have done, though, has been to send the TNLA letters telling them that the situation is becoming unbearable for civilians and asking them to please inform civilians about their actions in advance so that they can cope with that, or to inform them about potential dangers, where it is going to be dangerous for them and when, or the use of landmines so that they can be aware. I think this is a way in which civil society can influence them.[40]

The transformations characterizing borderlands—and more generally, Myanmar—civil societies in the last thirty years have proliferated into rebel movements, allowing the circulation of discourses and practices linked to humanitarian and human rights logics. These permeations, which often make it extremely difficult, if not impossible, to draw clear lines between armed organizations and CSOs, have mostly unfolded via that political capital that is ethnicity. While in other ambits, such as humanitarian first response or land grabbing, ethnic CSO networks have produced interethnic territorial scales of action and practices, in the field of controlling the means of violence they have remained more limited to the tracks of ethnonationality and ERO territorial entities. Commenting on his own attempts at launching a cross-ethnic land-mine ban campaign in Myanmar, and identifying a focus group to kick it off, a humanitarian mine action researcher recalled:

> When I was talking to the TSYU about doing a campaign . . . because in a lot of countries we got campaigns to ban landmines and I am looking for a focus group that can really start this . . . they said . . . well, we would be interested in that but there are other communities in the area [northern Shan State] and they would not listen to us, because we are Ta'ang . . . and immediately I could see the limitations to what they could do based on ethnic identity just from that comment."[41]

What is interesting here is not the question of the impact of such influence—for actually, EROs do not adjust their behavior on the basis of a pure bargain of legitimacy with CSOs and other sections of society, nor do CSOs carry an intrinsic power to leverage armed actors—rather, the circulation of rationalities and techniques, as well as the production of different scales of resolution for governing military means.

FIGURE 14. ERO military post on top of a hill. Hu Sun, Kyaukme Township, Shan State, Myanmar, March 2019, author's fieldwork.

PLACES

In October 2019, during fieldwork in the northern parts of Kyaukme, the bustling trading town that gives the name also to the respective township and district military-government administration, an informant recounted the story of a woman who had recently fallen victim to a landmine explosion (Buscemi 2021a). On the mountains that edge Kyaukme to the north, the woman was walking on her everyday path to the local market. After the explosion, it was quite a while before anyone could find her. A local civil society organization (CSO) with connections in the woman's village had been alarmed and her transportation to the nearest hospital, a clinic about a four hours' walk away from the place of the explosion, had been arranged. To pay medical expenses, the village community started to collect money in collaboration with the CSO. The money would hopefully help her not only to face medical expenses but also to cope with unemployment thereafter. Next to the path she had taken, there was a tea plantation—quite an usual landscape at these latitudes, for the hilly slopes north of the town are dotted with tea bushes, an ecological trait they share with the broader region. So, as news spread in the following days, work in the tea fields dropped, some workers left or lost their jobs, and people stopped using that same path to go to the market. Another, more circuitous road started to be used, although it would take an additional hour and a half to reach the market from the woman's village.

Yet, the circumstances of the explosion remained unclear. Kyaukme township sits on a quite important logistic, economic, and military crossroads: it straddles the Mandalay-Muse highway, one of the main arteries of the country that connects it to China and carries the bulk of the overland trade of Myanmar. This is why these areas have been affected by the presence of multiple armed forces. The Sit-Tat maintains several important garrisons here, People's Militia forces have long been present in some villages, while many rebel movements operate on the hills—the PSLF/TNLA, the RCSS/SSA-S, and the SSPP/SSA-N have alternated control at different points in time, often in the same posts, camps, checkpoints. Most probably the landmine had been freshly laid, since other people used to take that path on a daily basis and no incident had occurred. Nobody knew of military activity in the area, so nobody could really tell whether one of the EROs operating here or the Sit-Tat had laid the landmine.

The story of the landmine explosion suggests how place—for example, the path, the hillside tea plantation, or the village area—is made and remade through the presence of weapons, as well as through the processes and practices of governing them (Agnew 2011, 325; Cresswell 2004, 82). Place can be defined as a geographical context for the mediation of sociopolitical and physical processes (Agnew 2011, 317): a series of relations that mediate social processes by operating through notions and actions of proximity, spatial embedding, areal and horizontal differentiation among cores and peripheries at different scales (Jessop et al. 2008, 393). It can be understood through the prism of three main politico-geographical dimensions (Cresswell 2004; Agnew 2011). First, place as *locations* and *localities*—meaning specific sites in space that mediate the "where" of activities and objects (Agnew 2011, 326; Cresswell 2004, 7, 51). Second, place as *locales*—meaning the actual shape of the material setting in which everyday action and life is conducted. And third, place as a sense of belonging, identification or, more generally, ideational linkages produced by, and productive of, certain sociospatial relations (a "sense of place"). In the accident, the armed assemblage revolving around the landmine reshaped the place of the explosion: the path along which it occurred stopped being used, the tea plantation around that location—a locale of work for many—was abandoned and fell into disrepair in the aftermath of the event, while the village community organized a fundraising campaign to help the woman as one of their fellows.

In this sense, the story of the landmine explosion provides an example of how the rearmament of the Ta'ang rebel movement and its wars against the Sit-Tat, its militias, and the RCSS/SSA-S have shaped not only political geographies of territory and scale but also political geographies of place. Throughout the process of disarmament and rearmament, similar to what the story of the landmine explosion shows about Kyaukme and its surroundings, Ta'ang areas have become a veritable kaleidoscope of forms of rule. A complex mosaic of governable places emerged (Korf et al. 2018), one made up of fragmented, nonhomogeneous, and overlapping zones under the influence of EROs, the Sit-Tat, and militias. All along that underbelly belt that is the Mandalay-Muse Union Highway, as well as the Shwe Gas and Oil Pipeline, large Sit-Tat bases have been established on both sides. North of the mountains overlooking major towns like Kyaukme, Hsipaw, Lashio, and Namtu, the RCSS/SSA-S and SSPP/SSA-N have cyclically been interspersed together with the PSLF/TNLA troops. South to Namkham and the Muse special economic border zone, bases and camps of the Sit-Tat and its affiliate militias—*in primis,* the Pansay militia—could be found next to ERO ones. In the area of Kutkai and Tarmoenye, militia areas mix with the presence of the KIO/A and the Kawngkha Kachin militia. And the colors of this rich image become even more blurred when one attempts to trace the boundary lines of these place-areas and to depict their intersections and overlaps. These lines, in fact, shift both spatially and temporally. Outside the main areas of Sit-Tat, militias, and ERO bases, armed actors' presence may be concomitant and/or alternated. Here checkpoints, other key locations, and the frontiers of rebel forces' areas turn into especially fluid spaces.

In what follows, this chapter explores how techniques and practices of governing weapons and armed collectives have molded place. In particular, I highlight four main sets of techniques and practices that, similar to other parts of these borderlands, turned Ta'ang regions into changing sets of "governable" places. These have to do with the creation and maintenance of fixed and mobile military installations, armed forces roving presence, landmines deployment and control (or lack of control), and the politics of governing the implications of their presence.

Mobile and Fixed Military Installations

Mountainous and rural village areas have been characterized by a constant shifting and alternating presence of different armed actors, especially EROs and the Sit-Tat. This has been part of armed actors' military tactics and techniques. As an SSPP/SSA-N official pointed out, if the SSPP/SSA-N's troops "go away it means that TNLA will come instead, so . . . in areas where the SSPP and TNLA are both present, we try to move together to balance out power. Not exactly in the same place but if they are there in that village, then we stay here close in this other village."[1] Units remaining outside villages normally move inside settlements to meet specific needs and usually sojourn for a few hours or a couple of days at most.

The presence and roaming movement of armed actors has unfolded through techniques and practices of controlling weapons and the means of violence. For example, the RCSS/SSA-S, like other EROs, has organized its troops in platoons of about thirty men further broken down into squads of five to ten soldiers who would spread and roam around. These at times would sojourn inside villages during the day and move out to the bush right after dusk.[2] Depending on the area, armed actors would alternate with each other, staying for longer periods only in villages with more stable security conditions and withdrawing to the bush at night. Such fluid presence at times has been accompanied by physical security installations like mobile checkpoints, posts, camps, and bases. ERO troops normally operated inside forests and on the mountaintops that surround village areas where they set up camps. Occasionally bases have been built inside villages as well. Many EROs have adopted a system of rotational shifts to ensure soldiers do not settle down in a given community. While security protocols adopted by EROs tend to relax as time goes by and troops familiarize themselves with local residents, the arrival of new platoons or squads usually coincides with stricter measures on local communities' mobility and livelihoods.[3]

All of this has contributed to mold the geographies of place of Ta'ang areas. Landmines, armed men's roaming presence, and physical military installations have altered the landscape in at least three ways. First, by reconfiguring the altitudinal and morphological top-down relations between lower valley, or slope areas, and hilltops. EROs have adopted a technique of landmine use that sees mines as part of the architecture of posts and bases, whereby they are used around installations in a defensive/offensive manner. Landmines are laid to create a perimeter around bases, camps, and posts located on hilltops and forested areas. This in turn makes it harder for villagers and farmers to navigate certain areas and creates a locational differentiation between higher and lower parts of villages and surrounding areas.

Second, the viability of different kinds of walking paths has been redefined by military techniques. Institutionalized military practices of laying mines for offensive purposes carve out a distinction between types of paths and roads.[4] Generally speaking there are two main kinds of routes in these mountainous areas: main roads—normally unpaved paths running up and down the hills; and field paths—narrower, less visible paths that zigzag throughout areas off the beaten tracks and lead to agricultural fields.[5] The latter are more often a location of landmine contamination, offering better cover because they are less known and cut through forested areas at times. Landmines are less commonly found on main roads.

Third, the accessibility of tea and other crop fields and forested areas has been altered in fluid and shifting ways. Mountain guides can be heard recalling how "beautiful [it] was to be free to go in whatever area you wanted, taking whatever route you knew. Since around four years ago [ca. 2015–2016], though, this started to change. Forested areas started to be suspicious in part because of landmines presence, [and] in part because you could never know who you would actually meet inside the area."[6] Plantations through which field paths run have also been carved out as a dangerous and uncertain place. This is particularly true for tea fields, which tend to be located on the slopes of the hills and offer direct covered access to military camps and posts on the hilltops. Shifting places generated by techniques of controlling the means of violence intertwine with land tenure practices. In fact, in Ta'ang areas, as in many parts of the borderlands, land is often managed via shared land ownership systems that include regimes of freehold land, community forest reserves, and customary tenure for rotational farming entailing a certain degree of sharing and exchanging livelihood areas.

Landmine contamination in these locations, however, is never patterned and does not demarcate mined fields or no-go areas. Rather, it is as shifting as armed actors' presence. The accordion-like movement of armed actors' presence and control has generated very fluid, shifting, and overlapping place-spaces. Here landmines used to delimit outposts or temporary encampments are often left in place once armed actors have moved away. This is especially true for areas contested by multiple armed actors,[7] where encounters, a roaming armed presence, and shifts in positions tend to leave landmines around. The landmine, as a technical object inherently linked to techniques of use and control, actively contributes to entrenching the sociospatial relations of place. This does not occur as a result of actors' agencies, discourses, and weapons' properties, or at least not only; but rather as a combination of technical objects and techniques diffused throughout the social body. Often many villagers do not know what a mine— industrially manufactured or handcrafted—looks like, or where it could be. Other times mines are laid in visible spots to deter movement. The absent presence, and the present absence, of landmines renders tea fields, hilltops, and forests identifiable locations even where and when armed actors have moved away: "One mine, in a farmer's hill, does not just stop that farmer; it probably stops every neighboring farmer as well. So, even a small number of mines can remove a huge amount of land from a community's economic base."[8]

Techniques of landmine use and armed forces management that attempt to govern armed actors' presence circulate throughout society and shape how dwellers understand and practice place, as location and locale, as well as forms of counterconduct. The village head of K.L., located north to Kyaukme, notes how "the RCSS . . . usually does not stop in the village for long periods, neither [do] they sleep here. Behind this house there is the mountain, we have tea plantations on the slopes of the mountain but after the plantations nobody goes further because there is RCSS there. They have their camps and place landmines there."[9] Mobility and livelihood are channeled not merely as a result of coercive restrictions but also through the socialization and institutionalization of military techniques. For example, porterage, forced (at times just short-lived) recruitment, or using local guides as minesweepers are all part of an array of practices of managing armed units locally and are often employed to various ends. Locals moving alone are especially targeted. Thus, besides avoiding certain locations, people try to move in groups rather than alone, so as to be less exposed to porterage and forced recruitment.[10] Similarly, migrant workers from central Myanmar that around April usually move to Ta'ang areas looking for jobs as tea-leaf pickers have been shifting location in relation to hostilities and armed actors.[11] As a Mong Tin village resident explains:

> In the last years we have not had any problem in relation to the ensuing conflict . . . so the tea-related activities have not been affected. But in other areas that have been more exposed to conflict people have been afraid to go and pick up the tea leaves because of the presence of landmines. Overall, it is difficult to say where conflict is in Shan State, because it is not like Rakhine state, where there is regular fighting and you know it that there is fighting: here you never know, it can happen here or there but you will never know. Why? Cause here they [EROs and Sit-Tat] are never face to face, but they instead move always around. At times they meet and fight, at times they obtain the info that their opponents are coming so they may try to avoid meeting and fighting.[12]

The lives of local villagers in especially contested areas have often been characterized by osmotic movements from village areas to Buddhist monasteries, from mountainous areas to larger towns and urban agglomerates' outskirts in relation not only to oscillations in armed hostilities but also to the presence of mines, explosive remnants, and roaming armed units.[13]

Other areas instead have been characterized by permanently built checkpoints, trenches, posts, camps, and bases. Although stable, these are often only temporarily and alternately manned by EROs, while others represent fixed military installations. In the borderlands, larger fixed military installations belong most often to the Sit-Tat. To illustrate the governable places emerging in locations characterized by the presence of fixed military installations, I take the village of N.O. as an example.[14] N.O., it should be noted, is located ten kilometers east of Myitkyina, in Kachin State. Although not in Ta'ang regions of northern Shan, nonetheless N.O.'s case vividly illustrates several aspects that have been observed also in Ta'ang areas. Similarities between different borderland locations have been especially pronounced due to the waves of militarization as part of the processes of frontierization and (re)territorialization of the borderlands analyzed in previous chapters and the construction of Sit-Tat military bases that characterized the 1990s and 2000s.

N.O. has been surrounded by two Sit-Tat military bases built in the last decades: an airfield and an artillery corps base (with annexed artillery firing range).[15] The main road leading to the village winds through the fences and for-tifications of the two bases. Or, better said, fences and fortifications cross through the village, with barracks and infrastructures just a few hundred meters from vil-lage houses and fields. The KIO/A, whose closest military presence is located far-ther north, in the mountainous areas of Namti Pum and Auche, has maintained a presence inside the village and frequently targeted the two bases. The building up of the two bases has run in parallel with a layered landmine and explosives con-tamination. First (in no order of importance), rockets, artillery shells, mortars, grenades, and landmines have remained behind after KIO/A's attacks to Sit-Tat positions. Second, the Sit-Tat operated in a similar manner, mining wood areas that offered cover to the KIO/A's troop movements and attacks. Third, the firing range has been used for artillery shooting trainings that have caused explosive remnants spilling over the base's perimeter into village areas. Fourth, construc-tion and fortification techniques have brought along a concerted use of fences, barbed wire, and mined perimeters around the two bases intersecting the village. Rather than trenching off the two bases and impeding life in the village, all of this has had major impacts on the areal differentiation and mobility of this place. Part of the woods—a key agricultural and livelihood communal asset—and vil-lage outer areas have become hardly accessible and increasingly dangerous in a combination of military occupation, landmine contamination, and explosive remnants of faulty artillery trajectories due to training errors. Fields adjacent to the bases and parts of the village have been annexed to them and/or abandoned. Human and animal bodies have also been differentiated and targeted. Both have often been affected by acoustic and air pollution linked to both air traffic, landing and taking off, and miscalculated trajectories of artillery shelling and explosions. This has impacted pregnancies of livestock, with hens delivering damaged or empty eggs, for example.

The Places of Armed Forces' Roving Presence

The rhythms of EROs' troop movements and presence in rural areas, for exam-ple, are not only informed by strategic mobility, armed forces management, patrolling, and landmines, but also by the cyclic recurrence of specific events, festivities, and connected celebrations. Weddings, religious occasions (like the Buddhist Thadingyut and Tazaungdaing), or the inauguration of communally meaningful sites are often accompanied by arms and armed-human presence. In this sense, governing the entanglements of arms, armed forces, and populations is informed also by techniques linked to rationalities of ethnonationality, which are terraformed via sociospatial relations of place.

In March 2019, the village of B.H.—a mixed Shan/Ta'ang village in Kyaukme township with a majoritarian Shan population—inaugurated a new stupa built in the village monastery.[16] Households had invited friends and relatives from nearby villages to come over and stay overnight for the celebrations, so a mix of Ta'ang and Shan families had poured into B.H. Around lunchtime of the inauguration day, an RCSS/SSA-S unit of around ten men showed up, coming from a nearby base not far from B.H. Establishing a security perimeter around the monastery and some adjoining houses, an officer and his soldiers were hosted by the household of a former RCSS/SSA-S member who had invited the commander to the stupa inauguration. As the units' rank-and-file soldiers rotated to sit at a bamboo dining table and the officer was entertained by the former RCSS/SSA-S member, another unit of three arrived. Differences in their demeanor were illustrative of different layers of conducting the conduct of guns and gun(wo)men via training techniques and the materiality of uniforms.

The first, larger RCSS/SSA-S unit had traveled from a local base, escorting a commander in a green service dress uniform and two officers in field uniform (carrying sidearms only) to take part to the inauguration. The escort carried clean and well-maintained, nonrusty weapons. Weapons that are always pointed downward, with fingers always off the firearms' trigger guards, and seemingly shielded behind the soldiers' bodies when they talked with civilians—either because they embraced the rifles behind their arms, or because they shifted them to their side. The second, smaller RCSS/SSA-S unit instead had been patrolling the area and, hearing about the festival, came to the village. These three rank-and-file wore camouflaged uniform pants with casual tops, or un-uniformed garments. They carried rusty and older rifles, some of which seemed refurbished with components from different types of rifles. These were leaned against the ground, laid on or kept dangling from their shoulders as they strolled through the crowd, hung out watching the celebrations being staged, or stopped by to perform donations at monastic donation stalls.

Techniques of military enlistment and recruitment also contribute to spatially differentiate communities, shaping or reinforcing inner places within villages along ethnonational lines. In Hu Kwet village, for example, a student visiting during university holidays noted:

> Since RCSS came here . . . tensions between the communities, Shan and Ta'ang, have increased. The issue is that some of our relatives joined with the TNLA while other people in the village from Shan families joined with RCSS. Since then, the tensions in the village have increased a bit. We did not have open clashes or violence but frictions have heightened let's say. Half of the village here, as in other ones actually, is Ta'ang—so

there is a Ta'ang part of the village and there is a Shan part of the village. So, when the TNLA was coming before 2015, the Shan families would run away from the village to Kyaukme. Now instead that the RCSS is here sometimes there is a bit of conflictual relations in the communities cause they [Shan villagers] tell us, "Oh, you Palaung, you allowed the TNLA in here" or "the head of the village is Palaung so he allows TNLA to be in the village." This overall has intensified the tensions and frictions between ethnic groups. There are people I know, Shan people, that now are not willing to have contacts anymore or that look down at me and Palaung in general etc.[17]

An acquaintance of the former RCSS/SSA-S member explained to me how he had sent his son to Loi Tai Leng to study English.[18] Remarking on the need for an ethnonational bond to underpin the connections between families, communities, and EROs, he differentiated between the RCSS/SSA-S and the SSPP/SSA-N (both Shan ethnonational rebel movements): "The RCSS, they say their policy is to destroy drugs production, trafficking, and consumption, to destroy opium plantations; while instead the SSPP wants to grow opium plantations. SSPP do not care about the Shan people; about the youth, for example, they do not do anything."[19]

Logics and techniques linked to ethnonationality can also be seen for what concerns Buddhist monasteries as distinctive places in relation to weapons and armed actors. EROs' armed forces normally refrain from entering the monasteries' buildings with weapons. Buddhist monasteries maintain a value as places to be respected out of religious and social practices that variously inform people's lives. More than a matter of monks' intrinsic authority, or rational interests of rebel movements in legitimizing their behavior alone, such practice is linked to religious and social ideas and practices of how one should conduct themselves as well as the weapons one is acting with according to Buddhist precepts and the teachings of the Buddhas. As a former TNLA soldier put it: "We would leave the weapons outside as well as we were leaving our shoes, in order to pay respect to the monk. We would organize sentry turns in order to enter the monastery in small groups, not to enter all together with weapons. Some of us would stay outside and guard the weapons. But this was exceptional I would say and would happen practically only at a monastery."[20] It is interesting to note how a parallel is drawn here with a Buddhist and cultural norm, meaning the removal of shoes on holy sites, that is paramount in Myanmar; one that has often been recalled to shed a light on the inextricable political and social value of discourse and practices of conducting oneself based on Buddhist *sāsana* (the teaching of the Buddha) (Turner 2014). By the same token, the abbot of a monastery on the mountains alongside the Kyaukme-Mogok road, talking about the presence of an RCSS/SSA-S' contingent, remarked:

> Sometimes they would sleep in that building over there [points to a low building in front of the monastery, inside the compound] which is . . . where visitors of the monastery . . . coming . . . for the celebration of the Tazaungdaing Festival can stay and sleep. They would leave weapons outside of the building where we are now because they think it is not respectful for the monk so they should enter and pay respect to the Buddha statues there. TNLA was not really present here but came from time to time and would sleep in the guest house building outside as well.[21]

Nonetheless this is only one dimension, for in fact one cannot discern a clear pattern whereby EROs, or armed actors in general, never encroach on Buddhist monasteries' premises or actual buildings. The recognition of monasteries as places of restraint is linked to practices of ethnonationality inasmuch as ERO units tend to engage more respectfully with monks of the same ethnicity.[22] Concerning schools, the logic of ethnonational bonds can at times turn the building from a useful and strategic edifice—being one of the few throughout rural mountainous areas to be made of concrete walls—to a place to be preserved.

Deploying Landmines and the Politics of Place

The use and management of landmines and their explosions—as well as the dangers associated with them—inform, and are informed by, different shifting and fluid ethnonational places. As noted in the previous chapters, when livestock accidentally alters the ecology of landmines and triggers them, the ERO units present in the area often request the owner to provide financial compensation.[23] More often this occurs when the farmer is of a different ethnicity than the armed actors. ERO units' mobility and techniques of landmines and explosions management informed by ethnonational logics generate, from the perspectives of farmers, different places of risk concerning the request to compensate EROs. Considering that armed actors so swiftly and unpredictably alternate, such places of risk and compensation are temporally and spatially fluid with the added layer of a politics of identity.

A politics of identity is activated also when landmine victims or people directly/indirectly prevented from accessing agricultural fields or other livelihood means move to urban areas.[24] In this sense one can glimpse the intersectional nature of such politics of identity, which does not have to do with fixed ethnic categories per se. People losing their livelihoods and deciding to migrate are most often people of ethnic minority background from the mountains and, as a Kyaukme resident put it, having "always been used to living . . . a mountain life," find it extremely dangerous and difficult "to move to the town and the town life."[25] Or at least this is how some urban dwellers frame them.

It would be erroneous to dismiss the role of landmines and armed forces in exacerbating the friction among ethnic communities in villages. At the same time, it would be erroneous to portray these tensions as a result of only landmine presence or of the tense relations between one ethnonational rebel movement and another ethnic population, or between different ethnic populations. Instead, landmines as technical objects contribute to make visible and more concrete demarcations or lines of division that previously were not being activated. For example, the impossibility to access one's own livelihood, to which the landmine contributes, makes one more wary of an ethnic distinction as well. Why can the "other" move freely to their fields, for example, while "I" am prevented by the realities of landmine presence and absence?

This has occurred especially in relation to a differential distribution of mines by the RCSS/SSA-S.[26] The Shan armed movement has tended to use landmines and booby traps made of bamboo spikes in predominantly Ta'ang villages or in areas of villages where Ta'ang live. (Often Ta'ang and Shan households can be found in distinct places of the village.[27]) Such techniques and the use of these technical objects are an integral dimension of military defensive architecture and the infrastructure of bases, camps, and trenches of the RCSS/SSA-S. But they are also connected to a certain deployment of ethnonational logics and the Shan ERO's use of strategies akin to the Sit-Tat's four cuts, which construct Ta'ang populations as potentially part of the opponent rebel organization and territory. Ta'ang communities are thus to be targeted if they resist reterritorialization, or if they constitute an obstacle to it by supporting the Ta'ang rebels.[28] It is neither the landmine technology nor the actions or interpretations of armed units alone that deterministically shapes ethnonational lines and frictions, but rather the ways in which the technical properties of the landmine, as part of military infrastructures (and how they shape spaces), combine with diffused techniques of managing and coping with landmines use, their explosions, and the consequences. These combinations generate different shifting governable places.

Temporal-Spatial Relations of Place and Humanitarian Arms Control Logics

Governable places are also shaped via techniques of governing guns and armed collectives that are informed by humanitarian arms and armed violence control rationalities. Armed actors have often located shelter places, camps and posts very close to civilian-inhabited areas, monastery compounds, or stupas. This is in part practiced in the widespread consideration—widespread especially among field units—that other rival rebel forces will not dare attacking these places and in turn jeopardize local populations.[29] Such a rationale, though, operates also in reverse, and underpins mobility rather than fixity: for, in fact, EROs also often move away from civilian areas in order to avoid confrontations that could jeopardize the latter.

Discourses and practices of preserving civilian populations also shape the political configuration of places as liminal spatialities in between different orders. Talking about the town of Namhsan, a TNLA member pointed out: "We hope that we can control Namhsan as well and, if we could do it, that would mean a lot but it is difficult too. Because the base of the regime is on the top of the city, which is very close and then there is the hospital close and if we go there it would be a huge problem for the civilians. . . ."[30]

Different places emerge out of complex temporal and spatial combinations throughout which the ruling of military means is exercised, or at least attempted. Techniques are not only linked to spatially shifting movements of armed actors but also to shifting times. This can be seen in the temporally contingent use and control of landmines around camps and posts. Even in places where the more stable presence of only one armed actor may render it a site of one predominant governable order and regime of control over violence, shifts may occur in time, as an SSPP/SSA-N officer described:

> The troops [SSPP] do not place landmines for a long time. They normally place them at night, around 6 p.m., and they pick them up in the morning. No landmine is left there in the early morning at 5 a.m. Soldiers are mandated to inform civilians. When we move up to a new camp in a guerrilla zone we inform civilians that after 6 p.m. they must not come up where the camp is located. We inform the head of village and he refers to the others.[31]

Yet, as we have seen, the mine cannot be disposed of at will, and instead acts on its own by making it difficult to remove and replace so smoothly as described here. Furthermore, like its spatialities, the temporalities of the landmine are not so well defined but maintain blurred boundaries that may expand or shrink in relation to various elements: for example, how long the armed actors have camped; in what manner has their governable order been integrated, embedded in place; how have techniques of landmine use and control been assimilated and practiced by units.

Dynamics of sociospatial relations of place are also produced via practices, techniques, and discourses of coping with landmine contamination. Since around

2014, INGOs in partnership with local NGOs and CSOs have been progressively expanding the provision of mine and explosive ordnance risk education workshops and courses.[32] During the latter, dwellers are instructed concerning explosive items' most common shapes and sites of risk—dangerous areas that may likely be contaminated with mines: bridges, electricity pylons or related installations, ditches, roadsides, livestock and animals' dead bodies, victims, damaged vehicles and wreckage, areas traversed by fences or walls. Not all of the shapes and sites fit, some may be pertinent to borderland contexts, others not, and adjustments with the local places need to be made. Workshops are normally composed of an awareness session, in which standardized guidelines and formats are followed and a map-drawing exercise in which local inhabitants are asked to map out their village area and trace dangerous sites. While people have room for reformulating the "standard" sites disseminated via risk education, the entanglement of techniques of coping produces areas of risk, uncertainty, and danger that border sites of (potential) contamination.

This in turn is aggravated not only by the temporal and spatial shifting of different armed actors but also by different layers of sense of place that are embedded in one another. Talking about the provision of humanitarian assistance, a Ta'ang Women Organization's member underlined the embeddedness of roadsides and paths in relation to landmines inside forested areas and remarked how one has to adjust their own self-conduct in those places: "Even we, when we have to go to the jungle and forested areas to reach remote villages, we are traumatized and are afraid to use the motorbike, since the road is not smooth and straight and we have to proceed like a snake."[33] Private individuals, religious associations, and CSOs have been at the forefront of a largely self-organized mechanism to respond to armed conflict outbreaks and landmine contamination accidents. As part of such a response, these actors have practiced the transportation of families and communities away from sites of armed clashes toward Buddhist monasteries that often serve as places of refuge for people fleeing their homes.[34]

Mines too—the materials and techniques that compose them—contribute to reproducing sociospatial relations of place. Regardless of the entity of contamination, which to date remains both unknown and impossible to estimate, there lays in the ground an incredible heterogeneity in types of explosive items, and especially materials composing them, that in turn holds the potential to generate a kaleidoscope of places. While it is unknown how many are industrially manufactured and how many are handcrafted, or crafted through factory-grade processes, the unstandardized nature of many explosive devices tends to defy detection. Improvised mines have all very different and minimal metal content, usually in the form of wires. At times they may be more visible given that they often remain aboveground, but their minimum and disparate metal content make it difficult for metal detectors to spot them, and eventually may slow down any demining process in the future. This, on the one hand, reminds us of the complex ecologies and assemblage processes that bring from commercial and military cast-off, surplus, materials to an explosive weapon; and, on the other hand, illustrates the juxtaposition of what seem to be distant places and temporalities in the past and the future with the complex realities of place. Such juxtaposition reveals that in the complex material histories and futures of the explosive weapon (together with techniques of use and control), and the ways in which it unfolds through and shapes place, one can see a variety of governable orders or ways of governing the relations between weapons, people, and space.

FIGURE 15. Ukrainian BTR-3 armored personnel carrier in Yangon, February 2021. Source: Sai Aung Main/AFP.

Epilogue

The hammock he is sitting on swings back and forth to the pace of his right thumb scrolling down what the world says about the war in Ukraine. Commander Maui's eyes stick to the tablet and indulge a little longer in that gesture at the core of his morning routine, picture after picture, video after video.[1] Sobbing social media sounds alternate with the voices of the radio station next to him crackling orders and communications for frontline troops in between the last shades of night and the first light of dawn. After the 5:30 a.m. military flag salute, he usually takes a moment to relax here, under the hut that hosts his operational command, stretched out in an olive-green hammock framed by two unexploded bombs dropped by Sit-Tat's Russian-made fighter jets. He decided to sling several shells from every corner of the roof, for people to constantly keep an eye on the munitions, whenever they would visit his base. "It is very easy—Maui said, when I asked him about the bombs—to forget the pervasiveness of the violence we face, and that brought us to pick up guns as last resort. So, we have to remind ourselves—he continued—of what bombs do to people. Because, even if one has been lucky enough not to be hit by one of them, the violence of the bombs and the regime continues to work." Before joining his comrades at the table for breakfast, this is a ritual he has kept for himself since I arrived for a three-week fieldwork stay during August 2023 at a former government primary school in Demoso that the Karenni Nationalities Defense Forces (KNDF) have transformed into a military base. Today, though, he breaks his silence—"You are interested in weapons, right?"—and turns the screen toward me, to show the picture his thumb got stuck on: "These are the APCs [armored personnel carriers] of the Sit-Tat marching down there, through the Loikaw-Hpruso road where we are fighting. Made in Ukraine. Why is it so that Ukraine gets so much attention, but nobody is interested in the war in Myanmar, even though the Sit-Tat got their weapons from Ukraine and from Russia? Is our war not global?"

Almost two years after the same BTR-3 Ukraine-made vehicles used by the army (figure 15) paraded on the streets of Myanmar's major cities to enforce the February 1 coup,[2] Maui has become the deputy commander-in-chief of the KNDF in Karenni State and one of the major figures in the armed struggle. An armed struggle that he had previously long deserted, for desertion had been his way to engage with the military state up to the 2021 events. As a peace activist, in his twenties he had committed to the Brazilian playwright Augusto Boal's theater of the oppressed practices to generate societal change concerning environmental degradation in the Mekong River region, and to spread the same methods back in Myanmar's rural communities too. Around 2019, Maui had quit his career in the NGO and peace activism circles and had opened an organic farm. But when the coup happened, he ran to Karenni State's capital, Loikaw, together with friends, determined to facilitate protests by people flooding the streets. Like in many other Myanmar cities, people had decided to oppose the military state with peaceful means. They knew, he recalled, about the power of refraining from the gun. They were conscious of the fact that military violence was at the very root of the military-state oppression they faced.

But as the Sit-Tat crushed the peaceful protesters, atrocity after atrocity, "enough was enough." At first, he took part in short bottom-up self-defense and military trainings popping up everywhere in Karenni. When mobilization activities gathered momentum, however, he and his comrades reached out to the PSLF/TNLA leadership for military support. The TNLA agreed to host and train him to become a sniper. In search of support to create an armed force, Maui had contacted an older generation of partisan fighters—fighters who, like him, had long posed to observers, both inside and outside Myanmar, an iteration of that question he was now turning to me, albeit from a different perspective: How global are wars at the margins? How global are Ta'ang wars, and why is it that the Kachin, Karen, and Arakan wars, for example, got a modicum of international attention, but nobody cares about the wars of the minorities' minorities?

Deserting Global Wars

Since the very first weeks of the 2022 Russian armed invasion of Ukraine, the Italian political philosopher Sandro Mezzadra argued that the war was as much a

European war as it was a global one. The global character of war, and the global character of the tendencies toward rearmament and militarization that *the* war had only accelerated, he reasoned, made of desertion the main—if not the only—revolutionary imperative toward a truly revolutionary future, one that would break with the dominant politics of soil and ethnoracial identity as a continuation of war by other means. In Mezzadra's opinion, the glaring pervasiveness of war, and the interdependencies that underpin it, required delineating new paradigms and practices of desertion, so as to quit war, weapons, and militarism on a new international scale.

Only a year into the Myanmar coup, nowhere was the global character of war clearer than from the "margins" and borderlands of the Southeast Asian country, as Commander Maui *indirectly* pointed out. He noted how armored vehicles produced in an invaded country, and fighter jets manufactured by its invader, were used by the army of yet another state to oppress civilian communities. If the histories of Ukraine, Russia, and Myanmar could not be disentangled, then a radical demilitarization of social relations in order to defeat war would entail the desertion of all materials, practices, techniques, and logics that underpin weapons and militarism on a planetary scale. Notwithstanding these global connections, the longest active wars on the planet, often dismissed as "low intensity," "forgotten," or "small" conflicts (Brenner and Han 2021), never rose to the scale of "global" wars from Western points of view. And, similarly, at another scale, in Myanmar too, some borderland conflicts have always been smaller than others and—up to the 2021 coup—would not even be considered "Myanmar" wars by some, thus making it difficult to unify antiregime struggles. In contrast, the recognition of the global character of the Myanmar wars would require us to desert weapons and war, to practice the complete dismantlement of all the technical objects, practices, techniques, and rationalities that codify violence into weaponry and armed assemblages. In this sense, according to Mezzadra, desertion—as a counterconduct to war, weapons, and militarism—would incubate and shape alternative social and political futures departing from the politics of soil and ethnoracial identity, yet only if performed by a new internationalism.

But while Mezzadra focused on the effects of successful international desertion, political futures are shaped also in the failure of international desertion strategies. That is to say, while he focused on how desertion would generate a revolutionary future from the perspective of the effects that such a practice would have if imagined and performed on a global scale, his remark enclosed another postdesertion scenario too—the scenario engendered by the political processes and political geographies that emerge when the confrontational space vacated by the deserter is appropriated and molded by armed resistance, if working out a (global) large-scale desertion strategy turns out to be impossible and armed struggle becomes the only feasible way.

In his book, *Disertate!* (*Desert!*), Franco Bifo Berardi has argued that in a situation of crisis—that is, in a situation in which, as Antonio Gramsci noted, the "old" is dying but the "new" cannot be born yet and a great variety of symptoms appear in between—oftentimes the most effective political behavior in order to resist is to desert the politico-military battle. In different public talks he gave to launch the book, Bifo drew on the experiences of his father during the Second World War to articulate this point. After the September 8, 1943, armistice, his father deserted the Royal Italian Army. He fled toward the Apennines and, almost by chance, like many others, joined the partisan formations operating there, eventually becoming a member of the Communist Party. While Berardi uses his father's experience as an example of how deserting the battlefield in that specific historical moment had become the sole effective and meaningful behavior to refuse the horror of the Fascist and Nazi regimes, and the warfare paradigm inherent in them, it is interesting that what had started as a desertion actually became a defection. His father eventually changed side. He became a partisan, Communist fighter. Rather than deserting, he passed from one war paradigm to another, notwithstanding the abyss that separated them.

The act of deserting in this sense, with its ontological refusal of the established order, and its paradigms, generates a political and social space in which other forces thrive before the new can actually be born. So, when deserting cannot be carried out as a viable revolutionary practice across scales—that is to say, a practice of life that shapes subjectivities and communities defined more by the renunciation of violence than by the embrace of it (through militarism) in order to define and defend the political community—how is the space of resistance left void by the act of desertion reconfigured? In other words, how does the governing of military means, amidst the interregnum of rebellion and in a situation in which annihilating violence forecloses the possibility to quit war, shape the politics (and political geographies) to come?

These are particularly important questions in the wake of the coup in Myanmar and considering its long-standing histories of refusal, runaway, fugitive, and anarchist politics (Leach 1959, 1960; Scott 2009; Sadan 2013a). When the Sit-Tat will be defeated—because it is not a question of if, rather one of when and how—how will the militarism that stood at the core of about seven decades of armed conflict and politics on all sides be transformed? How will the void left by the Sit-Tat be filled, and what repercussions will be produced by armed resistance during the interregnum?

In this epilogue I do not formulate clear-cut answers. Rather, I wish to draw the lines of two conceptual approaches that the book generated, and that I believe allow us to foreground these issues and to reflect upon them. These approaches provide us with a prism to read the ways in which the politics of governing weapons and armed collectives, the governmentality of military means, shapes the political geographies of rule in borderlands and frontiers at the margins. I do so by drawing on the cycles of war that characterized the stories of the Ta'ang rebel polity at the core of the book.

Blunt Rebel Rule

Throughout the processes of ceasefire, disarmament, and rearmament of the Ta'ang rebel movement, like for many other armed organizations, the "old" sent the "new" to fight. A recurrent dynamic in Myanmar's conflicts, one that emerged most recently in the aftermath of the coup—when, for example, the PSLF/TNLA saw an influx of new volunteers both willing to join the ranks of the Ta'ang armed force and to obtain training to create People's Defense Forces units, Local Defense Forces, or other new armed movements. In the case of the creation of the TNLA, though, the "old" that sent the "new" to fight had initially refused war, weapons, and militarism. The PSLF leadership had originally refused to continue the armed struggle and had shifted to peaceful resistance, fleeing to the Thai-Myanmar border. Here, in these frontier assemblages that had long been molded by the processes of frontierization-cum-territorialization of the Sit-Tat, rebel counterconducts and missionary projects, neoliberal development, and humanitarianism, the rationalities at the nexus of war, political geographies of vital space, and identity continued to work and morph. For its part, the PSLF ended up reappropriating the paradigms of the political community, the state, and the nation.

Endorsing the *taingyintha* rationality, the PSLF positively drew the boundaries of a polity with its biological and geographic body—especially via the logic of ethnonationality, narcotics eradication, and humanitarian security. Yet the PSLF/TNLA reformulated the *taingyintha* rationality through complex dynamics of class relations as well as political and military diplomacy with other rebel movements (especially the UWSP/A eventually). Certain elite sections of the Ta'ang communities promoted a single Ta'ang identity as a unitary population body that, for example, aimed to draw in populations and areas outside the self-administration zone enshrined by the 2008 constitution. At the same time, the leadership of the movement worked to align its political vision toward that of the UWSP/A, thinking about the political geographies of the *taingyintha* rationality in terms of the proportionality between ethnonational population density and politico-administrative boundaries of autonomy, or performing a war on drugs framed as a countergenocide campaign.

The re-creation of a war machine by the PSLF redefined the void left by the initial desertion and refusal of violence of 1992. The impossibility to work out a transborderland peaceful resistance strategy, among other dimensions, brought the PSLF to fall back into armed struggle. The processes that brought about the formation and consolidation of the TNLA today impact what can be thought of as an interregnum of Myanmar revolutionary politics where the "old" of the military-state polity is dying but the "new" of the revolutionary movements cannot be born yet.

Still, through the TNLA, the PSLF performed its own revolutionary political practice. The revolutionary aspects of the PSLF politics resided in how, by forming the TNLA, it reproposed and repurposed the *taingyintha* rationality but without clearly drawing territorial borders. The movement placed politico-territorial borders under scrutiny. It questioned and reformulated them. Via the constitution and governing of its armed force, the PSLF reformulated the borders of "Palaungness" and the Palaung SAZ. It moved out of the boundaries of the *taingyintha*-imposed identity grid thanks to a new political geography. It broke inner borders of autonomy and managed to consolidate autonomous political relations with other rebel movements and regional political actors that allowed the movement to rescale Ta'ang Land. At the same time, the PSLF/TNLA essentially played with the vocabulary of the Taingyintha paradigm and refused to draw and fix the borders of Ta'ang Land.

This was an inherently bio-geo-political process that, although peculiar in its trajectory, illuminates broader considerations on the political geographies of rebel rule at the margins. Avoiding the so-called territorial trap that often has informed understandings of rebel organizations and other armed actors as sorts of "states-within-the-state," as sorts of proto-states with their bounded territories, the book has delved into the ways in which governmental actors and their individual and communal subjects reproduce governable spaces in Myanmar. They reproduce ragged and unstable forms of rule and order that hardly instantiate well-oiled disciplinary and biopolitical modes of governing and yet present their own rationalities of governing populations, which are terraformed and spatialized in specific places and times (Watts 2003; Korf, Hagmann and Engeler 2010).

Myanmar, it has been convincingly argued (Callahan 2003; Sadan 2013b; Prasse-Freeman 2023c; Campbell and Prasse-Freeman 2021; Yuzana Khine Zaw 2022), presents a society saturated in structural violence, that is compounded by forms of necroeconomy and a biopolitical system of power that nonetheless remains indifferent to the promotion of life (Mbembe 2019; Prasse-Freeman 2021b, 2023c; Yuzana Khine Zaw 2022; McCarthy 2023). A politico-economic apparatus in which savage extractive capitalism and accumulation by dispossession functioning regardless of the promotion of life (rather, even better with a certain amount of death) combine with absence of governmental care for populations on the part of the state (and at times likewise on the part of other governmental actors). In Myanmar a blunt biopolitical regime of power operates blunt massifications and divisions of populations, for the sake of governing aggregate subjects, via the deployment of violence as a key governing technique to "making live by making die" (Prasse-Freeman 2023c, 438; Prasse-Freeman and Ong 2021). Despite the usefulness of governmentality approaches to look at rebel rule (Hoffmann and Verweijen 2018), and aspirations and attempts of many long-standing rebel movements in Myanmar (Brenner and Tazzioli 2020), not all rebel organizations can be framed as promoters of life at an individual as well as aggregate level, for they do not all appear to deploy the full array of governmental techniques needed to know, and henceforth manage, populations at a distance in order to foster life at a large aggregate scale.

The case of the PSLF/TNLA is instructive here. In fact, especially up to the 2021 coup, the Ta'ang ERO has been seldom understood as a large-scale service provider per se but, from the perspectives of many, after the creation of its armed branch in 2009, it has become something of an enabler of a vital space for Ta'ang communities and civil society organizations in a context of relentless violence orchestrated by the military-state apparatuses. By reformulating the assemblages of the means of violence, as we have come to understand them through the lenses offered by this research, rebel formations and other politico-armed actors insert themselves in this blunt biopolitical apparatus and operate their own massifications and divisions of populations and (governable) spaces (Prasse-Freeman and Ong 2021).

Reproducing and governing armed assemblages, which as we have seen is a costly and constant endeavor to be reposed in specific time/space conjunctures, these different governmental actors "reach out and cleave" the polity apart (Prasse-Freeman 2023c, 10). In the case of the Ta'ang politico-armed movement, the possibility to constitute an armed assemblage force (the TNLA) and practice the governing of military means contributed to solidify a Ta'ang polity. Many in Shan State did not (and in some circumstances still do not) know whether they identify as Ta'ang of Rumai or Rukin communities, but the armed branch of the rebel political movement greatly contributed to constitute a Ta'ang polity across sociogeographic divisions—for example, between Ta'ang in northern and southern areas of Shan State, or Ta'ang people subject to Kachin or Shan EROs' influence. Governing military means allowed the political movement to counter a sociopolitical erosion of Ta'ang communities at regional scales operated by Kachin and Shan armed actors, while also working against processes of blunt biopolitical integration through violence operated by the Sit-Tat at the scale of the Myanmar nation-state (which subsumed the "Palaung" under the Shan as a *taingyintha*).

When, in the aftermath of the February 1, 2021, coup, the "old" once again sent the "new" to fight—as in the case of the KNDF in Karenni State, a newly formed rebel movement that received key support from the PSLF/TNLA—how did blunt rebel rule morph? The newly born rebel movement carefully crafted its governmentality of military means and war so as to overcome the *taingyintha* rationality. Its name provides some evidence in this sense. As Commander Maui once put it, explaining why they decided to use the words Karenni Nationalities Defense Forces (Karenni a-myo-tha ka-kwe-ye tat[3]): "We are the new generation and we want to overcome that ideology that makes indigenous people as built into the land and predating the arrival of the British. We wanted to recognize every national community by that name [KNDF], rather than tracing a dividing line."

The main purpose in using the word *nationalities* (a-myo-tha-mya) in their name was to create an armed force that would embody the defense of everybody living in the political geography of Karenni, all the nationalities to be found there. The primary intent was to distance itself from one of the main logics that has characterized the understanding of the Myanmar polity by the military state. Thus, organizing its armed force, the rebel movement moved beyond the practice of representing an ethnonational population and generating hierarchies among ethnonational communities based on a connection with the land, and wished instead to embody all the nationalities living in Karenni. In this way, the KNDF has been working to disentangle citizenship from *taingyintha*. It managed to intercept younger generations who had mobilized in the wake of the coup by constituting Local Defense Forces who did not feel represented by the idea of Karenni as a state for the Karenni indigenous people only (or mainly). The KNDF's definition of the polity allowed the newly established revolutionary movement to avoid generating friction and conflict with older EROs—such as the KNPP, its breakaway factions,[4] and other rebel groups.[5] The KNDF'S polity in this sense was framed to become a sort of a military and politico-geographical umbrella that could mediate between different Karenni EROs, their visions and conflicts.

A-myo-tha was used with a strong politico-geographical connotation. This word in fact implies a link between groups of people with their social ways of life and a circumscribed land on which they live together. Still acknowledging the heritage of the Karenni indigenous communities, the idea of the nationalities living in Karenni, and of the KNDF as an embodiment of them, aimed to break with the *taingyintha* and its link between soil and blood, while producing a political community based on the positive definition of the traits shared in common with its subjects. At a minimum, such traits include living within the political boundaries of Karenni State. Yet, the issue of the nexus between the military paradigm, identity, and bordered political spaces remained—for, what would the political boundaries of Karenni State be? In this regard, to date, the KNDF has had to accommodate political stances that politicize (for example) the Pekon Lake Basin as a natural and historical space of Kayan and other Karenni people, arguing for Moebye and Pekon townships to be included inside the borders of Karenni State rather than Shan State.

Weapons as Metaresources

The governmentality of military means that inflected blunt rebel rule in the case of the Ta'ang rebel movement was never the result of political actors' agencies and projects alone but a combination of technologies, techniques, and rationalities to manage armed assemblages. In this sense, the book foregrounded in particular the roles of weapon technologies and related techniques and practices of control. It has shown that weapons, when inserted into assemblages of the means of violence, contribute to reproduce and materialize spaces of surveillance and control, both through the processes of monitoring and acquisition of targets and through the different forms of management of individuals and space that underlie the use and control of the weapon. Like other technologies, weapons help to enhance forms of belonging and exclusion. They contribute to entrench the rationalities of control by which they are inflected, and to make more apparent and concrete the massification and division of populations and governable space.

Regarding the linkages between Nats worship and arms control proposed in chapter 5, for example, the infantry rifle constitutes a metric and interface with its own agency of calibrating what counts as acceptable death and injury—an agency cemented by technical properties, rationalities, and techniques (Shah 2017b)—which is remolded by the entities collectively playing around it. This extends to rebel armed forces as armed assemblages: they are at one time tools but also metrics of violence and control. In other words, the rationalities of control, techniques, and technologies that are put in place to create and govern collective ensembles of violence contribute to configure the metrics and parameters for what constitutes legitimate/illegitimate violence and control. Such metrics and parameters, on the one hand, come from long-term sociopolitical processes and, on the other, provide an interface for subaltern actors to confront armed actors and their violence, to contest and renegotiate the rationalities of control attached to the assemblages of the means of violence. The rationalities of narcotics eradication, ethnonationality, and humanitarian security, when cojoined with technologies and techniques, materialize an ontology of acceptable violence and control that is at the same time a platform for political subaltern confrontation.

Weapons in turn are metaresources. They configure a technical interface that provides certain parameters to discern legitimate, less legitimate, and illegitimate modes of violence, thus providing also to subaltern actors a domain to confront the power effects of relations of authority. But they are not only metrics to calibrate killing, injury, control, (il)legitimate lethal or nonlethal force (Shah 2017a, 2017b). They are also material resources that distribute other resources and reproduce space. Control over weapons is mutually interrelated with the production of authority and power as well as with the social and spatial distribution of resources, territory, and political arrangements (Buscemi 2019). Like (and together with) other technical objects, weapons act by making a difference, by reproducing certain techniques of control, and by becoming themselves an "arena" to be managed. Of such geographical and metaresource(ful) character of weapons and related networks of military means, the book supplied ample evidence.

In relation to the production of sociospatial relations, the book moved beyond conceptual approaches that, as elaborated in the introduction, characterize weapons as autonomous factors in determining social and political relations (substantivist positions), or suggest the irrelevance of weapons and their subordination to interests and rational agencies, intersubjective relations and norms (instrumentalist ones). Contrary to instrumentalist or substantivist visions that would connect weapons acquisition to the creation of rebel territory, or would expunge weapons entirely from the picture, the prism of analysis offered here sheds light on the precariousness of territorialization processes and the role of governing assemblages of the means of violence in reproducing territory as well as networks, scales, and places.

If we were to engage the case of the Ta'ang rebel movement and Ta'ang areas of the Shan State borderlands through purely substantive readings, the PSLF/TNLA would appear as having yet to reach a stage where its weapons and military capabilities allow it to consolidate a territory called Ta'ang Land. While, if we were to think of weapons as pure instruments, the Ta'ang movement would be the result of negotiations—among elites and civilians of the Ta'ang polity as well as among elites of different armed actors—that allowed the rearmament of PSLF and the consolidation of some form of territorial control. The rich empirical material mobilized by this research instead detailed how the PSLF/TNLA's networked ensembles of the means of violence simultaneously operate materially and symbolically to stabilize and circulate practices of violence and control that are intertwined with a (discursive and material) calculation-cum-production of

Ta'ang Land as a sociopolitical and territorial entity. It also showed that fluid combinations of weapons as technical objects that come from somewhere, rationalities of controlling military means, and techniques of arms control reconfigure the scalar arrangements of territorial arenas of authority over violence at the edge of the state. In this sense, governing armed assemblages both requires and reproduces spatialized configurative effects of political orders most immediately discernible in terms of localities, locales, and sense of place; scalar arrangements of territorial authority; and territory. Such a prism provides a more nuanced picture of how control of military means unfolds and how it links to the production of governable orders and spaces in Ta'ang areas of Shan State.

Another insight that the book generated—one that tends to remain hidden via substantive or instrumentalist frameworks—concerns the relevance of weapons acquisition processes, and in particular the means of weapons production, for structuring political orders at the margins. The means of weapons production beyond the state (most conspicuously arranged and managed by the KIO/A and UWSP/A but also by the RCSS/SSA-S and, to a certain extent, other EROs when it comes to weapons repair, refurbishment, reassemblage, and forms of craft manufacturing or technological creolization) have contributed to create constellations of politico-armed movements. The KIO/A and UWSP/A, in particular, have emerged as those maintaining the means of weapons production. For other EROs, the way they relate to military-state authorities is in part also a function of how they relate to those two EROs. Similar dynamics emerged concerning the regulation of access to weapons and control over arms acquisition processes. In these constellations, the position of armed movements in the Myanmar political system is in part linked to how they relate to other armed actors with means of weapons production.

I have further unpacked and qualified the dominant argument that the modalities of weapon acquisition and the feasibility of weapon access by rebel movements and other armed actors are mostly a function of state governance capacities and "external" conditions and support (Bourne 2007; Krause 2009; Marsh 2012). Without discrediting the role of state sovereignty, the book provided ample ground for conceiving weapon acquisition and feasibility of access as partly shaped by diffused techniques and rationalities of control over the means of violence that travel together with weapons. For example, the acquisition of weapons and support for forming and strengthening the TNLA that the Ta'ang movement of the PSLF received by Wa rebel movements (WNO/A and UWSP/UWSA) has necessarily to be read, at least in part, through the logics of ethnonationality and narcotics eradication as sets of rationalities, techniques, and practices of conducting oneself and populations that circulated throughout societies in the borderlands. This allows us not to reduce the acquisition of weapons and the formation of armed forces to the results of strategic calculus, support, alliances, structural factors, and actors' interactions, and to understand it instead as coproduced by diffused rationalities and logics shared by different governmental actors. If long-term processes of military regimes' control over production and acquisition of weapons, strategies, ideologies, and alignment between politico-armed movements were relevant to shape rebel access to weapons, logics that inform the governing of populations and are spread throughout governable orders in the borderlands played a role too. In the case of the TNLA, not only was rearmament connected to a series of conjunctures that made it possible to organize and manage arms acquisition protractedly, but it was also imbued with rationalities of control over weapons and the means of violence circulating throughout Ta'ang sociopolitical communities. Far from being a purely "internal" matter to the PSLF, the rearmament entailed navigating different changing communities of Ta'ang societies in different spatial junctures (for example, youth and civil society movements, higher and lower ranks of the PSLO/A, different components in the leadership of the new political front to be armed, former members now active in militias, community defense militias, local communities and leaders, women associations). This was a long, historically embedded process wrought in part through rationalities of means of violence control.

The point is that weapons accessibility and sociotechnical webs of military means (re)produce not only opportunities—for example, potential opportunities to exert violence—but also rationalities and techniques of control, violence, and war. Rationalities of ethnonationality and stateness, narcotics eradication, and humanitarian security spread throughout long histories of weapon flows, arms acquisition, and armed collective (trans)formation and control. Governable orders as sets of norms, rules, regulations, and practices neither follow nor preexist weapons and the means of violence but evolve together with them. In this sense weapons and the means of violence become an arena to reproduce or reformulate certain governable orders. Hidden in the techniques and micropractices inherently associated with weapons, one can often glimpse logics and modalities of injuring, killing, and controlling people and spaces. One could raise here the example of drug addicts pressed into military and "social" services with the PSLF/TNLA. People with drug-related issues have been subject to diversified techniques of weapon use and management, their bodies marked by head shaving while other recruits are not subject to that, and their being pressed into the armed force underpinned by, and reproductive of, rationalities and techniques of conduct linked to narcotics eradication that have been diffused in and through the family, communities, and civil society organizations. The governmentality of military means produced differentiated subjectivities through the conduct of which, by contrast, the political body and space of the Ta'ang polity could be envisaged and practiced.

Coda

It is not possible to dismantle a war machine (for example, the Sit-Tat) through weapons and military means without reproposing, albeit in a very different guise, or partially at least, the very war paradigms and the social and spatial identity differentiations that inform it. This is the core provocation that I wish to read behind the picture of the BTR-3 that Maui showed, behind that question he posed—*our war is not global?* Even though he, like the PSLF, has chosen *armed* revolution against the military-state regime, his question gestures toward a critical deconstruction of militarism. To me, by asking "is our war not global enough?" he not only hints at global military supply chains and networks of war but also at how war is consubstantial to the production of alterity and political order. If war can be waged from the center to the margins without seeing it as part of a broader global issue, this implies that the center is built upon and constantly shaped by war. On the one hand, the question points to the global interconnections of war. It is not possible to disentangle that weapon made in Ukraine (the BTR-3), and used against a form of (post)colonial occupation (i.e. that of the Russian army), from its deployment in another form of colonization, that carried out by the Sit-Tat at its polity's margins in Myanmar's borderlands. And this is the more explicit provocation enclosed in Maui's question. But, on the other hand, because Maui and the KNDF for the time being—like the PSLF did in 2009—have embraced the gun and chosen armed revolutionary struggle, they have had to replicate the war paradigm. They once again shape the positive boundaries of the political and geographical community of the polity and its enemy. Thus they reproduce, once and again, logics of exclusion and the co-constitutive relationship between violence, war, and the political community.

This is a provocation that opens a different perspective on the role of weapons and military means in such a relationship (i.e., in the co-constitutive relationship between violence and the political community). From Ta'ang Land to other regions of the Myanmar borderlands, from Palestine to Israel, from Ukraine to Russia, the occurrence of war foregrounds the issue of who actually stands behind violence. Is it weapons or humans? Be it concerning the legitimacy of providing military support to one or another party to a conflict, or be it concerning the ethical and legal dilemmas foreshadowed by weapons with some degree of autonomous functioning (including landmines, which featured prominently in this book), the occurrence of war raises the question of whether it is by controlling, restraining, and legitimately deploying weapons or human action that eventually armed violence can be limited, brought back into acceptable dimensions. Yet, this way of approaching the dilemmas of war hides what I think is a more pervasive and deeper role of weapons and military means that I hope this text managed to illuminate. As the cycles of disarmament and rearmament experienced by the Ta'ang rebel movement(s) in the context of the longest ongoing set of armed conflicts on the planet show, "becoming and being" a weapon unfolds processes of sociopolitical and geographical identity marking and community bordering.

Weapons, military means, and cycles of (re)armament are at the very base of processes and tendencies toward the production of ethnoracial boundaries, borders, and territorialization. This conclusion does not dispute the legitimacy that for many the idea and practice of fighting back militarily can have, in a situation in which violence and death saturate political space to the extent that deserting war completely becomes unviable. In a situation in which one's own extended families and communities have been bombed, shelled, slaughtered, and torched since unmemorable (but clearly memorable) times. (At least since the Sit-Tat, or the colonial power before it, was able to operate via automatic firepower, and mortars, and bomb shells and ultimately planes.) It does not translate into a value judgment on armed resistance struggles, or in a general call for disarmament that has become anathema to EROs in Myanmar. Rather, such a conclusion points to the possibilities for a demilitarization of rebel polities beyond revolutionary moments, and to the transformative potential of desertion practices broadly understood. That is to say, political practices and approaches that recognize the co-constitutive relation between violence, war, and political communities, to make rebel polities constantly aware of the dangerous drifts and shapes this relation can take. I have always been impressed, for example, by the way in which the Zionist nation- and state-building process of Israel is often taken as a model by some among EROs in Myanmar, notwithstanding the analogies and connections it bears with the nation- and state-building trajectories of the Sit-Tat polity. Albeit purely anecdotal, and without necessarily implying any political endorsement for the slaughters in Palestine, this parallel reminds us of the constant challenge to imagine and practice futures in which living together is decoupled from that co-constitutive relationship between violence, war, and the political community.

INTERVIEWS AND FIELDWORK METHODS TABLE

Interviews and Fieldwork Methods Table

INTERVIEWEE (PRECEDED BY REFERENCE NUMBER, REF.N.XXX)	MODE AND DATE (MM/DD/YYYY)	PLACE	FORMAT	RECORDING TECHNIQUE	CONFIDENTIALITY REQUESTED OR REQUIRED
1. Researchers (2) with fieldwork experience in N. Shan	9/14/2018	Yangon	Semistructured	Concurrent notes and supplementary notes within 1 hour	Required
2. INGO researcher	9/24/2018	Yangon	Unstructured	Concurrent notes and supplementary notes right after	Required
3. Journalist/consultant with fieldwork experience in Myanmar's borderlands	9/26/2018	Bangkok	Semistructured	Concurrent notes and supplementary notes within 2 hours	Requested
4. Independent researcher/consultant with fieldwork experience	Conducted via Skype 10/24/2018	Mae Sot	Semistructured	Concurrent notes and supplementary notes right after	Requested
5. Mountain guide and resident (Kyaukme township)	9/21/2018	Kyaukme township	Unstructured	Concurrent notes and supplementary notes the same day	Required
6. Humanitarian mine action workers (2) with field experience	2/26/2019	Yangon	Unstructured	Concurrent notes and supplementary notes within 1 hour	Requested
7. Mine action consultant	2/27/2019	Yangon	Semistructured	Concurrent notes and supplementary notes right after	Requested
8. Independent consultant with field experience	2/28/2019	Yangon	Semistructured	Concurrent notes and supplementary notes right after	Requested
9. Mine action expert with field experience	3/4/2019	Yangon	Semistructured	Concurrent notes and supplementary notes same day	Requested
10. Academic researcher (conversation)	3/4/2019	Yangon	Unstructured	Concurrent notes and supplementary notes same day	Required
11. ICRC's weapon contamination and armed and security forces units	3/6/2019	Yangon	Semistructured	Concurrent notes and supplementary notes right after	Requested

	Date	Location	Type	Notes	Status
12. Mine action expert with field experience	3/8/2019	Yangon	Unstructured	Concurrent notes and supplementary notes within 1 hour	Required
13. Mine action worker with field experience (Namkham born and Lashio raised) (conversation)	3/8/2019	Yangon	Unstructured	Concurrent notes and supplementary notes right after	Required
14. Independent consultant with field experience	3/12/2019	Yangon	Unstructured	Concurrent notes and supplementary notes right after	Required
15. Fieldwork note (community mapping and risk assessment activity)	Participated in person/ 3/14/2019	Myitkyina township	/	Concurrent notes and observation and supplementary notes the same day	Required
16. Local mine action worker (conversation)	3/15/2019	Myitkyina township	/	Supplementary notes right after	Requested
17. Humanitarian worker (conversation)	3/15/2019	Myitkyina township	/	Supplementary notes right after	Requested
18. Local humanitarian worker	3/14/2019	Myitkyina township	Unstructured	Concurrent notes and observation and supplementary notes the same day	Required
19. Fieldwork note (mine risk education session)	3/15/2019	Myitkyina township	/	Concurrent notes and observation and supplementary notes the same day	Required
20. Field note (meeting with landmine victim)	3/15/2019	Myitkyina township	Unstructured	Concurrent notes and observation and supplementary notes the same day	Required
21. Mountain guide and inhabitant (Kyaukme township)	3/17/2019	Kyaukme township	Unstructured	Concurrent notes and supplementary notes the same day	Required

(Continued)

(Continued)

INTERVIEWEE (PRECEDED BY REFERENCE NUMBER, REF.N.XXX)	MODE AND DATE (MM/DD/YYYY)	PLACE	FORMAT	RECORDING TECHNIQUE	CONFIDENTIALITY REQUESTED OR REQUIRED
22. Ta'ang CSO active in mine action (victims assistance) and humanitarian work	3/18/2019	Kyaukme township	Semistructured	Concurrent notes and observation and supplementary notes the same day	Required
23. Village dweller/resident (Mine Din, Kyaukme)	3/18/2019	Mine Din, Kyaukme township	Unstructured	Concurrent notes and observation and supplementary notes the same day	Required
24. Head of village of Hu Kwet	3/18/2019	Hu Kwet, Kyaukme township	Unstructured	Concurrent notes and supplementary notes the same day	Required
25. Field note	3/19/2019	Mine Din and Hu Kwet, Kyaukme township	Unstructured	Concurrent notes and supplementary notes the same day	Required
26. Village inhabitant (Bang Hone, Kyaukme township)	3/19/2019	Bang Hone, Kyaukme township	Unstructured	Concurrent notes and supplementary notes the same day	Required
27. Field note (observations from Bang Hone festival)	3/19/2019	Bang Hone, Kyaukme township	/	Concurrent notes and supplementary notes the same day	Required
28. Head of village of Hone Sar	3/20/2019	Hone Sar, Kyaukme township	Unstructured	Concurrent notes and supplementary notes within 1 hour	Required
29. Head of village of Kyein Low	3/20/2019	Kyein Low, Kyaukme township	Unstructured	Concurrent notes and supplementary notes the same day	Required
30. RCSS recruits (4) based at Hu Sun, Kyaukme township	3/20/2019	Hu Sun, Kyaukme township	Unstructured	Concurrent notes and supplementary notes the same day	Required
31. Hunter and village inhabitant (Hi Pu, Kyaukme township)	3/21/2019	Hi Pu, Kyaukme township	Unstructured	Concurrent notes and supplementary notes the same day	Required

	Date	Location	Type	Notes	Status
32. Field note (Kyu Shaw, Kyaukme township)	3/22/2019	Kyu Shaw, Kyaukme township	/	Concurrent notes and supplementary notes the same day	Required
33. Deputy head of village of Kyu Shaw, Kyaukme township	3/22/2019	Kyu Shaw, Kyaukme township	Unstructured	Concurrent notes and supplementary notes the same day	Required
34. TNLA former combatant	3/23/2019	Kyu Shaw, Kyaukme township	Semistructured	Concurrent notes and supplementary notes the same day	Required
35. Mine action expert with field experience	3/24/2019	Hsipaw township	Semistructured	Concurrent notes and supplementary notes within 1 hour	Required
36. Humanitarian mine action worker	4/1/2019	Loikaw	Unstructured	Concurrent notes and supplementary the same day	Required
37. Humanitarian mine action worker	4/2/2019	Loikaw	Unstructured	Concurrent notes and supplementary the same day	Required
38. Geneva Call	3/23/2019	Lashio township	Unstructured	Concurrent notes and supplementary notes the same day	Requested
39. Mine action expert with fieldwork experience	3/28/2019	Yangon	Semistructured	Concurrent notes and supplementary notes the same day	Required
40. Karenni CSO working on mine risk education	4/3/2019	Loikaw	Semistructured	Concurrent notes and supplementary notes the same day	Required
41. Field note	3/22/2019	Lashio township	/	Concurrent notes and supplementary notes the same day	/
42. Pa'O CSO working on peace building in N. Shan	4/6/2019	Taunggyi	Semistructured	Concurrent notes and supplementary notes the same day	Requested
43. Field note Naung Pyaet, Kyaukme township	September 2018	Naung Pyaet, Kyaukme township	/	Supplementary notes the same day	Required

(Continued)

(*Continued*)

INTERVIEWEE (PRECEDED BY REFERENCE NUMBER, REF.N.XXX)	MODE AND DATE (MM/DD/YYYY)	PLACE	FORMAT	RECORDING TECHNIQUE	CONFIDENTIALITY REQUESTED OR REQUIRED
44. Local journalist with field experience	4/8/2019	Yangon	Semistructured	Concurrent notes and supplementary notes the same day	Required
45. Field note	4/10/2019	/	/	Supplementary notes the same day	/
46. Humanitarian mine action NGO	4/23/2019	Yangon	Unstructured	Supplementary notes the same day	Required
47. Local humanitarian mine action worker	4/28/2019	Lashio township	Unstructured	Concurrent notes and supplementary notes the same day	Required
48. Journalist and consultant with field experience	4/24/2019	Bangkok	Semistructured	Concurrent notes and supplementary notes the same day	Requested
49. Ta'ang CSO operating in N. Shan	4/29/2019	Lashio	Semistructured	Concurrent notes and supplementary notes the same day	Required
50. The Border Consortium—NGO researcher with field experience	5/7/2019	Yangon	Semistructured	Concurrent notes and supplementary notes the same day	Required
51. Local humanitarian mine action worker	4/30/2019	Lashio	Semistructured	Concurrent notes and supplementary notes within 2 hours	Required
52. Local humanitarian worker with mine action INGO	5/2/2019	Lashio	Semistructured	Concurrent notes and supplementary notes the same day	Required
53. Humanitarian mine action worker and Lashio resident	5/2/2019	Lashio	unstructured	Concurrent notes and supplementary notes within 3 hours	Required
54. Informant working for weapon tracing organization with field experience in Kachin and Karen	Conducted via phone call 5/6/2019	Yangon	Semistructured	Concurrent notes and supplementary notes within 1 hour	Requested

#	Description	Date	Location	Interview type	Notes	Status
55	Member of Union Level Joint Cease-fire Monitoring Committee (JMC)	5/14/2019	Yangon	Semistructured	Concurrent notes and supplementary notes within 2 hours	Requested
56	Investigative journalist with specialized field experience in Myanmar rebel movements	5/18/2019	Mudon township	Semistructured	Concurrent notes and supplementary notes the same day	Required
57	Field note	5/18/2019	Mudon township	/	Concurrent notes and supplementary notes the same day	/
58	Ta'ang CSO	4/30/2019	Lashio township	Semistructured	Concurrent notes and supplementary notes the same day	Requested
59	Field note, observation to mine risk education and community assessment session	6/5/2019	Kutkai and Lashio townships	/	Concurrent notes and supplementary notes the same day	Required
60	Ta'ang CSO	6/5/2019	Lashio township	Semistructured	Concurrent notes and supplementary notes the same day	Required
61	Kachin Baptist Church (KBC) Pastor	6/8/2019	Yangon	Semistructured	Concurrent notes and supplementary notes the day after	Required
62	Head of village	3/18/2019	To Hu Kwet, Kyaukme township	Semistructured	Concurrent notes and supplementary notes the same day	Required
63	Academic researcher with field experience in Kachin	Conducted via Skype/ 5/3/2019	Yangon	Semistructured	Concurrent notes and supplementary notes within 1 hour	Required
64	Journalist and consultant with specialized field experience in Myanmar's conflicts	9/22/2019	Bangkok	Semistructured	Concurrent notes and supplementary notes the same day	Requested
65	Local journalist on Myanmar's conflicts with field experience	9/24/2019	Chiang Mai	Semistructured	Concurrent notes and supplementary notes within 1 hour	Requested

(Continued)

(Continued)

INTERVIEWEE (PRECEDED BY REFERENCE NUMBER, REF.N.XXX)	MODE AND DATE (MM/DD/YYYY)	PLACE	FORMAT	RECORDING TECHNIQUE	CONFIDENTIALITY REQUESTED OR REQUIRED
66. Volunteer of a multiethnic humanitarian movement	9/25/2019	Chiang Mai	Semistructured	Concurrent notes and supplementary notes the same day	Requested
67. Small arms and light weapons expert with multiethnic humanitarian movement	9/26/2019	Chiang Mai	Semistructured	Concurrent notes and supplementary notes the same day	Requested
68. Small arms and light weapons expert with multiethnic humanitarian movement	9/24/2019	Chiang Mai	Semistructured	Concurrent notes and supplementary notes the same day	Requested
69. Security and conflict analyst based at research institute	10/1/2019	Yangon	Unstructured	Concurrent notes and supplementary notes the same day	Required
70. Local journalist specializing in Myanmar's conflicts with field experience	10/3/2019	Mudon township	Semistructured	Concurrent notes and supplementary notes the same day	Required
71. Mountain guide, inhabitant of Kyaukme (conversation)	10/6/2019	Kyaukme township	unstructured	Concurrent notes	Required
72. Mountain guide, Hsipaw resident	10/6/2019	Hsipaw	unstructured	Concurrent notes and supplementary notes the same day	Required
73. Mountain guide, Kyaukme resident	10/7/2019	Kyaukme	Unstructured	Concurrent notes and supplementary notes the same day	Required
74. Former SSPP/SSA combatant (sergeant)	10/7/2019	Kyaukme township	Unstructured	Concurrent notes and supplementary notes the same day	Required
75. Head of CSO working on narcotics eradication and mine action	10/8/2019	Kyaukme	Semistructured	Concurrent notes and supplementary notes the same day	Required

76. Shan CSO	10/8/2019	Kyaukme township	Unstructured	Concurrent notes and supplementary notes the same day	Required
77. Abbot of a monastery in north Kyaukme township	10/8/2019	Kyaukme township	Unstructured	Concurrent notes and supplementary notes the same day	Required
78. Mine Din (Mong Tin) resident	10/8/2019	Mine Din (Mong Tin), Kyaukme township	Unstructured	Concurrent notes and supplementary notes the same day	Required
79. Head of village, Hu Kwet	10/10/2019	Hu Kwet, Kyaukme township	Unstructured	Concurrent notes and supplementary notes the same day	Required
80. Ta'ang National Party member and village resident	10/9/2019	Hu Kwet, Kyaukme township	Unstructured	Concurrent notes and supplementary notes the same day	Required
81. Pyi Thu Sit village community militia (5)	10/10/2019	Pang kunn, Kyaukme township	Unstructured	Concurrent notes and supplementary notes the same day	Required
82. Village head	9/10/2019	Kyaukme-Mogok road	Unstructured	Concurrent notes and supplementary notes the same day	Required
83. Mountain guide, Hsipaw resident	10/12/2019	Hsipaw	Semistructured	Concurrent notes and supplementary notes the same day	Required
84. Mountain guide, Lashio resident	10/15/2019	Lashio	Semistructured	Concurrent notes and supplementary notes the same day	Required
85. SSPP/SSA combatant (sergeant major)	10/16/2019	Lashio	Unstructured	Concurrent notes and supplementary notes within 3 hours	Required
86. Ta'ang CSO	10/16/2019	Lashio	Unstructured	Concurrent notes and supplementary notes the same day	Required
87. Ta'ang CSO	10/17/2019	Lashio	Semistructured	Concurrent notes and supplementary notes the same day	Required

(Continued)

(*Continued*)

INTERVIEWEE (PRECEDED BY REFERENCE NUMBER, REF.N.XXX)	MODE AND DATE (MM/DD/YYYY)	PLACE	FORMAT	RECORDING TECHNIQUE	CONFIDENTIALITY REQUESTED OR REQUIRED
88. TNLA officer and TNLA Local Guerilla Force (LGF) members (2)	11/1/2019	Namtu	Unstructured	Concurrent notes and supplementary notes the same day	Required
88bis. TNLA second lieutenant and major	11/4/2019	TNLA temporary mobile HQ, Namhsan township	Unstructured	Concurrent notes and supplementary notes the same day	Required
89. TNLA officer	11/4/2019	TNLA temporary mobile HQ, Namhsan township	Unstructured	Concurrent notes and supplementary notes the same day	Required
90. TNLA lieutenant, General Administrative Department (GAD) chief	11/5/2019	TNLA temporary mobile HQ, Namhsan township	Semistructured	Audio-recorded	Required
91. TNLA second lieutenant	11/2/2019	TNLA temporary mobile HQ, Namhsan township	Semistructured	Audio-recorded	Required
92. TNLA major and (third) secretary general	11/3/2019	TNLA temporary mobile HQ, Namhsan township	Semistructured	Audio-recorded	Name Disclosable
93. TNLA major and (third) secretary general	11/4/2019	TNLA temporary mobile HQ, Namhsan township	Semistructured	Audio-recorded	//
94. TNLA major and (third) secretary general	11/4/2019	TNLA temporary mobile HQ, Namhsan township	Semistructured	Audio-recorded	//
95. TNLA major and (third) secretary general	11/5/2019	TNLA temporary mobile HQ, Namhsan township	Semistructured	Audio-recorded	//
96. TNLA major and (third) secretary general	11/5/2019	TNLA temporary mobile HQ, Namhsan township	Semistructured	Audio-recorded	//
97. PSLO/A last vice chairman and PSLO/A last chairman (Tar Khun Yee and Tar Aik Mone)	12/3/2019	Lashio	Semistructured	Audio-recorded	//

98. Local translator/interpreter working with INGOs (original of a village in Mrauk U township, Rakhine)	9/30/2019	Yangon	Semistructured	Concurrent notes and supplementary notes the same day	Required
99. CSO working on education and social cohesion in Rakhine	12/9/2019	Sittwe	Semistructured	Concurrent notes and supplementary notes the same day	Required
100. CSO working on education and social cohesion in Rakhine (2 informants from Rathedaung and Mrauk-U)	12/10/2019	Sittwe	Semistructured	Concurrent notes and supplementary notes the same day	Required
101. TNLA officer	11/6/2019	Pansali, Namhsan township	Semistructured	Concurrent notes and supplementary notes the same day	Required
102. Former NUPA/AA combatant and head of humanitarian CSO	12/11/2019	Sittwe	Semistructured	Concurrent notes and supplementary notes the same day	Required
103. Landmine and cluster munition monitor researcher specializing in Myanmar and with field experience	12/17/2019	Yangon	Semistructured	Audio-recorded	Requested
104. Field note from observation of a mountain guide association	10/6/2019	Hsipaw	/	Concurrent notes and supplementary notes the same day	/
105. Field note	10/8/2019	Hu Kwet, Kyaukme township	/	Concurrent notes and supplementary notes the same day	/
106. Kyaukme businessman (conversation)	10/10/2019	Kyaukme	Unstructured	Supplementary notes the same day	Required
107. Field note (Kyaukme golf club and "Ta'ang neighborhood")	10/10/2019	Kyaukme	/	Supplementary notes the same day	/
108. Field note	10/12/2019	Mann Loi, Hsipaw township	/	Supplementary notes the same day	Required

(Continued)

(Continued)

INTERVIEWEE (PRECEDED BY REFERENCE NUMBER, REF.N.XXX)	MODE AND DATE (MM/DD/YYYY)	PLACE	FORMAT	RECORDING TECHNIQUE	CONFIDENTIALITY REQUESTED OR REQUIRED
109. Field note	10/13/2019	Pan Kam, Hsipaw	/	Supplementary notes the same day	/
110. TNLA soldier and officer	11/1/2019	Lashio-Namtu-Namhsan township	Unstructured	Supplementary notes the same day	Required
111. TNLA officer	11/1/2019	Lashio-Namtu-Namhsan township	Unstructured	Supplementary notes the same day	Required
112. TNLA sub-lieutenant and officer	11/1/2019	TNLA temporary mobile HQ, Namhsan township	Unstructured	Supplementary notes the day after	Required
113. Third secretary general PSLF-TNLA	11/2/2019	TNLA temporary mobile HQ, Namhsan township	Unstructured	Concurrent notes and supplementary notes the same day	Required
114. Third secretary general PSLF-TNLA (conversation)	11/2/2019	TNLA temporary mobile HQ, Namhsan township	Unstructured	Supplementary notes the same day	Required
115. TNLA sub-lieutenant and officer	11/2/2019	TNLA temporary mobile HQ, Namhsan township	Unstructured	Supplementary notes the same day	Required
116. Field note (workshop with TNLA cadets)	11/3/2019	TNLA temporary mobile HQ, Namhsan township	Unstructured	Supplementary notes the same day	Required
117. Field note (sport activity inside camp)	11/4/2019	TNLA temporary mobile HQ, Namhsan township	/	Supplementary notes the same day	Required
118. Field note	11/5/2019	TNLA temporary mobile HQ, Namhsan township	/	Supplementary notes the same day	/
119. Focus group (TNLA women cadets/lower-rank officers)	11/4/2019	TNLA temporary mobile HQ, Namhsan township	Semistructured	Concurrent notes and supplementary notes the same day	Required

	Date	Location	Structure	Notes	Disclosure
120. Field note (wedding party)	11/3/2019	Man Lon, Namhsan township	/	Concurrent notes and supplementary notes the same day	Required
121. Field note	11/4/2019	Namhsan	/	Supplementary notes the same day	/
122. TNLA officer (conversation)	11/4/2019	Namhsan township	/	Supplementary notes the same day	/
123. TNLA special forces/commando brigade's commander	11/5/2019	Namhsan township	Unstructured	Supplementary notes the same day	Required
124. Major General Tar Ho Plan	11/5/2019	TNLA temporary mobile HQ, Namhsan township	Semistructured	Concurrent notes and supplementary notes the same day	Name disclosable
125. TNLA Major, 3rd secretary general	11/5/2019	TNLA temporary mobile HQ, Namhsan township	Unstructured	Supplementary notes the same day	Required
126. TNLA captain	11/2/2019	Namhsan township	Unstructured	Concurrent notes and supplementary notes the same day	Required
127. Field note (weapons management inside camp)	11/5/2019	Namhsan township	/	Concurrent notes and supplementary notes the same day	Required
128. Field note	11/10/2019	/	/	/	/
129. TNLA officers and field note	11/6/2019	Tar Khun Aye memorial, Pansali, Namhsan township	Unstructured	Concurrent notes and supplementary notes the same day	Required
130. TNLA officers	11/6/2019	Pansali, Namhsan township	Unstructured	Concurrent notes and supplementary notes the same day	Required
131. Former head of village	11/7/2019	Pansali, Namhsan township	Unstructured	Concurrent notes and supplementary notes the same day	Required

(Continued)

(Continued)

INTERVIEWEE (PRECEDED BY REFERENCE NUMBER, REF.N.XXX)	MODE AND DATE (MM/DD/YYYY)	PLACE	FORMAT	RECORDING TECHNIQUE	CONFIDENTIALITY REQUESTED OR REQUIRED
132. Former head of Pyi Thu Sit village community militia	11/7/2019	Khung Ka Mai/Vain Mail/ Myo Thit, Namhsan township	Unstructured	Concurrent notes and supplementary notes the same day	Required
133. TNLA officer	11/7/2019	Khung Ka Mai/Vain Mail/ Myo Thit, Namhsan township	Unstructured	Concurrent notes and supplementary notes the same day	Required
134. Sao Hsur Hten, SSPP/SSA's political patron	11/11/2019	Lashio	Semistructured	Concurrent notes and supplementary notes the same day	Required
135. Shan CBO	11/12/2019	Lashio	Semistructured	Concurrent notes and supplementary notes the same day	Required
136. Ta'ang researcher descendant of last Tawngpeng Sawbwa	11/13/2019	Lashio	Unstructured	Concurrent notes and supplementary notes the same day	Required
137. Sub-lieutenant SSPP/SSA	11/14/2019	Lashio-Mongyai-Kyethi (Keshi)-Wan Hai	Unstructured	Concurrent notes and supplementary notes the same day	Required
138. Sub-lieutenant SSPP/SSA	11/15/2019	Wan Hai, SSPP/SSA HQ	Unstructured	Concurrent notes and supplementary notes the same day	Required
139. SSPP/SSA sub-lieutenant and major	11/16/2019	Wan Hai, SSPP/SSA HQ	Unstructured	Concurrent notes and supplementary notes the same day	Required
140. SSPP/SSA foreign office and education department head	11/17/2019	Wan Hai, SSPP/SSA HQ and Haipa village	Unstructured	Concurrent notes and supplementary notes the same day	Required
141. SSPP/SSA major	11/18/2019	Wan Hai, SSPP/SSA HQ	Unstructured	Concurrent notes and supplementary notes the same day	Required

142. Local humanitarian worker operating in Rathedaung, Bothidaung, Mrauk U (especially)	12/10/2019	Sittwe	Unstructured		Concurrent notes and supplementary notes the same day	Required
143. Local humanitarian worker based in Sittwe working throughout Rakhine	Email correspondence 2/19/2020		Semistructured	/		Required
144. TNLA secretary general	Private correspondence (Signal) 3/29/2020		Unstructured	/		Required
145. Humanitarian worker and researcher specializing in N. Shan conflicts	Private correspondence 3/29/2020		Unstructured	/		Required
146. SSPP/SSA officer and foreign affairs representative	11/19/2019	Wan Hai, SSPP/SSA HQ	Semistructured	Concurrent notes and supplementary notes within 2 hours	Required and requested

Notes

INTRODUCTION

1. The Myanmar state armed forces refer to themselves as Tatmadaw using a formal royal register. The suffix -*daw* aims to confer to the armed forces an elevated institutional status through reference to the Bamar dynasties' kingdoms. Among ordinary people, though, such a term is often avoided and the word *Sit-Tat*—which translates as "military force" or "armed group"—is used instead, in order to refuse the legitimacy of the state's armed forces. Throughout the book I privilege the term Sit-Tat to reflect the sentiment expressed by the majority of the population in Myanmar who, especially after the latest coup d'etat on February 1, 2021, refuses to recognize the Tatmadaw as the Myanmar state's armed forces. I use the term *Tatmadaw* when interlocutors used it in speaking about the Sit-Tat or when referring to the military state's own political and military processes.

2. Throughout the book I use the acronym ERO (ethnic revolutionary organization) rather than EAO (ethnic armed organization) to reflect a recent terminological shift endorsed by the majority of the rebel armed movements in Myanmar after the coup. "EAO" is perceived to be a remnant of the military-bureaucratic jargon of the highly disputed peace process that led to the 2015 National Ceasefire Agreement. After the first of February coup, many among the resistance started to argue in favor of a new terminology that would stress the revolutionary character of rebel movements to break away from the military-state-imposed terminology.

3. Ref.N.24. Throughout the text I have protected informants by anonymizing names and circumstantial details that may be used in reconstructing their identities. I provide only a general description of the type of interlocutor followed by a reference number (Ref.N.XXX) that guides the reader to the table in the appendix, where more interview details—date, place, format, recording technique, and confidentiality status—are provided.

4. See, for instance, A. Davis, "China's Mobile Missiles on the Loose in Myanmar," *Asia Times*, November 28, 2019, https://asiatimes.com/2019/11/chinas-mobile-missiles-on-the-loose-in-myanmar/.

5. See, for example, Aung Zaw, "Myanmar's Generals Make a Show of Displeasure at China's Arming of Rebels," *Irrawaddy*, November 26, 2019, https://www.irrawaddy.com/opinion/commentary/myanmars-generals-make-show-displeasurechinas-arming-rebels.html; and Myat Thura, "Tatmadaw Seizes Huge Caches of Chinese Made Weapons in Shan State," *Myanmar Times*, November 25, 2019, https://www.mmtimes.com/news/tatmadaw-seizes-huge-cache-chinese-made-weapons-shan-state.html.

6. Ref.N.111.

7. For a critique on democratization as a fetishized Western lens that misunderstands Myanmar's resistance, see Brenner 2024.

8. My claim that the PSLF/TNLA or other EROs in Myanmar do not maintain large-scale biopolitical governing apparatuses should be qualified. I do not claim that ties of mutual civilian-rebel support do not exist, or that the movement does not perform governance functions. It also should not be interpreted to suggest that state governing apparatuses do it more, or better, than the rebel movements. And it is not even to say that the rebel movement in question does not exert territorial control, or that its rule is

coercive only. This claim instead points to the fact that the PSLF/TNLA (for example) does not govern life in a more traditional biopolitical sense: it is not capable of taking care of and enhancing life on a large aggregate scale and in capillary ways. That of the PSLF/TNLA instantiates a peculiar form of biopolitics, as chapter 1 will explain. The overall argument of the book should also be framed in the precoup context and considered in perspective. That is, it should be understood against the background of what, in the Ta'ang rebel movement, like in others in Myanmar, is thought of as the "revolutionary period." The Ta'ang Political Consultative Committee, a civilian consultative committee created after the coup, has started to formalize a draft for a constitution and an interim collective agreement that regulates a Ta'ang civilian government arrangement for the revolutionary period only in 2024. This has occurred after the PSLF/TNLA enlarged its areas of influence and control in northern Shan State as a result of so-called Operation 1027 launched in late October 2023. Before the coup and these events, all forms of governing activities remained under the PSLF/TNLA—at least officially, as in practice they would be carried out by a variety of actors, among which civil society organizations featured prominently.

9. For an exception see Hoffman and Verweijen 2019.

10. For a full conceptual critique see Bourne 2012.

11. The literature in peace and conflict studies cannot be reduced only to substantivist and/or instrumentalist frameworks. Many scholars have approached weapons in terms of tools providing opportunities and situated them in an overall conceptual frame primarily concerned with the willingness and opportunities of actors (Most and Starr 1989; Sislin and Pearson 2001; Bara 2015; Marsh 2020). In this sense, the dualism between materiality and sociopolitical relations is overcome, although there remains the risk to grant a privileged role to human agency and intentions. In the case of a land mine, for example, there might be a risk to obscure the agency of the weapon that (due to the logics and technical features inscribed in it) continues to operate and make a difference in sociopolitical relations even when actors' willingness and opportunity do not materialize.

12. In Waterman and Worrall's 2020 review of the literature on ordering processes in armed conflict and civil wars, besides other traditionally more prominent cases, Myanmar's borderlands have been given much more space (2020).

13. Throughout the research for this book I have looked for other examples of rebel movements in Myanmar that have experienced some form of disarmament and later rearmed and remobilized themselves. Yet, the only instance of official disarmament of which I am aware is the PSLO/A and the subsequent rearmament into the PSLF/TNLA.

1. FRONTIERS OF VIOLENCE

1. Ta-ya-daw-siq' [တရားသောစစ်] in Burmese.

2. The central areas of the Dry Zone and the Anyatha (Aung-Thwin and Aung-Thwin 2012).

3. Pan is a pseudonym.

4. Ta'ang activist, interview by the author, June 2022, via Signal.

5. For a discussion of the nexuses of violence, care, and biopolitics, see Prasse-Freeman 2023a, 24–27, 65–66.

6. Several authors outside the framework of biopolitics have emphasized the centrality of mass violence, military means, and other forms of violence as a key governing mechanism in Myanmar. See, for example, Aung-Thwin 1985; Callahan 2002, 2003; Fink 2009; Tharaphi Than 2015; Thawnghmung 2019; MacLean 2022.

7. For a discussion on the paramount role of boundaries, borders, and frontiers in biopolitical forms of rule, see Esposito 2010 and 2011b. Esposito remarks how boundaries,

borders, and frontiers—while very ancient mechanisms to separate and protect what lays within from what lays without, and to selectively exclude the latter—assume two specific forms that are exploited by biopolitics: individual property (both as juridical-economic mechanism and as "positive" features that delineate the subject) and territory (as a technology to calculate the space of the capitalist and colonial nation-state with its borders and frontiers).

8. Many scholars have accounted for the circulation of these political ideas in Myanmar. For example, see Maung Maung 1980; Aung-Thwin 2005; Osada 2011; Nehginpao Kipgen 2016; Ferguson 2015, 2021; Boshier 2017; Thant Myint U 2019; Campbell and Prasse-Freeman 2021; Thawnghmung 2022. For a comprehensive overview and analysis of the links between governmentality and biopolitics in Myanmar see Prasse-Freeman 2023a and 2023b.

9. Kyaukme resident, personal communication, December 2019.

10. The concept of *Taingyintha* cannot be equated exclusively with or reduced to the terms *race, ethnicity,* or *indigeneity* alone (Cheesman 2017). It has to be considered as a state political/legal idea and mechanism developed hand in hand with what Elliott Prasse-Freeman has understood as the environments of a state-led capitalist economic model in a socialist regime (1948–1988) and neoliberal foreign capital investment and accumulation by dispossession (1988–present) (Prasse-Freeman 2021; Campbell and Prasse-Freeman 2021).

2. FRONTIER HISTORIES OF WEAPON FLOWS

1. Like many villages in Ta'ang areas of northern Shan State, Myo Thit has three different names: Myo Thit (Bamar), Vain Moe (Shan), and Kung Ka Mai (Ta'ang).

2. Interview with TNLA officer and field notes, Ref.N.133.

3. Ref.N.133.

4. Interview with TNLA sub-lieutenant, Ref.N.91.

5. Myanmar Constitution, clause 338, https://www.myanmar-law-library.org/IMG/pdf/constitution_de_2008.pdf.

6. Interview with the former head of Pyi Thu Sit village community militia, Ref.N.132.

7. Some *sao hpa* were hereditary rulers, while others were nominated by Burmese kings and operated in the form of a garrison under agents.

8. Especially the *sao hpa* of Hsenwi, Mongmit, Kengtung, and Yawnghwe were envisioning the possibility to reshape the status of the principalities into a confederation of independent states or a league of princes under the suzerainty of a new Burmese king proclaimed by the princes themselves (Yawnghwe 2010, 70–71).

9. Gunpowder flows became especially important around the turn of the twentieth century, when the technological sophistication of newer rifles required fine machine-made powder and impeded the use of craft, locally manufactured powder (Tagliacozzo 2005, 292–294).

10. Explosives and dynamite assumed a great technological and geopolitical relevance as they became more commercially widespread and accessible (Tagliacozzo 2005, 292–294).

11. Based on archival research and citing H. Maxim's biography (1915), Thant Myint U identifies the 1885 expedition as the first time Maxim guns were deployed. See also That Myint U's Twitter post (@thantmyintu), November 21, 2020, https://twitter.com/thant myintu/status/1330130613098197007. For examples from other colonized areas, such as Benin, Ghana, or South Africa, see Hicks (2020). Dan Hicks has underlined how militarized colonial expeditions, and the weapons they brought along, at one time molded and

were molded by new forms of racism and technologies of classifying, dividing, managing human bodies, and space.

12. For an excursus on the trajectories of the administration of the frontier areas, see Tun (2009, 223–226). Ministerial Burma comprised Tenasserim, Arakan, Pegu, and Irrawaddy divisions; frontier areas comprised the Shan plateau principalities, the Chin Hills, and the Kachin tracts.

13. Myanmar firearms regulations have remained a patchwork of different laws and provisions until today. The Shan State Arms Order is still in effect and used for prosecution of arms trafficking; see Htun Htun (2020). In Myanmar firearms manufacture, trade, possession, and transport is regulated by the 1878 Indian Arms Act, the Kachin Hill Tribes Regulation of 1895, the 1924 Shan States Arms Order, and the 1948 Chin State Arms Provisions. See Myat Thura (2020).

14. In particular, the Allies supplied the Burma National Army (BNA) and the Anti-Fascists Organization's units, and weapons were left over by British and Indian troops of the British Burma Army first and by Japanese forces later on.

15. The Arms (Temporary Amendment) Act of 1947. Burma Act No. LXIV of 1947, Rangoon, October 16, 1947. The amendment added the death penalty and transportation for life to the existing imprisonment and pecuniary punishments in case an individual was found in possession of characteristic World War II weapons such as "machine-gun, bren-gun, tommy-gun, sten-gun or rifle or American carbine, or ammunition for any of the said guns, or a hand-grenade, or any other arms of the description which the Governor may, by notification, declare in this behalf." The Arms (Temporary Amendment) Act of 1947, Burma Act No. LXIV of 1947, clause 2, https://www.burmalibrary.org/sites/burmalibrary.org/files/obl/docs25/1947-collected_laws%2Bamendments-en-tu.pdf. After independence, the 1947 Amendment Act was repealed in 1949, thus removing the harsher penalties.

16. Communist formations started to fight against the Anti-Fascist People's Freedom League (AFPFL) government and quickly fractured into two factions, the Red and White Flag communists. The People's Volunteer Organization (PVO)—a militia of the AFPFL created by Aung San to pressure the British to grant independence—rebelled after his death, in part joining with the communist movement. At the same time the Burma Rifles (ethnic Bamar units) were affected by mutinies while Rakhine State was hit by a wave of Muslim insurgency. In addition, armed insurrection started among the Karen, with the Karen National Defense Organization (KNDO) leading, which triggered other Karen mutinies from the Burma Army. The same was true also for the mutiny led by the Kachin Captain Naw Seng followed by units of the Kachin Rifles. In Shan State, besides the mutinies commanded by Naw Seng to the north (in the so-called Kachin substate—areas of Hsenwi, Kutkai, and parts of Tawngpeng), the first Pa-O rebel movement (allied with the Karen), the PVOs, and White Flag communists were active. These for some time were based in major Shan State towns like Lashio, Hsenwi, Kutkai, Muse, Taunggyi and in areas bordering Burma proper, such as Lawksawk, Muang Kung, Hsipaw, and Namhsan.

17. Such as 77 mm recoilless rifles that had been released only eighteen months before.

18. For an in depth historical reconstruction see Gibson and Chen (2011, 154).

19. The program was named after a king warrior from the second century AD (see Buchanan 2016). The creation of this militia program was connected to foreign assistance received by the central government and the Sit-Tat in these years. The program was partially inspired by Israel's settlement defense system studied by Sit-Tat officers during a visit to the country.

20. Such contestations were in part linked to broader demands concerning the redistribution of natural resources and land. These were epitomized by the scandal of the

management of electricity and the Lawpita hydroelectric power plant located east of Demoso, Karenni State. Here electricity was produced and subsequently distributed to Mandalay and the central areas of Burma. While passing through Shan State, often along highly militarized itineraries, the power grid failed to serve ethnic minority areas.

21. In the early 1950s, for example, the Directorate of Defense Industries (Karkweye Pyitsu Setyoun—ကာကွယ်ရေး ပစ္စည်းစက်ရုံ; Ka-Pa-Sa) released its first locally produced firearm: the BA-52/Ne Win Sten. The Ne Win Sten was a Myanmar copy of the Italian TZ45 submachine gun. According to Maung Aung Myoe, by the late 1950s, the contract with Heckler & Koch granted Ka-Pa-Sa factories with the capability to produce different firearms: "BA-63 Automatic Rifle (G-3A2), BA-72 Assault Rifle (G-3K), BA-64 Light Machine Gun (G-4), BA-100 (G-3A3ZF) and 7.62-mm and 9-mm ammunition" (Maung Aung Myoe 2009, 106).

22. Carbines and grenade launchers, as well as mortars, were imported from the US; rocket launchers, recoilless guns, artillery, signal/wireless machinery from the US, UK, Federal Republic of Germany, Yugoslavia, and Israel (Maung Aung Myoe 2009, 107).

23. The weaponry acquired through ex-KMT consisted of (1) supplies—including landmines (Taylor 1973, 59–60)—that the US and Taiwan supplied to the ex-KMT in connection with its deployment in Laos; and (2) weapons transferred by Thai police and military forces.

24. Maha San was the son of the last *sao hpa* of Vingngun.

25. In these years of recurrent splits and breakaways, munition supplies and military storages could often become a key component of alliance deals. The reshuffling and reorganization of armed forces was often to be operated through harmonization of rules and regulations in command and control, which circulated via adaptations of US Army manuals, uniforms, leadership courses for officers, and territorial reorganization of battalions (Yawnghwe 2010, 18, 27, 31). Similar activities occurred for civilian administration.

26. In particular, after rounds of negotiations, the SSA had managed to bring on board the Shan State Independence Army (SSIA), Shan National United Front, and Jimmy Yang's Kokang Resistance Force (Smith 1999, 220).

27. Before the PNF there had been the Palaung Nationalities United League (PNUL), the first political ethnic movement, then abolished in 1962 by the Revolutionary Council military regime. The PNF had been initially set up as a group subordinate to the SSIA (Lintner 1999, 490).

28. Interview with the TNLA special forces brigade's commander, Ref.N.123.

29. Headquarters of the Third KMT.

30. For a detailed analysis of the history of militia programs, see Buchanan 2016.

31. Interview with local humanitarian mine action worker, Ref.N.51.

32. For example, Browning machine guns, M-70 grenade launchers, and 57 mm recoilless rifles; and, since the mid-1960s, also M16 rifles.

33. Smith specifies the provision of AK-pattern assault rifles, mortars, antiaircraft guns, field machine guns, and ammunition.

34. Interview with the TNLA special forces brigade's commander, Ref.N.123.

35. Chao Nor Far, Khun Li, and Khun Aye. Interview with TNLA officers and field note from the Tar Khun Aye memorial, Ref.N.129.

36. Smith, in his book *Burma: Insurgency and the Politics of Ethnicity*, talks about Wang Mai as the headquarters of the Fifth Battalion, but in interviews I was referred to Vain Moe/Myo Thit/Khun Ka Mai. Though plausible, I was unable to crosscheck whether "Wang Mai" and Vain Moe (the Shan name for Myo Thit/Khun Ka Mai village) are the same place but just differently transliterated. Among PSLF/TNLA's members,

the story of the fracture of the PNF is recounted by emphasizing the role of the Sit-Tat in fueling internal disagreement between the two factions and remarking on the decision of the Sixth Battalion to ally with the Kachin movement so as to remind both soldiers and the grassroots groups of the importance of unity and cooperation in the rebel movement.

37. Interview with the TNLA special forces brigade's commander, Ref.N.123; and interview with TNLA officers and field note from the Tar Khun Aye memorial, Ref.N.129.

38. In 1986, under the leadership of Mai Aik Mong, the name was changed again into Palaung State Liberation Party, although the rebel movement is normally referred to as PSLO/A.

39. Interview with the PSLF/TNLA commander in chief Major General Tar Ho Plan, Ref.N.124.

40. Interview with Tar Ho Plan, Ref.N.124.

41. Interview with SSPP/SSA officer and foreign affairs representative, Ref.N.146.

42. Interview with SSPP/SSA officer and foreign affairs representative, Ref.N.146; and Interview with Sao Hsur Hten, SSPP/SSA's political patron, Ref.N.134.

43. Sakhong notes that such policies have been rooted in the anticolonialists' motto of "amyo, batha, thatana" [where Myanmar-lumyo (amyo) stands for ethnicity/race; Myanmar-batha-ska (batha) stands for Myanmar language; and Myanmar-thatana of Buddha-batha (thatana) stands for Buddhism as religion]. Myanmarization policies have aimed to create a Myanmar Naing-ngan (kingdom) as opposed to Pyi-daung Suh ("Nations Together," nations coming together to build a state). The main policies were embodied by (1) U Nu's government (1948–1962): Myanmar-thatana of Buddhism, that is, religious homogenization policy; (2) Ne Win's regime (1962–1988): Myanmar Batha-Ska, linguistic homogenization; and (3) Saw Maung and Than Shwe's government (1989–2010): Myanmar-lumyo or ethnic assimilation (see also the change of country name from Burma to Myanmar) (Sakhong 2014, 2–8).

44. The US increased its assistance through a formal antinarcotics program, which involved the supply of helicopters, small arms, and light weapons to the Sit-Tat, ostensibly for antinarcotics operations (Lintner 1999, 312).

45. At this stage, with the Vietnam-US conflict well underway, there was a boom in opiates consumption and the golden triangle was emerging as one of the first producer regions worldwide.

46. Especially in the southern and eastern Shan State borderlands.

47. In 1978 invading Vietnamese troops brought down the regime of the Khmer Rouge, who settled on the Thai-Cambodian border. At this point, Thailand started to support different Khmer Rouge and other Cambodian factions along the Thai-Cambodian border (Pongpaichit et al. 1998, 130–133) At the same time, Chinese arms were delivered to the Khmer Rouge at the Thai-Cambodian borderlands via Laos, by port in southern Cambodia, and through airlifts, although overland routes in Thailand became privileged (Pongpaichit et al. 1998, 130–133). Moreover, the Vietnam-supported armed actors were supplied with weaponry that Vietnam received from the USSR.

48. The US scaled up its financial assistance for arms acquisitions and increased the provision of military advisors and trainers (from 30 to 120 between 1979 and 1980) to support such plans after the Vietnamese invasion of Cambodia (Pongpaichit et al. 1998, 154).

49. In particular in Chantaburi, Sa Kaew, Korat, Chonburi—but also Bangkok.

50. Interview with TNLA former combatant, Ref.N.34; and interview with humanitarian mine action worker, Ref.N.36.

51. Interview with Karenni CSO working on mine risk education, Ref.N.40.

52. Or the Type-59 Chinese-produced copy. Interview with TNLA former combatant, Ref.N.34; humanitarian mine action worker, Ref.N.36; and SSPP/SSA foreign office and education department head, Ref.N.140.

53. With this I do not want to assume a deterministic diffusion of weapon technologies and techniques of control but only to remark on their messy spread. For example, at times anecdotes are recalled concerning the haphazard acquisition, use, and control of weapons by armed actors in these borderlands. In one case an interlocutor used to recount the episode of a rebel movement's unit that, having acquired an antipersonnel Taiwan-produced landmine leaked from a stockpile of the Thai army, proceeded to plant it with the trigger mechanism upside-down (private correspondence with humanitarian mine action worker, October 15, 2020).

54. CPB was itself profiled according to ethnic lines despite ideological stances rejecting ethnonationalism. For an analysis of the factors leading to the breakup of the CPB, see Smith 1999, 374–383; Kramer 2007, 16–18; and Lintner 1990.

55. Initially named Burma National United Front.

56. For a detailed navigation of the events that characterized the ceasefires subsequently signed by CPB's breakaway factions, see Min Zaw Oo and Win Min (2007).

57. In 1988, after the dramatic events of the pro-democracy protests erupted on August 8, the Tatmadaw operated a self-inflicted coup d'état, deposed Ne Win and his Burma Socialist Programme Party, and installed a restyled military regime under the leadership of General Saw Maung and the newly created State Law and Order Restoration Council.

58. From the taxation of farmers to the provision of armed escort services to convoys, the setting up of tollgates at key junctures, or the organization of refineries, warehouses, or rest hubs.

59. Among such implications were rampant HIV/AIDS rates, the impact of drug abuse on local communities, and the creation of a system of co-optation and coercion of armed actors through which state authority has been reproduced in the borderlands (Meehan 2011).

60. Kramer identifies the roots of this shift in the approaches implemented across the ex-CPB areas spanning the Mong La, Kokang, and Wa regions, where opium bans were adopted in 1997, 2003, and 2005, respectively (Kramer 2007, 24).

61. M-150 energy drink produced by Osotspa provided the nickname for the hand-crafted mine (M-150). Interview with mine action consultant, Ref.N.7.

62. Interviews with mine action researcher specialized in Myanmar and with field experience, Ref.N.103; mine action expert with fieldwork experience, Ref.N.39; mine action consultant, Ref.N.7; Karenni CSO working on mine risk education, Ref.N.40; volunteer of a multiethnic humanitarian movement, Ref.N.66; investigative journalist with field experience specializing in Myanmar rebel movements, Ref.N.56.

63. For what concerns the Thai-Burma border, especially logging concessions.

64. In this period, the Ka-Pa-Sa started manufacturing mortars, opened new defense factories for landmines, ammunition, and assault rifle manufacturing plants in Magway, with technological and engineering assistance from Singaporean defense industries and the South Korean company Daewoo (Maung Aung Myoe 2009, 107). In particular, the Tatmadaw secured vast military assistance from China (Smith 1999, 426; Maung Aung Myoe 2009, 107). The cruel crushing of the 1988 pro-democracy demonstrations had triggered a US and European Economic Community arms embargo on Myanmar, which forced the Sit-Tat to diversify imports previously covered by some European arms manufacturers. For example, in the early 1980s the Burma Army had acquired about 1,200 M2

Carl Gustaf antitank recoilless rifles from the Swedish manufacturer (Aung Maung Myoe 2009, 109). In the 1990s it would rely instead on Chinese-manufactured recoilless rifles (Aung Maung Myoe 2009, 109). It is interesting to note that Conflict Armament Research (CAR) documented a Carl Gustaf rifle during tracing activities carried out in KIO/A areas (Laiza and Mae Ja Yang) in January 2015 (CAR iTrace portal at https://www.conflictarm. com/itrace/). Nonetheless, the recoilless rifle documented was a later model produced in Sweden in 2003 (a Carl Gustaf 3 by Saab Bofors Dynamics, which had been seized by the KIA from the Tatmadaw). It has been argued that the weapon was probably acquired by the Sit-Tat through third-country retransfer from India.

65. Interview with NGO researcher with field experience, Ref.N.50.

66. Interview with mine action expert with fieldwork experience, Ref.N.39.

67. Interview with mine action expert with fieldwork experience, Ref.N.39.

68. Interview with journalist/consultant with fieldwork experience in Myanmar's borderlands, Ref.N.3.

69. The Mong Tai Army emerged in 1985 as a merger between Khun Sa's Shan United Army and Moh Heng's Shan Revolutionary United Army.

70. Interviews with independent consultant with field experience, Ref.N.8; and SSPP/ SSA foreign office and education department head, Ref.N.140.

71. What today is known as Revolutionary Council of Shan State/Shan State Army-South.

72. Toward the end of the 1990s UWSA substituted earlier Type-56 rifles with batches of Type-81 decommissioned in Yunnan.

73. Interviews with Kachin Baptist Church (KBC) pastor, Ref.N.61; and small arms and light weapons expert with multiethnic humanitarian movement, Ref.N.67–68.

74. Interviews with investigative journalist with field experience specialized on Myanmar rebel movements, Ref.N.56; journalist/consultant with fieldwork experience in Myanmar's borderlands, Ref.N.3; and SSPP/SSA foreign office and education department head, Ref.N.140. Three different arms-control programs were adopted in Cambodia in the period 1999–2007: (1) Cambodian government arms control endeavors in 1999 and before March 2000; (2) EU-ASAC (European Union Assistance on Curbing Small Arms and Light Weapons in Cambodia) assistance programs from March 2000 to June 2006; and (3) since 2003, the Japanese Assistance Team for Small Arms Management in Cambodia, mostly working with the ministry of interior (as compared to EU-ASAC, which was mostly working with the ministry of defense) (Roberts 2008, 103–4).

75. Interviews with journalist/consultant with fieldwork experience in Myanmar's borderlands, Ref.N.3; Kachin Baptist Church (KBC) pastor, Ref.N.61; small arms and light weapons expert with multiethnic humanitarian movement, Ref.N.67; and TNLA soldier and officer, Ref.N.110.

76. China National Precision Machinery Import-Export Corporation was responsible for the FN6 MANPADS export. >

77. In particular KIO/A, SSA-N (which for the UWSA represents an important buffer on the western banks of the Salween river) MNDAA, Arakan Army, TNLA.

78. Among these in particular are heavy machine guns, mortars, rocket-propelled grenades (RPGs), recoilless rifles, more recently produced assault rifles, antimateriel rifles, MANPADS, artillery, and antitank missiles. Interview with local journalist specializing in Myanmar's conflicts with field experience, Ref.N.70. Especially Type-81 (both of Chinese and Wa manufacture), 107 mm rockets, Type-69 RPGs, grenades, M99-Zijiang sniper rifles, and FN6 MANPADS. Further confirmation of such trends was brought by recent Tatmadaw seizures. Two in particular are indicative: one carried out in Homein village, Namhsan township in 2019 (Irrawaddy 2019), another on January 15, 2020, in Hsenwi, near Namkhaik village (Irrawaddy 2020a).

3. DISARMING AND REARMING TA'ANG LAND

1. Interview with Major General Tar Ho Plan, Ref.N.124.

2. Ref.N.124.

3. Interview with TNLA major and (third) secretary general, Ref.N.92. In a context of relatively low population density, sixty-two villages in Namkham and forty in Namhsan were relocated to sites closer to towns during the beginning of the preparations for the tea-harvesting season of 1991—between February/March and end of April, when the best tea (so-called shwe-phi-oo in Burmese, or "spring tea") is harvested (PWO 2006, 22).

4. Interview with TNLA major and (third) secretary general, Ref.N.92.

5. Interview with TNLA major and (third) secretary general, Ref.N.92. The current secretary general of the PSLF/TNLA, Tar Bong Kyaw, also experienced displacement. A native of Myo Thit/Khun Kha Mai village, between Namhsan and Kyaukme townships, he moved to Kyaukme in these years.

6. Interview with Sao Hsur Hten, SSPP/SSA's political patron, Ref.N.134.

7. Interview with Major General Tar Ho Plan, Ref.N.124.

8. Renamed Kachin Defense Army, also known as Kawngkha militia.

9. KIO/A agreed to a ceasefire with the Tatmadaw only in 1994.

10. Interview with PSLO/A last vice chairman and PSLO/A last chairman (Tar Khun Yee and Tar Aik Mone), Ref.N.97.

11. Interview with PSLO/A last vice chairman and PSLO/A last chairman (Tar Khun Yee and Tar Aik Mone), Ref.N.97.

12. Ref.N.97.

13. Ref.N.97.

14. Ref.N.97.

15. Ref.N.97; interview with Ta'ang CSO, Ref.N.87.

16. Interview with TNLA major and (third) secretary general, Ref.N.92.

17. This is not something peculiar to PSLO/A's experiences. Many other EROs at this stage, and in part also today, were agreeing to ceasefires on word.

18. Interview with TNLA major and (third) secretary general, Ref.N.92.

19. Minor defections had already occurred in November 1989 when a faction of the PSLO/A under Kyaw Yin—second from the right in figure 3—had temporarily broken away and sided with a faction of the SSPP/SSA. The latter faction had remained close to the CPB and decided to split from the SSPP/SSA and join the wave of ceasefires (Smith 1999, 379). Kyaw Yin's defectors would later join the PSLO/A's ceasefire.

20. Interview with the TNLA major and (third) secretary general, Ref.N.92.

21. Ref.N.92.

22. Ref.N.92. Mai Aik Pan would later become an important figure following the founding of the Ta'ang student and youth association (the PYNG, today renamed Ta'ang Student and Youth Union) and his arrest in October 2001. Captured while carrying out intelligence-gathering activities in Myawaddy, he subsequently died under suspect circumstances in Mawlamyine prison in July 2002 (Kyaw Zwa Moe 2002). Tin Moung instead was a doctor born in Namhsan that had later moved to Yangon before fleeing to Manerplaw (Meehan 2016a, 370).

23. Interview with TNLA major and (third) secretary general, Ref.N.92.

24. Interview with PSLO/A last last chairman (Tar Aik Mone), Ref.N.97.

25. Interview with TNLA major and (third) secretary general, Ref.N.92.

26. Interview with Ta'ang CSO, Ref.N.87.

27. In bamaza ငြိမ်းချမ်းရေးတောင်. Field note, Kyaukme, 10.10.2019.

28. Interview with Ta'ang CSO, Ref.N.87.

29. Ref.N.87. Located in the homonymous village of Pansay, south to Namkham, this militia had played a role as antirebel (Ta-Ka-Sa-Pha) militia in the Tatmadaw offensives

against the CPB in the late 1980s (TWO 2006, 24–26). In the following decades the military regime assigned the area of Namkham to the militia; here the latter was entrusted with the management of security, the collection of taxes or extortion fees, the provision of logistics, and stable conditions for economic activities.

30. Ref.N.87.

31. As Patrick Meehan and Ta'ang researchers have noted (2016, 381; TWO 2006, 24–26), the crisis of the tea industry was to be ascribed to various drivers. It was connected to a combination of the creation of a cartel of regime-affiliated companies managing tea distribution to the national markets, the introduction of a quota set by the military government, increased taxes and bribing practices, as well as arbitrary confiscations at armed checkpoints. See also the work of the Ta'ang (Palaung) Working Group, a coordination platform among TSYO, PWO, and PSLF (Ta'ang [Palaung] Working Group 2011).

32. The history of the Bawdwin mine stretches back to at least the fifteenth century, when Chinese workers began to mine the area for silver to be marketed back in China or to Bengal for the purpose of minting silver coins (see Van Schendel 2020, 43). For a streamlined historical trajectory of the Bawdwin mine, see Kyaw Lin Htoon 2018.

33. Interview with Ta'ang CSO, Ref.N.87.

34. The Shweli river runs from the border crossing of Ruili-Muse southwest along the border into Namkham and farther south where it becomes a tributary of the Irrawaddy River. Particularly important for Ta'ang communities, among which it is known as "Ohn-mtamao," the Shweli River was designated for the construction of three dam and hydropower plant projects (PYNG 2007). Shweli Dam 1 was the first build-operate-transfer agreement in Myanmar between regime-affiliated holdings and foreign companies (PYNG 2007, 43), to which two more would have followed in the 2000s.

35. "Prescribing Duties and Rights of the Central Committee for the Management of Culturable Land, Fallow Land and Waste Land"; see Meehan 2016a, 379.

36. PSLO/A had maintained five main bases: the headquarter were located in Manton town; two more bases were in Manton township, in Komone and Homein villages (with the Komone base being in charge of the areas in Namkham and Kutkai); a third in Kyaukme township, in Kun Kar village; and a fifth one in Loi Yar village, Momeik/Mongmit township. Interview with PSLO/A last vice chairman and PSLO/A last chairman (Tar Khun Yee and Tar Aik Mone), Ref.N.97.

37. As Tar Khun Yee and Aik Mone recall: "Just after we agreed to the ceasefire in 1991, the headquarters was in Manton and then at that time the Tatmadaw also came and set up their military base in Manton. At that time this did not occur also in the other places in which we had the military bases. Only in Manton at that time, because we had the headquarters in Manton and then they also came there to be based." Interview with PSLO/A last vice chairman and PSLO/A last chairman (Tar Khun Yee and Tar Aik Mone), Ref.N.97.

38. Interview with PSLO/A last vice chairman and PSLO/A last chairman (Tar Khun Yee and Tar Aik Mone), Ref.N.97. Excluding western parts of Namhsan.

39. Ref.N.97.

40. Women were stigmatized both as partners of drug addicts and as addicts themselves (PWO 2006).

41. Interviews with Ta'ang CSO, Ref.N.87; and PSLO/A last vice chairman and PSLO/A last chairman (Tar Khun Yee and Tar Aik Mone), Ref.N.97.

42. Interviews with TNLA sub- lieutenant and officer, Ref.N.115; and Ta'ang CSO, Ref.N.58. In the early 1990s at Manerplaw the NDF members organized to create an educational institution that, cooperating with the Thai NGO Santiprachatam, would have provided advanced education to youth in order to bring them to a university level and

allow for possible admissions abroad. The curriculum of the institution included political science–related subjects, political ecology, public and international law, humanitarian law, and human rights classes but also seminars conducted by outside organizations, such as the International Commission of Jurists (Ref.N.115 and 58). In this sense, Manerplaw was on the one hand a gateway between different forms of armed-nonarmed political struggle, and on the other hand "a gateway between the leadership of many of the ethnic armed groups and external entities through this federal university set up under the auspices of the national democratic front" (Interview with mine action expert with field experience, Ref.N.103).

43. Interviews with Mine action expert with fieldwork experience, Ref.N.39; and The Border Consortium—NGO researcher with field experience, Ref.N.50.

44. The Network for Human Rights Documentation Burma (ND-Burma) is a coordination and collaboration platform for multiethnic CSOs and CBOs working on human rights violations documentation formalised in 2004. Interview with TNLA sub-lieutenant and officer, Ref.N.115.

45. Ref.N.115.

46. Interview with Kachin Baptist Church (KBC) pastor, Ref.N.61.

47. Interview with TNLA major and (third) secretary general, Ref.N.92.

48. General Khin Nyunt was the former chief of the military intelligence and SPDC's prime minister from August 2003 to October 2004.

49. Interview with PSLO/A last vice chairman and PSLO/A last chairman (Tar Khun Yee and Tar Aik Mone), Ref.N.97.

50. Interview with TNLA major and (third) secretary general, Ref.N.92

51. Interview with PSLO/A last vice chairman and PSLO/A last chairman (Tar Khun Yee and Tar Aik Mone), Ref.N.97.

52. Interview with TNLA major and (third) secretary general, Ref.N.92.

53. Note that while Meehan (2016a) indicates April 29 as the date of disarmament, PWO's report states the handover occurred on April 21, 2005 (PWO 2006, 23).

54. Interview with PSLO/A last vice chairman and PSLO/A last chairman (Tar Khun Yee and Tar Aik Mone), Ref.N.97; and Major General Tar Ho Plan, Ref.N.124. The disarmament procedure resulted in the consignment of "-420 assorted arms including one 81-mm mortar, five 60-mm mortars, six 2' mortars, one BA-103, one 57-mm recoilless gun, one RPG launcher, one launcher made at Homein, one 7B launcher, two M-7 launchers, one M-72 launcher, one HK-33, eight M-79 launchers, one MG-42 sub-machine gun, 10 M-23 China-made sub-machine guns, three .303 Bren guns, seven BA-64s, 17 BA-63s, nine BA-72s, one BA-94, two BA-52s, 37 carbines, 56 M-22s, M-20s, three .38 revolvers, one .225 gun, two revolvers, two mines, 18 grenades, 57 magazines, five walkie-talkies, 313 shells and grenades and 19,906 rounds of ammunition" (Keenan 2013, 122).

55. Including then SPDC first secretary Thein Sein. Interview with TNLA second lieutenant, Ref.N.91.

56. Interview with TNLA major and (third) secretary general, Ref.N.92. Viss is a Burmese unit of measure for weight. One viss corresponds to 1.63293 kilograms (3.6 pounds).

57. Interview with PSLO/A last vice chairman and PSLO/A last chairman (Tar Khun Yee and Tar Aik Mone), Ref.N.97.

58. Ref.N.97. And interview with TNLA major and (third) secretary general, Ref.N.92.

59. Interview with Ta'ang CSO, Ref.N.87.

60. Interview with PSLO/A last vice chairman and PSLO/A last chairman (Tar Khun Yee and Tar Aik Mone), Ref.N.97.

61. PSLO was allowed to form the so-called Palaung Ethnic Nationalities Group, which the military regime admitted to the national convention process. The Ta'ang National

Party (TNP) was instead allowed to form only later after the adoption of the 2008 constitution and in view of the 2010 elections. TNP was formed in 2010.

62. Ref.N.97.

63. Interview with Major General Tar Ho Plan, Ref.N.124.

64. This occurred especially in Manton, Namhsan, Namtu, eastern Kutkhai, and lower Namkham (see also Meehan 2016a, 374–375).

65. In 2016 Palaung researchers estimated the presence of fifteen militias active in Palaung areas of northern Shan for a total of approximately twelve hundred troops (TWO 2016, 52). The townships in which these groups are active include Namkham, Kutkhai, Muse, Namsan, Namtu, Mantong.

66. For example, for what concerned enrollment registration, grades, or exams. This is a widespread practice in many borderlands areas. Teachers give children of so-called *taingyinthar* (national races) ethnic minorities Burmese "school names," which are used for official paperwork throughout their education and beyond. For examples of such practice in Rakhine state see Balčaitė (2020).

67. Interview with TNLA major and (third) secretary general, Ref.N.92. Until very recently, the civic education curriculum of Myanmar state education system's primary schools included the teaching of songs reciting "we hate mixed blood, it will make our race extinct" (Balčaitė 2020).

68. Interviews with TNLA major and (third) secretary general, Ref.N.92–96.

69. Interview with Ta'ang CSO, Ref.N.58. Shweli Dam 1 in Man Tat village area, Namkham township; Shweli Dam 2 in Manton township; Shweli Dam 3, with the involvement of Électricité de France, in Momeik/Mongmit township (Kean 2018). Since 2009 additional hydropower projects entailing the construction of further dams along the Namtu/Myitnge River have also been planned. See Shan Human Rights Foundation et al. (2016).

70. Interviews with TNLA major and (third) secretary general, Ref.N.92–96; Ta'ang CSO, Ref.N.87; PSLO/A last vice chairman and PSLO/A last chairman (Tar Khun Yee and Tar Aik Mone), Ref.N.97; SSPP/SSA officer and foreign affairs representative, Ref.N.146.

71. This occurred also in the areas of Kawngkha, and Kutkai township, under the influence of the Kawngkha militia—the former Kachin Defense Army and breakaway faction of the former Fourth Brigade of KIO/A.

72. Interview with Ta'ang CSO, Ref.N.87.

73. Interview with Deputy head of village of Kyu Shaw, Kyaukme township, Ref.N.33.

74. Interview with TNLA major and (third) secretary general, ref.N.92.

75. Interview with TNLA second lieutenant, Ref.N.91.

76. Interview with TNLA major and (third) secretary general, Ref.N.93.

77. Interview with TNLA major and (third) secretary general, Ref.N.92.

78. Ref.N.92.

79. Interview with TNLA major and (third) secretary general, Ref.N.92.

80. Interview with Ta'ang CSO, Ref.N.37.

81. Myanmar 2008 Union constitution, art. 56.

82. Interview with Ta'ang CSO, Ref.N.37.

83. Less important, but still relevant, behind this choice there was also an issue of branding: during 2008 another ERO in Shan State had emerged with the very same acronym, the Pa-O National Liberation Army (PNLA), formed by Khun Thurein.

84. Distinctions in colors used to identify subgroups are said to be related to the colors of women's clothing.

85. Interview with TNLA officers, Ref.N.130.

86. PYNG was renamed Ta'ang Students and Youth Organization in December 2008, while PWO became the Ta'ang Women's Organization after 2012.

87. Interview with Ta'ang CSO, Ref.N.87. In 2018 Buddhist monastic institutions based in Mandalay carried out research suggesting that among the Ta'ang there are twenty communities (previously usually thought to be thirteen): Rumai; Katul; Tanma; Pannim; Rukin; Rugwan; Rukhau; Rukhpo; Rumà; Rumao; Rulain; Rubrang; Rukhot; Runà; Rulien; Rutra; Rubrang; Rukaw; Ngonrouk; Guanhai (Interview with Ta'ang CSO, Ref.N.87). These, the Ta'ang CSOs stress, are to be seen as an unicum rather than subgroups. The idea of subgroups is often reluctantly employed, avoided, or accompanied by caveats stressing unity of the ethnic nationality.

88. For a deeper account of how the household is an inherently gendered, militarized, and political space in and beyond Myanmar, see Hedström 2025.

89. Interview with investigative journalist with field experience specializing in Myanmar rebel movements, Ref.N.56.

90. Interview with investigative journalist, Ref.N.56.

91. Interview with investigative journalist, Ref.N.56.

92. In particular, the Human Rights Education Institute of Burma, founded in 2000, cooperated with local CSOs and CBOs to establish the documentation network ND-Burma. In the ambit of ND-Burma trainings were delivered to members of the CSOs and CBOs involved. As part of such trainings ND-Burma organized internship exchanges. Through one of these Tar Bong Kyaw spent a period in Johannesburg at the Center for the Study of Violence and Reconciliation (personal conversation with humanitarian worker, April 12, 2020).

93. One can see how the roots of the movement at this stage were branching out through different terrains. There were lived experiences of violence, oppression, and dislocation in northern and southern Shan State, study and activism in the realities of the refugee camps, as well as migration and marginalization in the informal/formal economic milieus of the Thai borderlands. Some of today's leaders, like the three current secretary generals [Tar Bong Kyaw, Tar Ohnm Ta Mao, and Tar Parn La], were important figures in the PYNG throughout the late 1990s and early 2000s, all of them living and working at the Thai border in Chiang Mai, Mae Sot, and Mae Hong Son. Field note, Ref.N.118.

94. Interview with PSLO/A last vice chairman and PSLO/A last chairman (Tar Khun Yee and Tar Aik Mone), Ref.N.97. Particularly, the Lone Kan/Lom Kum (Kutkai township) mililtitia and the Nonsai, Namhsan militia (TWO 2016, 52; Meehan 2016a, 375).

95. With the unfolding of Tatmadaw's BGF and PMF programs, the militia constituted by PSLO in Manton had been transformed into the M4 Manton PMF. With the formation of TNLA, the Manton militia—which is now headed by Tar Khun Li, the son of Tar Aik Mone, the last chairman of PSLO/A—saw its ranks reduced by half, from approximately one hundred to between thirty and fifty. Also, the Lom Kum Palaung Militia in Kutkhai, composed of former PSLA members, was disbanded in 2009. Interview with TNLA major and (third) secretary general, Ref.N.93.

96. Myanmar 2008 Union Constitution, art. 338.

97. The program was instituted after the promulgation of the 2008 constitution. It foresaw the transformation of ceasefire EROs into BGFs named Home Guard Forces in areas not adjoining a border) or PMFs. The designated EROs would be transformed into battalions of 326 soldiers in total, for what concerned the BGF program. Among these there would be eighteen officers and three commanders, two from the ERO and one from the Tatmadaw. Additionally, some key departments, like the General Staff Office and the Quartermaster Office, would be directed by the Tatmadaw and twenty-seven noncommissioned officers from the Myanmar Army would also be included. BGFs would have been inserted in the Tatmadaw structures also concerning personnel management and supplies (see Keenan 2014a, 186–191). By contrast, the PMF program entailed the transformation

of EROs or previous militia forces in Tatmadaw-controlled militias but without being integrated in the structure of the Tatmadaw, without being restructured with the integration of Tatmadaw officers, and being logistically and materially supported by the Tatmadaw. In addition it also required a downsizing of personnel (Buchanan 2016).

98. Myanmar Peace and Democracy Front (MPDF), a coalition that was composed of MNDAA, UWSA, NDAA, and KIO/A (Keenan 2014b).

99. Faced with Tatmadaw demands, before the outbreak of conflict three brigades of the SSPP/SSA (the headquarter brigade, brigade 3, and 7) had already decided to transform into Home Guard Forces—. Interview with SSPP/SSA officer and foreign affairs representative, Ref.N.146.

100. It should be noted that parallel to TNLA formation, another ERO was being constituted that was supported by KIO/A—the Arakan Army (AA). Emerging from Arakan youth involved in the All Arakan Student and Youth Congress in Mae Sot, the AA was established at the end of 2008 and took up base in Laiza, where KIO/A had agreed to host the nascent rebel movement. Interview with CSO working on education and social cohesion in Rakhine, Ref.N.99.

101. Interview with investigative journalist with field experience specializing in Myanmar rebel movements, Ref.N.56.

102. Interview with TNLA major and (third) secretary general, Ref.N.92–96.

103. Interview with Kachin Baptist Church (KBC) pastor, Ref.N.61.

104. Ref.N.61.

105. The Shwe Gas and Oil Pipeline connects gas fields in the Bay of Bengal off the Arakan/Rakhine coasts to the port area of Khyaukpyu and all the way to China (Van Schendel 2020, 46, 52). Gas and oil pipeline tracts in Myanmar have been operative since 2014. The tracts passing through Shan State traverse Ta'ang areas across Hsipaw, Namtu, Manton, Namkham townships (TSYO 2012).

106. By early 2015 TNLA was estimated at 3,000–3,500 combatants, while by early 2018 it had practically doubled, reaching 6,000 troops (Davis 2019). In 2019 it was understood to oscillate between lowest estimates of 5,000 and a highest one of 7,000. In the aftermath of the coup its ranks have significantly increased and are deemed between 10,000 and 15,000 soldiers.

107. Interviews with TNLA major and (third) secretary general, Ref.N.93; and Kachin Baptist Church (KBC) pastor, Ref.N.61.

108. Interview with TNLA former combatant, Ref.N.34.

109. Thein Sein's peace initiative, which preceded the starting of the negotiations for the Nationwide Ceasefire Agreement (NCA), led to the signing of fourteen bilateral ceasefires between 2011 and late 2013 (Buchanan 2016, 21).

110. Sixteen EROs formed the Nationwide Ceasefire Coordination Team to discuss the drafting of the NCA.

111. Interview with TNLA major and (third) secretary general, Ref.N.93.

112. Interviews with Head of CSO working on narcotics eradication and mine action, Ref.N.75; investigative journalist with field experience specializing in Myanmar rebel movements, Ref.N.56.

113. "PSLF/TNLA's & RCSS/SSA-S Conflict or Myanmar Proxy Army RCSS/SSA-S's War of Aggression to Ta'ang Region and Brutal Human Rights Violation," PSLF/TNLA briefing paper, obtained by the author through private correspondence with PSLF/TNLA.

114. The Ta'ang village of Tho San/Tort-san is often used as an example of such "Shannization" processes by different Ta'ang communities, which point out how throughout the ceasefire years the main language spoken in the village changed from Ta'ang to Shan (Interview with Head of CSO working on narcotics eradication and mine action, Ref.N.75; Interview with Ta'ang CSO, Ref.N.58).

115. Interviews with TNLA major and (third) secretary general, Ref.N.92–96. RCSS for its part had previously maintained small encampments and toeholds in Namkahm (near Maiwee/Minewee village not far from the Shwe Gas and Oil Pipeline) and Muse townships (about a hundred troops of the RCSS/SSA-S Task Force 701) (Davis 2019, 5; Keenan 2014a, 122).

116. The Shanni, "Red Shan" (in Shan, "Tai-leng"), are an ethnic population related to the Shan.

117. As a sub-lieutenant of TNLA recalls, this has to be carried out rather discreetly when trainings are organized in the actual Wa region: "[I]n theory they [UWSA] do not allow others to stay in. Even us, when we are there we have to wear their uniforms to join the training. Because at times they also have Burmese troops on their territory visiting and when they go and look around . . . they would know we are cooperating and that would not be safe." (Ref.N.91).

118. This occurred especially via two platforms. First, at the end of 2016, together with KIO/A, MNDAA, and AA, the PSLF/TNLA formed the "Northern Alliance," a military coalition of EROs that carried out operations especially along the China-Myanmar border (in particular, Pang Hseng, Muse, Namkham, Kutkhai, Kokang), but recently also launched an unprecedented attack on the Defense Services Technological Academy in Pyin Oo Lwin (Frontier 2019; Tønnesson et al. 2019). As of late, the alliance has renamed itself "Three Brotherhood Alliance" when acting without KIO/A. Second, after the signing of the NCA by eight EROs and the Tatmadaw in October 2015, UWSA was involved in promotion of a series of summits held between 2016 and 2017 out of which emerged the Federal Political Negotiating and Consultative Committee (FPNCC). FPNCC constituted a political coordination platform for EROs rejecting the NCA and proposing the initiation of a new peace process while calling for preemptive constitutional reforms guaranteeing genuine federal autonomy. The FPNCC brought together the following EROs: UWSP/A; Myanmar National Truth and Justice Party (MNTJP)/MNDAA; National Democratic Alliance Army/Peace and Solidarity Committee (NDAA/PSC); KIO/A; SSPP/SSA-N; United League of Arakan/AA (ULA/AA); PSLF/TNLA.

119. This alliance allowed the Ta'ang rebel movement to acquire munitions while sending members to the UWSA-maintained de facto autonomous areas east to the Salween River for training. Weapons were often consigned to the units trained and transported on foot at the end of the training. While training areas could be reached using cars, usually the way back was on foot in order to transport weapons and military supplies (Ref.N.91).

120. Interview with TNLA major and (third) secretary general, Ref.N.93.

121. Ref.N.93. UWSA's Southern Command is located approximately in the areas of Mong Yawn (renamed Yong Pang) and Yong Kha, located at the Thai-Shan border, west of Tachilek and just opposite to Chiang Rai.

4. TERRITORIES IN TA'ANG LAND

1. Field note, Ref.N.117.

2. Interview with TNLA captain, Ref.N.126.

3. The three Tawngpeng *sao hpa*'s relatives were Tar Khun Li, Khun Aye, and Chao Nor Far.

4. Interview with TNLA officers and field note (Tar Khun Aye memorial), Ref.N.129.

5. See PSLF/TNLA, "Press Release on Successful Celebration of Ta-ang National Resistance Golden Jubilee," January 18, 2013, http://en.pslftnla.org/press-release-on-successful-celebration-of-ta-ang-national-resistance-golden-jubilee/.

6. Interview with head of village, Ref.N.24.

7. The equivalent of what is the Dha among Bamar.

8. Interview with TNLA major and (third) secretary general, Ref.N.92.

9. Interviews with TNLA major and (third) secretary general, Ref.N.92–96. See also this interview with PSLF/TNLA's chairman Tar Aik Phone/Bong: "Ta'ang (Palaung) Leader Tar Aik Bong: 'Without Proper Political Solutions, There Will Be No Lasting Peace,'" *BurmaLink*, November 11, 2014, https://www.burmalink.org/taang-palaung-leader-tar-aik-bong-without-proper-political-solutions-will-lasting-peace/.

10. Interview with Ta'ang CSO, Ref.N.60.

11. Interview with TNLA former combatant, Ref.N.34.

12. Interviews with TNLA major and (third) secretary general, Ref.N.92–96.

13. Interview with TNLA major and (third) secretary general, Ref.N.92.

14. Field note, Ref.N.41; interviews with head of CSO working on narcotics eradication and mine action, Ref.N.75; Ta'ang National Party member and village resident, Ref.N.80; Ta'ang CSO, Ref.N.87.

15. Interview with head of CSO working on narcotics eradication and mine action, Ref.N.75.

16. Field note, Ref.N.41.

17. Interview with RCSS recruits (4) based at Hu Sun, Kyaukme township, Ref.N.30.

18. Ref.N.30.

19. Ref.N.30.

20. Interview with TNLA major and (third) secretary general, Ref.N.92.

21. Interview with head of CSO working on narcotics eradication and mine action, Ref.N.75.

22. Interview with TNLA major and (third) secretary general, Ref.N.92.

23. Interviews with TNLA major and (third) secretary general, Ref.N.92–96.

24. Interview with TNLA major and (third) secretary general, Ref.N.94.

25. Ref.N.94.

26. Field note, Ref.N.109.

27. Interview with TNLA second lieutenant, Ref.N.91.

28. B. 112—Kutkhai township; B. 256—Namhsan township; B. 478—Nam Kham township; B. 367—Man Tong township; B. 717—Moe Mit and Kyaukme; while B. 101 and B. 527 as special battalions (Ref.N.91).

29. Interview with TNLA major and (third) secretary general, Ref.N.92.

30. Interviews with investigative journalist with field experience specializing in Myanmar rebel movements, Ref.N.56; and TNLA major and (third) secretary general, Ref.N.92–96.

31. Interview with Ta'ang CSO, Ref.N.87.

32. Interview with head of village, Ref.N.28.

33. Ref.N.28.

34. Field note, Ref.N.121.

35. Interview with TNLA officer, Ref.N.101.

36. Interview with TNLA commander-in-chief Major General Tar Ho Plan, Ref.N.124.

37. Interview with TNLA second lieutenant, Ref.N.91.

38. Interview with TNLA major and (third) secretary general, Ref.N.92–96.

39. Interview with TNLA major and (third) secretary general, Ref.N.94.

40. Interview with TNLA major and (third) secretary general, Ref.N.92.

41. Interview with TNLA major and (third) secretary general, Ref.N.93.

42. Mong Kung in Shan, a southern Shan State town and township inhabited by considerable Ta'ang communities. Interview with Ta'ang National Party member and village resident, Ref.N.80.

43. Interview with TNLA major and (third) secretary general, Ref.N.96.
44. Interview with TNLA captain, Ref.N.126.
45. Interview with TNLA officer, Ref.N.101.
46. Interview with TNLA captain, Ref.N.126.
47. Interview with TNLA officer and TNLA Local Guerrilla Force (LGF) members (2), Ref.N.88.
48. Interview with PSLO/A last vice chairman and PSLO/A last chairman (Tar Khun Yee and Tar Aik Mone), Ref.N.97.
49. Interview with SSPP/SSA officer and foreign affairs representative, Ref.N.146.
50. Ref.N.146.
51. Series of attacks staged by the Brotherhood Alliance (BA)—comprising TNLA, MNDAA, and AA—and related Tatmadaw counteroffensives occurred in the second half of August 2019. Interview with Ta'ang CSO, Ref.N.87.
52. This may sound somewhat ironic given that on the one hand the state of Israel has been one of the major exporters of weapons, military knowledge, techniques, and the like to the Tatmadaw; and that, on the other hand, Israel itself has transformed Palestinian spaces into a battleground for the territorialization of its state authority (Weizman 2007).
53. Ref.N.87.
54. Interview with TNLA lieutenant, General Administrative Department (GAD) chief, Ref.N.90. GAD articulated Ta'ang Land as follows: Moe Goes district (townships: Owm Yai; Main Ngaw; Moe Goes; Mao Vai; Owm Paw); Owm Ta Mao district (Am Chri; Owm Po; Owm Ve; Ru Gaw); Owm Khron district (Mu Kyal; Kut Khtai; Main Hawm; Main Kyi); Owm Ngyi district (Owm Ngyi; Nam Phat Kham; Main Baw); Larshio district (Main Yaw; Kyan Yan; Kan Main; Nawn Lain)
55. Especially in the townships of Kyaukme, Hsipaw, Namtu, and Lashio.
56. Interview with Shan CSO, Ref.N.76.
57. Interviews with mountain guide, Hsipaw resident, Ref.N.72; and investigative journalist with field experience specializing in Myanmar rebel movements, Ref.N.56.
58. Interview with head of village, Ref.N.29.
59. Interview with Ta'ang CSO, Ref.N.87.
60. Field note, Ref.N.108.
61. Field note, Ref.N.108.
62. For example, see this Geneva Call statement in which the RCSS/SSA-S is referred to as the Shan State Army: Geneva Call, "Burma/Myanmar: Geneva Call urges an end to mine use in northern Shan State," Geneva Call, July 14, 2016, https://www.genevacall.org/burmamyanmar-geneva-call-urges-end-mine-use-northern-shan-state/.
63. Interview with Shan CSO, Ref.N.76.
64. Interview with Shan CBO, Ref.N.135.
65. Interview with village head, Ref.N.82.
66. Interview with deputy head of village, Ref.N.33.
67. Interview with TNLA officer, Ref.N.101.
68. Interviews with TNLA major and (third) secretary general, Ref.N.92–96.
69. Interviews with TNLA major and (third) secretary general, Ref.N.92–96.
70. Interview with TNLA former combatant, Ref.N.34.
71. Interview with TNLA captain, Ref.N.126.
72. Ref.N.126.
73. Interview with investigative journalist with field experience specializing in Myanmar rebel movements, Ref.N.56.
74. Ref.N.56.
75. Interview with TNLA officer and TNLA Local Guerrilla Force (LGF) members (2), Ref.N.88.

76. Interview with SSPP/SSA officer and foreign affairs representative, Ref.N.146.

77. Ref.N.146.

78. Interviews with TNLA major and (third) secretary general, Ref.N.92–96.

79. Ref.N.92–96.

80. Interview with TNLA second lieutenant, Ref.N.91.

81. Interview with TNLA officer and TNLA Local Guerrilla Force (LGF) members (2), Ref.N.88.

82. Interview with SSPP/SSA foreign office and education department head, Ref.N.140.

83. Interview with head of village, Ref.N.29.

84. Interview with head of village, Ref.N.62.

85. Interview with head of village, Ref.N.79.

86. Interview with SSPP/SSA foreign office and education department head, Ref.N.140.

87. Interviews with mountain guide and inhabitant, Ref.N.21; and head of CSO working on narcotics eradication and mine action, Ref.N.75.

88. The head of a CSO based in Kyaukme raises this point while talking about episodes of compensation requests: "They, the armed groups, see landmines as a strategic property. If an animal or sometimes even humans step on the landmine, they have to pay back to the ERO the price of the landmine. I have heard it can be something like 300,000 kyat and in addition, if it was an animal that stepped on the landmine, the meat of the cow or buffalo has to be donated to the armed group." (Ref.N.75)

89. Interview with mountain guide, Hsipaw resident, Ref.N.83.

90. Interview with mine action expert with field experience, Ref.N.12.

91. Interview with Ta'ang CSO operating in norther Shan, Ref.N.49.

92. Interview with deputy head of village, Ref.N.33.

93. Interview with Shan CSO, Ref.N.76.

94. Interview with TNLA major and (third) secretary general, Ref.N.96.

95. Interview with head of CSO working on narcotics eradication and mine action, Ref.N.75.

96. Interviews with Shan CSO, Ref.N.76; and mountain guide, Hsipaw resident, Ref.N.83.

97. Interviews with Ta'ang CSO, Ref.N.58; and mine action researcher specializing in Myanmar and with field experience, Ref.N.103.

98. Interview with SSPP/SSA officer and foreign affairs representative, Ref.N.146.

99. Interview with Ta'ang CSO, Ref.N.86.

100. Ref.N.86.

101. Interview with Shan CBO, Ref.N.135.

102. Interview with Ta'ang CSO, Ref.N.58.

103. Interview with TNLA major and (third) secretary general, Ref.N.95.

104. Interview with Ta'ang CSO, Ref.N.58.

105. Interview with Shan CSO, Ref.N.76.

106. Ref.N.76.

107. Interview with village head, Ref.N.82.

108. Interview with Ta'ang CSO, Ref.N.58.

109. Ref.N.58.

110. Interview with TNLA officer, Ref N.101.

111. Private correspondence with PSLF/TNLA second lieutenant, October 12, 2020.

112. Interview with Ta'ang CSO, Ref.N.87. Interview with TNLA officer, Ref.N.101.

113. Ref.N.101.

114. Ref.N.101.

115. Interview with TNLA major and (third) secretary general, Ref.N.94.

116. Interview with investigative journalist with field experience specializing in Myanmar rebel movements, Ref.N.56.

117. Interview with head of village, Ref.N.28.

118. Interview with Ta'ang CSO, Ref.N.60.

5. NETWORKS (WEAPON BIOGRAPHIES)

1. Woman's skirt (longyi) in Myanmar (ထဘီ/ထမိန်). The longyi is a sheet of cloth widely worn as a skirt in Myanmar.

2. Personal conversations, Signal, December 2022.

3. Interviews with Tar Parn La, TNLA major and (third) secretary general, Ref.N.93–94.

4. Personal conversation with landmines and explosives expert with extensive experience in Myanmar, July 30, 2020.

5. In particular, the most widespread types were the POMZ-2 and POMZ-2M stake-mounted antipersonnel fragmentation mines, designed and produced in the USSR, and the Chinese Type-58 and -59, modeled on the POMZ-2 and PMN Soviet-designed land-mines. Type-69 and -72A, as well as Soviet PMN-1, PMD-6 landmines also circulated, according to the Landmine Monitor (2004, 938).

6. The Federal People's Republic of Yugoslavia first and the Socialist Federal Republic of Yugoslavia in the 1960s. Particular types of mines included US M14, M16A1, M18; and Indian/British LTM-73, LTM-76.

7. Stockholm International Peace Research Institute (SIPRI) Databases—Arms Embargoes: EU arms embargo on Myanmar, SIPRI, accessed November 2, 2024, https://www.sipri.org/databases/embargoes/eu_arms_embargoes/myanmar.

8. It has been reported how some models of Italian-manufactured landmines have ended up in Myanmar (Selth 2000a; Landmine Monitor 2004, 2019). In particular the Italian VS 1.6. mine has been documented after 1988, while the VS-50 and Valmera 69 were allegedly supplied.

9. Five different types of landmines have been and are produced at these Ka-Pa-Sa's sites, which basically consist of antipersonnel landmines modeled on Chinese and US types. The MM1, based on the Type-59 Chinese stake-mounted fragmentation landmine; the MM2, based on the Type-58 Chinese blast mine; the MM3, a bounding mine; the MM5, based on the Claymore type; and the MM6, a copy of the US M14 plastic mine (minimum metal content mine) (Landmine Monitor 2019, 3).

10. Interview with Tar Khun Yee and Tar Aik Mone, PSLO/A last vice chairman and chairman, Ref.N.97.

11. Interview with SSPP/SSA-N officer and foreign affairs representative, Ref.N.146.

12. Interviews with Tar Parn La, TNLA major and (third) secretary general, Ref.N.92–96.

13. Ref.N.92–96; interview with mine action researcher specializing in Myanmar, Ref.N.103.

14. Ref.N.103. Also, Andrew Selth (2000, 17) mentions specifically that various rebel movements by the 2000 had gained long-standing familiarity in particular with gunpowder (black powder), dynamite, gelignite, TNT, RDX, amatol, C-4, and nitro-methane compounds.

15. In particular, Laiza has represented a key place for the development of both the AA and PSLF/TNLA. Panghsang, the capital of the Wa SAD, at times hosts trainings of UWSA's allied EROs (interview with TNLA sub-lieutenant, Ref.N.91). Moreover, some areas on the eastern banks of the Salween River, in proximity to the Wa SAD, have been

indicated as hubs in which military assistance activities have been provided to other EROs, in particular, a training area south to the town of Kunlong and the road connecting Hseni with Chinshwehaw and the Chinese border; and another area southeast to Mong Hsu, on the western banks of the Salween River. Interviews with TNLA sub-lieutenant, Ref.N.91; journalist and consultant with field experience, Ref.N.48; SSPP/SSA-N foreign office and education department head, Ref.N.140.

16. European Union Assistance on Curbing Small Arms and Light Weapons in Cambodia (EUASAC)

17. Interviews with Tar Parn La, TNLA major and (third) secretary general, Ref.N.92–96; interview with mine action researcher specializing in Myanmar and with field experience, Ref.N.103; interview with SSPP/SSA-N officer and foreign affairs representative, Ref.N.146; interview with former NUPA/AA combatant and head of humanitarian CSO, Ref.N.102.

18. Recall the March 2020 Tatmadaw crackdown on the Kawngkha militia in the area of Kutkai, northern Shan State, as an example (see L. Weng, "Myanmar Army Seizes Shan State Militia Chiefs over Drugs Bust," *Irrawaddy*, March 29, 2020, https://www.irrawaddy. com/news/burma/myanmar-army-seizes-shan-state-militia-chiefs-drugs-bust.html). Also known as the Kachin Defense Army, the Kawngkha militia is a splinter armed organization that served as the former KIA's Fourth Brigade and whose leader, Mahtu Naw, agreed to a ceasefire with the military regime in 1990. In 2020, the Tatmdaw severely cracked down on the militia, seizing its headquarters at Lwekham village, Kutkai, and arresting part of the leadership. Apparently one of the underlying factors for such harsh actions against what in fact is a military-backed militia was related to allegations that the narcotics production and smuggling of the Kawngkha militia were in part funded/backed by Kokang smugglers and the Myanmar National Democratic Alliance Army (MNDAA) (private correspondence with TNLA secretary general, March 29, 2020, and humanitarian worker and researcher specialized in northern Shan conflicts, March 29, 2020). Moreover, the militia was believed to be involved in weapons smuggling, operating as a bridge for flows across the Chinese border (private correspondence with humanitarian worker, March 29, 2020). In general, the area of the Kawngkha is geographically key as it lies in the midst of different overlapping EROs' areas of influence (e.g., KIO/A, MNDAA, TNLA) and represents a nodal point serving as a buffer between such areas, urbanized centers and key transportation links.

19. For an example, think of the charity Shadows of Hope, https://www.shadow sofhope.org/. Among others one can recall also the case of the Italian Fascist political movement Casapound. See L'Espresso, "Casapound, Guerra in Birmania," *L'Espresso*, November 5, 2012, https://lespresso.it/c/politica/2012/11/4/casapound-guerra-in-birm ania/33739.

20. Interviews with the sub-lieutenant SSPP/SSA-N (who participated in the trainings), Ref.N.137–139.

21. Myay hmyouq mine—မြေမြှုပ်မိုင်း

22. Interviews with Tar Parn La, TNLA major and (third) secretary general, Ref.N.92–96.

23. Interview with Tar Parn La, TNLA major and (third) secretary general, Ref.N.93.

24. Email correspondence with mine action expert working in Myanmar, October 13, 2020.

25. Martin Smith specifies the provision of AK-style assault rifles, mortars, antiaircraft guns, field machine guns, and ammunition (1999). The program aimed to reinforce the North-Eastern Command of CPB under Naw Seng's command.

26. See L. Weng, "Armed Insurgents in Burma Face Shortage of Ammunition," *The Irrawaddy*, December 22, 2008, http://www.irrawaddy.org/highlight.php?art_id=14829.

The Chittagong arms seizure of 2004 in Bangladesh, in which ten truckloads were captured at a jetty of the state-owned Chittagong Urea Fertilizer Limited on the Karnaphuli River, is taken as a turning point concerning weapon flows originating from surpluses in Southeast Asia and the acquisition of Chinese-manufactured weapons by armed actors in the India-Bangladesh borderlands. For example, the United Liberation Front of Asom in Bangladesh and India has since reportedly changed weapons acquisition processes, obtaining arms produced at UWSA factories through Chinese license agreements and production support. See S. Bhaumik, "Where Do 'Chinese' Guns Arming Rebels Really Come From?" *BBC News*, August 3, 2010, https://www.bbc.com/news/world-south-asia-10626034.

27. I follow the Small Arms Survey *Introductory Guide to the Identification of Small Arms, Light Weapons, and Associated Ammunition* in referring to small arms types (see Jensen-Jones and Schroeder 2018). For example, I use the term "AK-type" here because not all the rifles in the picture could be identified as a specific model.

The conclusions I draw here are based on interviews with journalist/consultant with fieldwork experience in Myanmar's borderlands, Ref.N.3; Kachin Baptist Church (KBC) pastor, Ref.N.61; small arms and light weapons expert with multiethnic humanitarian movement, Ref.N.68. I also draw on the *Small Arms & Light Weapons, Vehicles & Aircraft Recognition Guide*. The latter consists of an unpublished technical document drafted by the Free Burma Rangers in order to recognize and record armed actor's behavior during operations.

28. In 2008 Weng reported that a munition factory was "situated in Kunma, the hometown of UWSA Chairman Bao You-xiang, in the Wa hills, 125 kilometers (78 miles) north of the group's headquarters at Panghsang on the Chinese border" (Weng 2008).

29. Interview with Kachin Baptist Church (KBC) pastor, Ref.N.61.

30. Ref.N.61.

31. Especially the KIO/A's allies, in particular TNLA, AA, and MNDAA.

32. Interviews with mountain guide and resident of Kyaukme, Ref.N.5, 21, 73.

33. Interview with local investigative journalist with field experience specializing in Myanmar rebel movements, Ref.N.56; personal conversation with academic researcher, Ref.N.10. See also Weng 2015.

34. Interview with mine action researcher specializing in Myanmar and with field experience, Ref.N.103.

35. Interview with SSPP/SSA officer and foreign affairs representative, Ref.N.146.

36. Interview with mine action researcher specializing in Myanmar and with field experience, Ref.N.103.

37. Interview with small arms and light weapons expert of a multiethnic humanitarian movement, Ref.N.67.

38. Fifty rifles, thirty handguns, and two thousand to thirty thousand rounds of ammunition (depending on the type) per year. This limit may be circumvented by registering family members as dealers and requesting personal sale licenses so that import lots could be cumulated (see Carter 2019). For an overview of Thai firearms legislation see Suriyavorapunt 2018.

39. Firearms, Ammunition, Explosives, Fireworks, and the Equivalent of Firearms Act, B.E.2490 (1947), Section 5, http://www.vertic.org/media/National%20Legislation/Thailand/TH_Firearms_Ammunition_Act.pdf; Arms Control Act, B.E.2530 (1987), Section 6, http://www.vertic.org/media/National%20Legislation/Thailand/TH_Arms_Control_Act.pdf.

40. Arms Control Act, B.E.2530 (1987), Section 6, http://www.vertic.org/media/National%20Legislation/Thailand/TH_Arms_Control_Act.pdf.

41. Interview with mine action researcher specializing in Myanmar and with field experience, Ref.N.103.

42. Interviews with NGO researcher with field experience, Ref.N.50; and investigative journalist with field experience specializing in Myanmar rebel movements, Ref.N.56.

43. For the following reconstruction I draw on work done by Miles Vining (2020) for Armament Research Services on this issue and triangulate through interviews with small arms expert working with the Free Burma Rangers.

44. In particular, M16/A1 and M16/A2 variants, which were most widespread throughout surpluses of stockpiles supplied by the US as part of different forms of military assistance to armed actors in Vietnam, Cambodian, and Laotian armed conflicts. The RCSS and KNU/KNLA nowadays, just to mention two, are the EROs most usually seen with M16-patterne rifles, although surely not the only.

45. Interview with village inhabitant, Ref.N.26.

46. Vining provides a more technical explanation of the Yat-Thai rifle. Basically, gunsmiths repurpose and adjust the receiver, action, gas block and tube system, rear and front sights, and trunnion of an AK-pattern rifle. On this "baseline" firearm a barrel of an M-16 or Tatmadaw-produced rifle, which are produced to fire 5.56x45mm ammunition, is mounted. Lastly, magazines of either M16-pattern or Tatmadaw-produced rifles are modified so as to fit with the AK-type receiver and feed the Yat-Thai assemblage with 5.56x45 ammunition produced by the Tatmadaw without affecting the baseline components of the weapon.

47. Interview with hunter and village inhabitant, Ref.N.31.

48. Interview with hunter and village inhabitant, Ref.N.31.

49. Interview with hunter and village inhabitant, Ref.N.31.

50. This pattern has emerged clearly in the aftermath of the February 1 coup. Resistance formations in Chin and Sagaing were among the first to mobilize units armed with hunting muskets. Interestingly, these were areas without consistent ERO operational presence. Craft manufacture has been a major supply mechanism, and resistance forces in Sagaing and Chin have been able to produce bolt-action rifles, mortars, and at times Sten guns, although access to semiautomatic/automatic firearms has not been straightforward and, different from the southeastern borderlands, gunsmithing seems to have been less focused on semiautomatic/automatic weapons refurbishment.

51. Interviews with small arms and light weapons expert with multiethnic humanitarian movement, Ref.N.67.

52. In particular, lower receivers of M16-type rifles or others subject to usury.

53. Local feudal hereditary rulers exercising sovereignty over princedoms in today's Shan State throughout precolonial and colonial times.

54. Interviews with Pyi Thu Sit village community militia, Ref.N.81.

55. Interviews with Pyi Thu Sit village community militia, Ref.N.81.

56. In Bama sa: စစ်ဆင်ရေး ကွပ်ကဲမှုဌာနချုပ်—Sit Sin Ye Koot Ke Mu Htar Na Chote, from which comes the abbreviation Sa-Ka-Kha.

57. Ref.N.81. See also Radio Free Asia, "Skirmishes Continue between Myanmar Army and Ethnic Militias in Shan State," December 16, 2016, https://www.refworld.org/docid/58f9ca011b.html.

58. Interviews with Pyi thu sit village community militia, Ref.N.81.

59. Ref.N.81.

60. Ref.N.81.

61. Ref.N.81.

62. Specifically the four villages of Pan Tha Pyay, Pang Kun (Haung), Pang Kun (Thit), and Pang Hpyet.

63. Ref.N.81.

6. SCALES

1. Interview with TNLA soldier and officer, Ref.N.110.

2. Interview with head of village, Ref.N.29.

3. Ref.N.29.

4. Interview with SSPP/SSA officer and foreign affairs representative, Ref.N.146.

5. Using the expression "Geneva Conventions," he is referring to deeds of commitment the PSLF/TNLA agreed to.

6. Interview with TNLA soldier and officer, Ref.N.110.

7. Interview with Ta'ang CSO, Ref.N.87.

8. For a thorough discussion of the highly debated and problematic aspects of the FBR's activities as well as their embeddedness into larger political and religious constellations at different local and global scales, see Horstmann 2018, 2019.

9. As we have seen, these techniques (1) have been informed by an ethnonational logic of identifying areas as white where and when a military base is constructed in an area and a Myanmarization process is being enacted in that area; as brown where and when ethnonational minority institutions and authorities, cultures, and practices still hold; and as black where and when no frontier has been established; and (2) they have been unfolding also via arms control and firing practices.

10. Interview with FBR volunteer, Ref.N.66.

11. Ref.N.66

12. Ref.N.66.

13. Interview with sub-lieutenant SSPP/SSA, Ref.N.138.

14. Ibid.

15. Interview with volunteer of a multiethnic humanitarian movement, Ref.N.66.

16. Interview with Pyi Thu Sit village community militia, Kyaukme township, Ref.N.81.

17. Tatmadaw's Military Operations Command 1 (MOC-1) based in Kyaukme.

18. Interview with Pyi Thu Sit village community militia, Kyaukme township, Ref.N.81.

19. Ref.N.81.

20. Ref.N.81.

21. Interviews with humanitarian mine action worker, Ref.N.37; local humanitarian mine action worker, Ref.N.47; humanitarian mine action worker and Lashio resident, Ref.N.53; Ta'ang CSO, Ref.N.86; Shan CBO, Ref.N.135; and SSPP/SSA officer and foreign affairs representative, Ref.N.146.

22. Until December 28, 2018, the GAD remained a prerogative of the Sit-Tat in connection with its purview over the Ministry of Home Affairs as enshrined in the 2008 constitution. At the end of 2018, under the NLD government, the GAD was reformed and brought under the Ministry of the Office of the Union Government. Although detached from the Sit-Tat-controlled Ministry of Home Affairs and placed under the structure of the semi-democratically elected government, its staff was not substantially altered from the previous military legacy. With the 2021 coup d'état, the GAD has been reintegrated under military control.

23. Interview with SSPP/SSA-N officer and foreign affairs representative, Ref.N.146.

24. Drafted during colonial times, the Unlawful Associations Act has been widely used to target political and/or armed rebel movements by accusing people of being in contact with or involved in their activities. To make just one among many possible examples, in June 2017 the journalists Aye Nai, Pyae Phone Naing, and Thein Zaw (known as Lawi Weng) were arrested and provisionally detained at Hsipaw prison while returning from a work trip to document a PSLF/TNLA drug-burning ceremony in the UN International Day against Drug Abuse and Illicit Trafficking. See also Cheesman 2015, 115.

25. Similar dynamics, albeit slightly less convoluted, occur where only one ERO or the Sit-Tat only is present. The ties of EROs often require that a connection be maintained in the village, either through village heads or informants, as well as in areas where a certain ERO does not maintain any presence.

26. Interview with deputy head of village, Ref.N.33.

27. Interview with Ta'ang CSO, Ref.N.53.

28. Ref.N.58.

29. Interview with Ta'ang CSO operating in N.Shan, Ref.N.49.

30. Interviews with local humanitarian mine action worker, Ref.N.51; and local humanitarian worker with mine action INGO, Ref.N.52.

31. Interview with head of CSO working on narcotics eradication and mine action, Ref.N.75.

32. Ref.N.75.

33. Particularly in Kyaukme, Hsipaw, and Namtu. See ACLED, "Understanding Inter-Ethnic Conflict in Myanmar," September 28, 2018, https://acleddata.com/2018/09/28/understanding-inter-ethnic-conflict-in-myanmar/; and Relief Web, "Myanmar: Temporary Displacement in Northern Shan (2019–2020) as of 31 December 2020," January 31, 2021, https://reliefweb.int/map/myanmar/myanmar-temporary-displacement-northern-shan-2019-2020-31-december-2020.

34. The RCSS/SSA-S is a party to the NCA and at times has been accused of maintaining collaborative (although not smooth) relations with the Sit-Tat and conflictual stances vis-à-vis the UWSA. The SSPP/SSA-N is a member of the Federal Political Negotiation and Consultative Committee (FPNCC) political umbrella, has maintained privileged relations with the UWSA, and has allied at times with the PSLF/TNLA.

35. Eleven News, "SNLD Requests SSPP and RCSS to Discuss a Ceasefire," November 28, 2018, https://elevenmyanmar.com/news/snld-requests-sspp-rcss-to-discuss-ceasefire.

36. In particular, the CSSU was composed of the following organizations: the RCSS/SSA-S; the Shan State Joint Action Committee comprising SNLD and SSPP/SSA-N; the Shan Nationalities Democratic Party; and thirteen Shan CSOs and CBOs, including lawyer professional networks.

37. Interview with SSPP/SSA officer and foreign affairs representative, Ref.N.146.

38. Ref.N.146.

39. An RCSS colonel, Col. Sai Phone Han, put it this way in an interview with the journalist Lawi Weng: "Everyone has their own boundaries. Problems will always break out when someone comes to attack [the other side's] land. We need to mark our boundary lines to be able to build trust in each other." L. Weng, "RCSS Invites Rival Shan Group to Join Ceasefire, Excludes TNLA," March 22, 2019, https://www.irrawaddy.com/news/burma/rcss-invites-rival-shan-group-to-join-ceasefire-excludes-tnla.html.

40. Interview with Ta'ang CSO, Ref.N.85.

41. Interview with mine action researcher specializing in Myanmar and with field experience, Ref.N.103.

7. PLACES

1. Interview with SSPP/SSA-N officer and foreign affairs representative, Ref.N.146.

2. Interview with mountain guide and resident of Kyaukme township, Ref.N.5.

3. Interview with mountain guide and Hsipaw resident, Ref.N.83.

4. Interviews with mountain guide and inhabitant of Kyaukme township, Ref.N.21; Shan CSO official, Ref.N.76; and field note, Ref.N.109.

5. Field note, Ref.N.109.

6. N.109.

7. For example the mountainous areas north to Kyaukme and Hsipaw; those in between Hsipaw and Namtu townships; areas straddling the union highway passing through Kutkai; and areas of southern Namkham.

8. Interview with mine action researcher specializing in Myanmar and with field experience, Ref.N.103.

9. Interview with head of village, Ref.N.29.

10. Interview with researchers with fieldwork experience in northern Shan State, Ref.N.1.

11. Interview with Mine Din (Mong Tin) resident, Ref.N.78.

12. Ref.N.78.

13. Interview with SSPP/SSA-N foreign office and education department head, Ref.N.140.

14. Field note, Ref.N.15.

15. The base belongs to the artillery corps under the Sit-Tat Northern Command headquartered in Myitkyina.

16. Field note, Ref.N.27.

17. Interview with Ta'ang National Party member and village resident, Ref.N.80.

18. Loi Tai Leng is RCSS/SSA-S' headquarters area on the Thai-Myanmar border.

19. Interview with village inhabitant (Bang Hone, Kyaukme township), Ref.N.26.

20. Interview with TNLA former combatant, Ref.N.34. For a parallel and an example from a different borderland in Myanmar, refer to the case of Thamanya Sayadaw, named U Winaya (1912–2003), the monk and countrywide prominent figure who had established a community on Thamanya Taung (Mount Thamanya) in Karen State in 1980. As he acquired a reputation for his spiritual capacities, the land on which he had established the community gradually turned into a pilgrimage site and was eventually donated to him and registered as *thathana myei* (religious land) with the Religious Ministry (Tosa 2009, 240). In this area, among other things, weapons and armed forces were prohibited. Such proscription was due to the awareness and importance of the Buddhist practice of a *hbeme detha* (a peaceful land free of danger) in a civil war landscape (Tosa 2009, 245).

21. Interview with abbot of a monastery in north Kyaukme township, Ref.N.77.

22. Interview with local humanitarian mine action worker, Ref.N.51.

23. Interview with mountain guide and Kyaukme township resident, Ref.N.21; and mountain guide, Hsipaw township resident, Ref.N.83.

24. Interview with head of CSO working on narcotics eradication and mine action, Ref.N.75.

25. Ref.N.75.

26. I corroborated this information through private correspondence with a Shwe Phee Myay News Agency's journalist, Signal, July 21, 2021. See also Shwe Phee Myay News Agency's reportage at https://www.facebook.com/451214215694715/videos/59934237 4368738.

27. It is important not to overgeneralize these distinctions, but in Ta'ang areas of northern Shan one usually encounters villages that host only Ta'ang communities, villages that host mixed Ta'ang-Shan communities but in distinct parts of the village, and villages that host mixed communities but in which households are not grouped together into Ta'ang and Shan parts of the village.

28. See also a reportage by Shwe Phee Myay News Agency about the arson of more than 110 Ta'ang houses in the villages of Panlong and Chaungsa, Namtu township (https://www.facebook.com/watch/?v=494641431659732). In the video, witnesses recall how RCSS/SSA-S troops threatened villagers to "choose whether to save your homes or your life" and drove out people before burning down the houses.

29. See Shwe Phee Myay News Agency's reportage at: https://www.facebook.com/watch/?v=494641431659732.

30. Interview with TNLA second lieutenant, Ref.N.91.

31. Interview with SSPP/SSA-N officer and foreign affairs representative, Ref.N.146.

32. See the Landmine & Cluster Munition Monitor's 2019 update at http://www.the-monitor.org/en-gb/reports/2019/myanmar_burma/view-all.aspx.

33. Meaning zigzagging from one roadside to the other along the road or path. Interview with Ta'ang CSO senior officer, Ref.N.87.

34. Interview with the head of a Shan CSO, Ref.N.76.

EPILOGUE

1. Maui is his nom de guerre. Maui has agreed to be identified in these terms in these paragraphs, which he personally reviewed. What follows draws from multiple interviews and conversations carried out in August 2023.

2. See Justice for Myanmar (2021), "Ukraine Is Arming the Myanmar Military," https://www.justiceformyanmar.org/stories.ukraine-is-arming-the-myanmar-military.

3. Karenni Nationalities Defense Forces; in Burmese: ကရင်နီအမျိုးသား ကာကွယ်ရေးတပ်.

4. The Karenni National People's Liberation Front and the Karenni National Solidarity Organization.

5. Like the Kayan New Land Party.

References

Abel, P. 2000. "Manufacturing Trends—Globalizing the Source." In *Running Guns: The Global Black Market in Small Arms*, edited by L. Lumpe, 81–104. London: Zed Books.

Abraham, I., and W. Van Schendel. 2005. *Illicit Flows and Criminal Things: States, Borders, and the Other Side of Globalization*. Bloomington: Indiana University Press.

Agamben, G. 2005. *State of Exception*. Chicago: University of Chicago Press.

Agnew, J. 2011. "Space and Place." In *Handbook of Geographical Knowledge*, edited by J. Agnew and D. Livingstone, 316–331. London: Sage.

Arjona, A. 2016. *Rebelocracy: Social Order in the Colombian Civil War*. Cambridge: Cambridge University Press.

Arjona, A., N. Kasfir, and Z. Mampilly. 2015. *Rebel Governance*. New York: Cambridge University Press.

Ashkenazi, M. 2012. "What Do the Natives Know? Societal Mechanisms for Controlling Small Arms." In *Small Arms, Crime and Conflict: Global Governance and the Threat of Armed Violence*, edited by O. Greene and N. Marsh, 228–247. New York: Routledge.

Atkinson, P., and M. Hammersley. 2007. *Ethnography: Principles in Practice*. New York: Routledge.

Aung-Thwin, M. 1985. "The British 'Pacification' of Burma: Order without Meaning." *Journal of Southeast Asian Studies* 16 (2): 245–261.

Aung-Thwin, M. 2005. *The Mists of Ramañña: The Legend That Was Lower Burma*. Honolulu: University of Hawai'i Press.

Aung-Thwin, M., and Ma Aung-Thwin. 2012. *A History of Myanmar since Ancient Times: Traditions and Transformations*. Chicago: University of Chicago Press.

Aung Zaw. 2022. "Give Myanmar's Resistance What It Needs to Shoot Down Junta Helicopters." *Irrawaddy*, October 5. https://www.irrawaddy.com/opinion/commentary/give-myanmars-resistance-what-it-needs-to-shoot-down-junta-helicopters.html.

Balčaitė, I. 2020. "Race in Myanmar: Rigid Hierarchies, Blurred Boundaries and the Human Cost of Racism." *TeaCircle*, November 12, 2020. https://teacircleoxford.com/2020/11/12/race-in-myanmar-rigid-hierarchies-blurred-boundaries-and-the-human-cost-of-racism/.

Ballvè, T. 2020. *The Frontier Effect: State Formation and Violence in Colombia*. Ithaca, NY: Cornell University Press.

Bara, C. 2015. "Incentives and Opportunities: A Complexity-Oriented Explanation of Violent Ethnic Conflict." *Journal of Peace Research* 51 (6): 696–710.

Bartolucci, V., and A. B. Kanneworff. 2012. "Armed Violence Taking Place within Societies: SALW and Armed Violence in Urban Areas." In *Small Arms, Crime and Conflict: Global Governance and the Threat of Armed Violence*, edited by O. Greene and N. Marsh, 122–137. New York: Routledge.

Battle, W. 2017. *World Tea Encyclopaedia: The World of Tea Explored and Explained from Bush to Brew*. London: Troubador Publishing.

Baud, M., and W. van Schendel. 1997. "Toward a Comparative History of Borderlands." *Journal of World History* 8 (2): 211–242.

Beier, M. J. 2011. "Dangerous Terrain: Re-Reading the Landmines Ban through the Social Worlds of the RMA." *Contemporary Security Policy* 32 (1): 159–175.

Bergner, G. 2019. "Introduction: The Plantation, the Postplantation, and the Afterlives of Slavery." *American Literature* 91 (3): 447–457.

Bolton, M. 2015. "From Minefields to Minespace: An Archeology of the Changing Architecture of Autonomous Killing in US Army Field Manuals on Landmines, Booby Traps and IEDs." *Political Geography* 46:41–53.

Bohwongprasert, Y. 2020. "Thailand's Other Pandemic." *Bangkok Post*, June 2, 2020. https://www.bangkokpost.com/life/social-and-lifestyle/1928136/thailands-other-pandemic.

Boshier, C. A. 2017. *Mapping Cultural Nationalism: The Scholars of the Burma Research Society 1910–1935*. NIAS Monographs 136. Copenhagen: NIAS Press.

Bourne, M. 2007. *Arming Conflict*. London: Palgrave Macmillan UK.

Bourne, M. 2012. "Guns Don't Kill People, Cyborgs Do: A Latourian Provocation for Transformatory Arms Control and Disarmament." *Global Change, Peace & Security* 24 (1): 141–163.

Bousquet, A., J. Grove, and N. Shah. 2017. "Becoming Weapon: An Opening Call to Arms." *Critical Studies on Security* 5 (1): 1–8.

Bousquet, A., J. Grove, and N. Shah. 2020. "Becoming War: Toward a Martial Empiricism." *Security Dialogue* 51 (2–3): 99–118.

Boutry, M. 2016. "Burman Territories and Borders in the Making of a Myanmar Nation State." In *Myanmar's Mountain and Maritime Borderscapes: Local Practices, Boundary-Making and Figured-Worlds,* edited by Su-Ann Oh, 99–121. Singapore: ISEAS-Yusof.

Brenner, D. 2017a. "Authority in Rebel Groups: Identity, Recognition and the Struggle over Legitimacy." *Contemporary Politics* 23 (4): 408–426.

Brenner, D. 2017b. "Inside the Karen Insurgency: Explaining Conflict and Conciliation in Myanmar's Changing Borderlands." *Asian Security* 14 (2): 83–99.

Brenner, D. 2019. *Rebel Politics: A Sociology of Armed Struggle in Myanmar's Borderlands*. Ithaca, NY: Cornell University Press.

Brenner, D. 2024. "Misunderstanding Myanmar through the Lens of Democracy." *International Affairs* 100 (2): 751–769.

Brenner, D., and E. Han. 2021. "Forgotten Conflicts: Producing Knowledge and Ignorance in Security Studies." *Journal of Global Security Studies* 7 (1): 1–17.

Brenner, D., and M. Tazzioli. 2022. "Defending Society, Building the Nation: Rebel Governance as Competing Biopolitics." *International Studies Quarterly* 66 (2).

Buchanan, J. 2016. "Militias in Myanmar." *The Asia Foundation*. https://asiafoundation.org/wp-content/uploads/2016/07/Militias-in-Myanmar.pdf.

Burawoy, M. 1998. "The Extended Case Method." *Sociological Theory* 16 (1): 4–33.

Buscemi, F. 2019. "Armed Political Orders through the Prism of Arms: Relations between Weapons and Insurgencies in Myanmar and Ukraine." *Interdisciplinary Political Studies* 5 (1): 189–231.

Buscemi, F. 2021a. "The Art of Arms (Not) Being Governed: Means of Violence and Shifting Territories in the Borderworlds of Myanmar." *Geopolitics* 28 (1): 282–309.

Buscemi, F. 2021b. "Ecologies of Dead and Alive Landmines in the Borderlands of Myanmar." *Italian Political Science Review* (July):217–235.

Buscemi, F. 2022. "'Blunt' Biopolitical Rebel Rule: On Weapons and Political Geography at the Edge of the State." *Small Wars & Insurgencies* 34 (1): 81–112.

Buscemi, F. 2024. "Powers of the Gun: Spirituality and Weapon-Human Encounters in Frontiers." *Geopolitics* 0 (0): 1–28.

Buscemi, F., and M. Proto. 2023. "Telluric Geographies of the Means of Violence: On Alterity, Weapons, and Space at the Margins." *Political Geography* 109: 103046.

Callahan, M. P. 2002. "State Formation in the Shadow of the Raj: Violence, Warfare and Politics in Colonial Burma." *Japanese Journal of Southeast Asian Studies* 39 (4): 513–536.

Callahan, M. P. 2003. *Making Enemies: War and State Building in Burma*. Ithaca, NY: Cornell University Press.

Campbell, S., and E. Prasse-Freeman. 2021. "Revisiting the Wages of Burman-ness: Contradictions of Privilege in Myanmar." *Journal of Contemporary Asia* 52 (2): 175–199.

Capie, D. 2013. "Arms Trafficking in Mainland Southeast Asia." In *An Atlas of Trafficking in Southeast Asia: The Illegal Trade in Arms, Drugs, People, Counterfeit Goods and Natural Resources in Mainland Southeast Asia*, edited by P. Chouvy, 89–112. London: I.B Tauris.

Carter, L. 2019. "Gunning for Firearms." *Bangkok Post*, June 9, 2019. https://www.bangkok post.com/thailand/special-reports/1691936/gunning-for-firearms)

de Castro, E. V. 2012. *Cosmological Perspectivism in Amazonia and Elsewhere: Four Lectures Given in the Department of Social Anthropology, Cambridge University, February–March 1998*. Hau Books, Masterclass Series, Vol. 1. https://monoskop. org/images/e/e5/Viveiros_de_Castro_Eduardo_Cosmological_Perspectivism_in_ Amazonia_and_Elsewhere.pdf.

Cathcart, G. S. 2016. "Landmines as a Form of Community Protection in Eastern Myanmar." In *Conflict in Myanmar*, edited by N. Cheesman and N. Farrelly, 121–136. Singapore: ISEAS.

Chandler, D., and J. Pugh. 2022. "Interstitial and Abyssal Geographies." *Political Geography* 98:1–9.

Chatterjee, P. 2004. *Politics of the Governed*. New York: Columbia University Press.

Cheesman, N. 2015. *Opposing the Rule of Law*. Cambridge: Cambridge University Press.

Cheesman, N. 2017. "How in Myanmar 'National Races' Came to Surpass Citizenship and Exclude Rohingya." *Journal of Contemporary Asia* 47 (3): 461–483.

Chit Hlaing. 2008. "Anthropological Communities of Interpretation for Burma: An Overview." *Journal of Southeast Asian Studies* 39 (2): 239–54.

Christina, F. 2009. *Living Silence*. London: Zed Books.

Chu May Paing. 2020. "In Need of Daughters of Good Lineage: Placing Gender in Myanmar's Buddhist Nationalist Discourse." *Journal of Southeast Asian Linguistics Society* 32 (4): 40–90.

Chu May Paing and Than Toe Aung. 2021. "Talking Back to White 'Burma Experts.'" *AGITATE Now!* https://agitatejournal.org/ talking-back-to-white-burma-experts-by-chu-may-paing-and-than-toe-aung/.

Cons, J., and M. Eilenberg. 2019. *Frontier Assemblages: The Emergent Politics of Resource Frontiers in Asia*. Oxford: Wiley.

Cons, J., and R. Sanyal. 2013. "Geographies at the Margins: Borders in South Asia—An Introduction." *Political Geography* 35:5–13.

Cresswell, T. 2004. *Place: A Short Introduction*. Malden, MA: Wiley-Blackwell.

Cunningham, K. G., and C. E. Loyle. 2021. "Introduction to the Special Feature on Dynamic Processes of Rebel Governance." *Journal of Conflict Resolution* 65 (1): 3–14.

Dan Seng Lawn. 2022. "Conflict and Development in the Myanmar-China Border Region." *XCEPT*. https://www.xcept-research.org/wp-content/uploads/2022/05/ Report_Conflict-and-Development-in-the-Myanmar-China-Border-Region.pdf.

Dan, S. L., J. H. P. Maran, M. Sadan, P. Meehan, and J. Goodhand. 2021. "The Pat Jasan Drug Eradication Social Movement in Northern Myanmar, Part One: Origins & Reactions." *International Journal of Drug Policy* 89, 1031811

Das, V., and D. Poole. 2004. "Anthropology in the Margins of the State." *PoLAR: Political and Legal Anthropology Review* 30 (1): 140–144.

Davis, A. 2019. "Inter-Ethnic Tensions Drive Intensifying Conflicts in Myanmar's Shan State." *Jane's Terrorism & Insurgency Monitor*, January 17.

Dean, M. 2002. "Powers of Life and Death beyond Governmentality." *Cultural Values* 6 (1–2): 119–138.

Dean, M. 2010. *Governmentality: Power and Rule in Modern Society.* Los Angeles: Sage.

Debos, M. 2016. *Living by the Gun in Chad: Combatants, Impunity and State Formation.* London: Zed Books.

Deleuze, G., and F. Guattari. 1987. *A Thousand Plateaus: Capitalism and Schizophrenia.* Minneapolis: University of Minnesota Press.

Dunford, M. 2019. Indigeneity, Ethnopolitics, and Taingyinthar: Myanmar and the Global Indigenous Peoples' Movement. *Journal of Southeast Asian Studies* 50 (1): 51–67.

Dunford, M. 2024. *The Flavour of Empire: Tea Production and the Ta'ang World in Highland Southeast Asia.* PhD diss. Canberra: The Australian National University.

Duquet, N. 2009. "Arms Acquisition Patterns and the Dynamics of Armed Conflict: Lessons from the Niger Delta." *International Studies Perspectives* 10:169–185.

Egreteau, R. 2006. *Instability at the Gate: India's Troubled Northeast and Its External Connections.* New Delhi: Centre de Sciences Humaines.

Eilenberg, M. 2011. "Straddling the Border: A Marginal History of Guerrilla Warfare and 'Counter-Insurgency' in the Indonesian Borderlands." *Modern Asian Studies* 45 (6): 1423–1463.

Eilenberg, M. 2014. "Frontier Constellations." *Journal of Peasant Studies* 41 (2): 157–182.

Enomoto, T. 2018. "Giving Up the Gun: Overcoming Myths about Japanese Sword-Hunting and Firearms Control." *History of Global Arms Transfers* 6:44–59.

Esposito, R. 2008. *Bios: Biopolitics and Philosophy.* Minneapolis: University of Minnesota Press.

Esposito, R. 2010. *Communitas. The Origin and Destiny of Community.* Stanford, CA: Stanford University Press.

Esposito, R. 2011a. *Immunitas: The Protection and Negation of Life.* Polity.

Esposito, R. 2011b. *Dieci Pensieri Sulla Politica.* Bologna, Italy: Il Mulino.

Faxon, O. H. 2021. "After the Rice Frontier: Producing State and Ethnic Territory in Northwest Myanmar." *Geopolitics* 28 (1): 1–25.

Ferguson, J. M. 2015. "Who's Counting?: Ethnicity, Belonging, and the National Census in Burma/Myanmar." *Bijdragen tot de taal-, Land-en Volkenkunde/Journal of the Humanities and Social Sciences of Southeast Asia* 171 (1): 1–28.

Ferguson, J. M. 2018. "Buddhist Bomb Diversion and an American Airman Reincarnate: World War Folklore, Airmindedness, and Spiritual Air Defense in Shan State, Myanmar." *Cultural Geographies* 25 (3): 473–489.

Ferguson, J. M. 2021. *Repossessing Shanland: Myanmar, Thailand, and a Nation-State Deferred.* Madison: University of Wisconsin Press.

Foucault, M. 1977. *Discipline and Punish: The Birth of the Modern Prison.* New York: Vintage Books.

Foucault, M. 2003. *Society Must Be Defended: Lectures at the Collège de France, 1975–1976.* New York: Picador.

Foucault, M. 2007. *Security, Territory, Population: Lectures at the Collège de France, 1977–1978.* New York: Picador.

Foucault, M. 2008. *The Birth of Biopolitics*. New York: Picador.

Frontier Myanmar. 2019. "At Least 14 Dead in Unprecedented Northern Alliance Attacks." *Frontier Myanmar*. August 15. https://www.frontiermyanmar.net/en/at-least-14-dead-in-unprecedented-northern-alliance-attacks/.

Frontier Myanmar. 2021. "Rising Dragon: TNLA Declares 'Victory' in Northern Shan." *Frontier Myanmar*. February 4. https://www.frontiermyanmar.net/en/rising-dragon-tnla-declares-victory-in-northern-shan/.

Gelbort, J. 2018. "Implementation of Burma's Vacant, Fallow and Virgin Land Management Law: At Odds with the Nationwide Ceasefire Agreement and Peace Negotiations." *Transnational Institute*. December 10, 2018. https://www.tni.org/en/article/implementation-of-burmas-vacant-fallow-and-virgin-land-management-law.

Geneva Call. 2007. "Three Ethnic Armed Groups from Burma/Myanmar Commit to a Ban on Anti-personnel Mines." April 16, 2007. https://www.genevacall.org/three-ethnic-armed-groups-burmamyanmar-commit-ban-anti-personnel-mines/.

Geneva Call. 2019. "Deed of Commitment under Geneva Call for Adherence to a Total Ban on Anti-Personnel Mines and for Cooperation in Mine Action." https://www.genevacall.org/wp-content/uploads/2019/07/DoC-Banning-anti-personnel-mines.pdf.

Gibson, R. M., and W. Chen 2011. *The Secret Army Chiang Kai-Shek and the Drug Warlords of the Golden Triangle*. Singapore: Wiley.

Global Witness. 2021. "Jade and Conflict: Myanmar's Vicious Circle." *Global Witness*. https://www.globalwitness.org/en/campaigns/natural-resource-governance/jade-and-conflict-myanmars-vicious-circle/.

Glouftsios, G. 2020. "Governing Border Security Infrastructures: Maintaining Large-Scale Information Systems." *Security Dialogue* 52 (5): 452–470.

Goodhand, J. 2013. "Epilogue: The View from the Border." In *Violence on the Margins: States, Conflict, and Borderlands*, edited by B. Korf and T. Raeymaekers, 247–261. New York: Palgrave Macmillan.

Greene, O., and N. Marsh, eds. 2012. *Small Arms, Crime and Conflict: Global Governance and the Threat of Armed Violence*. New York: Routledge.

Gregory, D. 2008. "The Biopolitics of Baghdad: Counterinsurgency and the Counter-city." *Human Geography* 1 (1): 1–21.

Grove, J. 2016. "An Insurgency of Things: Foray into the World of Improvised Explosive Devices." *International Political Sociology* 10 (4): 332–351.

Hagmann, T., and B. Korf. 2012. "Agamben in the Ogaden: Violence and Sovereignty in the Ethiopian-Somali Frontier." *Political Geography* 31 (4): 205–214.

Hajdinjak, M. 2002. *Smuggling in Southeast Europe: The Yugoslav Wars and the Development of Regional Criminal Networks in the Balkans*. Sofia: Center for the Study of Democracy.

Han, E. 2019. *Asymmetrical Neighbors: Borderland State Building between China and Southeast Asia*. New York: Oxford University Press.

Harrison, A., and H. Kyed. 2019. "Ceasefire State-Making and Justice Provision by Ethnic Armed Groups in Southeast Myanmar." *Sojourn: Journal of Social Issues in Southeast Asia* 34 (2): 290–326.

Harrison, A. 2020. "Fish Caught in Clear Water: Encompassed State-Making in South-East Myanmar." *Territory, Politics, Governance* 9 (4): 533–552.

Hazama, I. 2018. "Ugandan Pastoralists' Everyday Histories of Gun Acquisition and State Violence." *History of Global Arms Transfers* 6:23–37.

Hazen, J. 2013. *What Rebels Want: Resources and Supply Networks in Wartime*. Ithaca, NY: Cornell University Press.

Hedström, J. 2020. "Militarized Social Reproduction: Women's Labor and Parastate Armed Conflict." *Critical Military Studies* 8 (1): 58–76.

Hedström, J. 2025. *Reproducing Revolution. Women's Labor and the War in Kachinland.* Ithaca: Cornell University Press.

Hendrickson, D., and K. Jolliffe. 2018. *Security Integration in Conflict-Affected Societies: Considerations for Myanmar.* Discussion paper. Saferworld.

Hicks, D. 2020. *The Brutish Museums: The Benin Bronzes, Colonial Violence, and Cultural Restitution.* London: Pluto Press.

Hoffmann, K., and J. Verweijen. 2018. "Rebel Rule: A Governmentality Perspective." *African Affairs* 118 (471): 352–374.

Hong, E. 2017. "Scaling Struggles over Land and Law: Autonomy, Investment, and Interlegality in Myanmar's Borderlands." *Geoforum* 82:225–236.

Howard, M. C., and W. Wattanapun. 2001. *The Palaung in Northern Thailand.* Chiang Mai, Thailand: Silkworm Books.

Horstmann, A. 2018. "Humanitarian Assistance and Protestant Proselytizing in the Borderlands of Myanmar: The Free Burma Rangers." In *Routledge Handbook of Asian Borderlands*, edited by A. Horstmann, M. Saxer, and A. Rippa, 349–360. Abingdon: Routledge.

Horstmann, A. 2019. "An American Hero: Faith-Based Emergency Health Care in Karen State, Myanmar and Beyond." *Religions*, 1–11.

Hsu Mu-Lung. 2019. "Making Merit, Making Civil Society: Free Funeral Service Societies and Merit-Making in Contemporary Myanmar." *Journal of Burma Studies* 23 (1): 1–36.

Htet Min Lwin and Thiha Wint Aung. 2024. "Beyond Operation 1027: A New Mandala Order Needed in Myanmar." *Irrawaddy*, January 4. https://www.irrawaddy.com/ opinion/guest-column/beyond-operation-1027-a-new-mandala-order-needed-in-myanmar.html.

Htun Htun. 2020. "Drug Squad Arrests Three, Seizes Assault Rifles, Narcotics in Myanmar's Shan State." *Irrawaddy*, August 7. https://www.irrawaddy.com/news/ burma/drug-squad-arrests-three-seizes-assault-rifles-narcotics-myanmars-shan-state.html.

Huang, R. L. 2016. "Myanmar's Way to Democracy and the Limits of the 2015 Elections." *Asian Journal of Political Science* 25 (1): 25–44.

IRN (Independent Research Network). 2022. "The Internal Struggle: The Fight for an Inclusive National Identity in Myanmar." Independent Research Network. https:// progressivevoicemyanmar.org/wp-content/uploads/2022/08/English_The-Fight-for-an-Inclusive-National-Identity-IRN.pdf.

Irrawaddy. 2019. "Myanmar's Military Seizes TNLA Arms Cache, Mostly Chinese-Made." *Irrawaddy*, November 25. https://www.irrawaddy.com/news/burma/myanmars-military-seizes-tnla-arms-cache-mostly-chinese-made.html.

Irrawaddy. 2020a. "Myanmar Military Finds Large Cache of Weapons in Cave in Northern Shan State." *Irrawaddy*, January 17. https://www.irrawaddy.com/news/burma/ myanmar-military-finds-large-cache-weapons-cave-northern-shan-state.html.

Irrawaddy. 2020b. "Weapons Seized in Mae Sot Destined for Myanmar's Rakhine State: Intelligence Sources." *Irrawaddy*, July 15, https://www.irrawaddy.com/opinion/ analysis/weapons-seized-mae-sot-destined-myanmars-rakhine-state-intelligence-sources.html.

Jentzsch, C., S. N. Kalyvas, and L. I. Schubiger. 2015. "Militias in Civil Wars." *Journal of Conflict Resolution* 59 (5): 755–769.

Jenzen-Jones, N. R., and M. Schroeder. 2018. *An Introductory Guide to the Identification of Small Arms, Light Weapons, and Associated Ammunition*. Geneva: Small Arms Survey.

Jessop, B., N. Brenner, and M. Jones. 2008. "Theorizing Sociospatial Relations." *Environment and Planning D: Society and Space* 26:389–401.

Jolliffe, K. 2015. *Ethnic Armed Conflict and Territorial Administration in Myanmar*. The Asia Foundation. https://asiafoundation.org/resources/pdfs/ConflictTerritorial AdministrationfullreportENG.pdf.

Jolliffe, K., J. Bainbridge, and I. Campbell. 2017. *Security Integration in Myanmar: Past Experiences and Future Visions*. Discussion paper. Saferworld.

Jones, L. 2014. "Explaining Myanmar's Regime Transition: The Periphery Is Central." *Democratization* 21 (5): 780–802.

Jones, A., and J. Clark. 2019. "Political Geography and Political Science: Common Territory?" *Geopolitics* 25 (2): 472–478.Joseph, J. 2010. "The Limits of Governmentality: Social Theory and the International." *European Journal of International Relations* 16 (2): 223–246.

Kaldor, M. 2012. *New and Old Wars: Organized Violence in a Global Era*. Cambridge: Polity Press.

Kasfir, N. 2015. "Rebel Governance—Constructing a Field of Inquiry: Definitions, Scope, Patterns, Order, Causes." In *Rebel Governance*, edited by A. Arjona, N. Kasfir, and Z. Mampilly, 21–46. New York: Cambridge University Press.

Kasfir, N., G. Frerks, and N. Terpstra. 2017. "Introduction: Armed Groups and Multi-layered Governance." *Civil Wars* 19 (3): 257–278.

Kean, T. 2018. "Hydropower Is Back." *Frontier Myanmar*, August 9. https://frontier myanmar.net/en/hydropower-is-back.

Keenan, P. 2013. *By Force of Arms: Ethnic Armed Groups in Burma*. Delhi: VIJ Books.

Keenan, P. 2014a. "The Border Guard Force: The Need to Reassess the Policy." In *Ending Ethnic Armed Conflict in Burma: A Complicated Peace Process; A Collection of BCES Analysis and Briefing Papers*, edited by L. Sakhong and P. Keenan, 186–191. Chiang Mai, Thailand: Wanida Press.

Keenan, P. 2014b. "The Kokang Clashes—What Next?" In *Ending Ethnic Armed Conflict in Burma: A Complicated Peace Process; A Collection of BCES Analysis and Briefing Papers*, edited by L. Sakhong and P. Keenan, 143–153. Chiang Mai, Thailand: Wanida Press.

Khun Moe Htun. 2018. *Living with Opium: Livelihood Strategies among Rural Highlanders in Southern Shan State, Myanmar*. Chiang Mai, Thailand: Chiang Mai University Press.

Klinke, I., and M. Bassin. 2018. "Introduction: Lebensraum and Its Discontents." *Journal of Historical Geography* 61: 53–58.

Ko Ko, N., and J. Braithwaite. 2020. "Baptist Policing in Burma: Swarming, Vigilantism or Community Self-help?" *Policing and Society* 30 (6): 688–703.

Korf, B., T. Hagmann, and M. Doevenspeck. 2013. "Geographies of Violence and Sovereignty: The African Frontier Revisited." In *Violence on the Margins: States, Conflict, Borderlands*, edited by B. Korf and T. Raeymaekers, 29–54. Basingstoke, UK: Palgrave.

Korf, B., T. Hagmann, and M. Engeler. 2010. "The Geography of Warscape." *Third World Quarterly* 31 (3): 385–399.

Korf, B., and T. Raeymaekers. 2013. *Violence on the Margins: States, Conflict, and Borderlands*. Basingstoke, UK: Palgrave.

Korf, B., T. Raeymaekers, C. Schetter, and J. M. Watts. 2018. "Geographies of Limited Statehood." In *The Oxford Handbook of Governance and Limited Statehood* edited by A. Draude, T. A. Börzel, and T. Risse, 167–189. Oxford: Oxford University Press.

Kozłowska, M., and M. Lubina. 2021. "The Burmese Road to Israeli-Style Cooperative Settlements: The Namsang Project, 1956–63." *Journal of Southeast Asian Studies* 52(4): 701–725.

Kramer, T. 2007. *The United Wa State Party: Narco-Army or Ethnic Nationalist Party?* Singapore: ISEAS Publishing and East-West Center.

Kramer, T. 2015. "The Current State of Counternarcotics Policy and Drug Reform Debates in Myanmar." *Journal of Drug Policy Analysis* 10 (1): 1–14.

Kramer, T., and K. Woods. 2012. *Financing Dispossession: China's Opium Substitution Programme in Northern Burma.* Amsterdam: Transnational Institute. https://www.tni.org/files/download/tni-financingdispossesion-web.pdf.

Krause, K. 2009. "War, Violence and the State." In *Securing Peace in a Globalized World*, edited by M. Brzoska and A. Krohn, 183–202. London: Palgrave Macmillan.

Kyed, H. M. and Gravers, M. 2015. "Integration and Power-Sharing: What Are the Future Options for Armed Non-State Actors in the Myanmar Peace Process?" *Stability: International Journal of Security & Development* 4 (1): 1–20.

Kyaw Lin Htoon. 2018. "Fluctuating Fortunes at the Bawdwin Mine." *Frontier Myanmar*, January 18. https://www.frontiermyanmar.net/en/fluctuating-fortunes-at-the-bawdwin-mine/.

Kyaw Zwa Moe. 2002. "Ethnic Leader Dies in Detention." *Irrawaddy*, August 28. https://www2.irrawaddy.com/opinion_story.php?art_id=2066.

Landmine & Cluster Munition Monitor. 2004. *2004 Myanmar/Burma Country Report.*

Landmine & Cluster Munition Monitor. 2019. *2019 Myanmar/Burma Country Report.*

Latour, B. 2005. *Reassembling the Social: An Introduction to Actor-Network Theory.* New York: Oxford University Press.

Law, J. 2002. *Aircraft Stories: Decentring the Object in Technoscience.* Durham: Duke University Press.

Law, J. 2007. "Actor Network Theory and Material Semiotics. Accessed April 25, 2007. http://www.heterogeneities.net/publications/Law2007ANTandMaterialSemiotics.pdf.

Law, J., and A. Mol. 2001. "Situating Technoscience: An Inquiry into Spatialities." *Environment and Planning D: Society and Space* 19 (5): 609–621.

Leach, E. 1959. *Political Systems of Highland Burma.* New York: Berg.

Leach, E. 1960. "The Frontiers of 'Burma.'" *Comparative Studies in Society and History* 3 (1): 49–68.

Lefebvre, H. 1991. *The Production of Space.* Oxford, UK: Blackwell.

Legg, S., and D. Heath. 2018. *South Asian Governmentalities: Michel Foucault and the Question of Postcolonial Orderings.* Cambridge: Cambridge University Press.

Lehman, F. K. 1967. "Ethnic Categories in Burma and the Theory of Social Systems." In *Southeast Asian Tribes, Minorities, and Nations*, edited by P. Kunstadter, 93–124Princeton, NJ: Princeton University Press.

Lemke, T. 2011. *Biopolitics: An Advanced Introduction.* New York: New York University Press.

Lintner, B. 1990. *The Rise and Fall of the Communist Party of Burma (CPB).* Ithaca, NY: Cornell University Press.

Lintner, B. 1999. *Burma in Revolt: Opium and Insurgency since 1948.* Chiang Mai, Thailand: Silkworm Books.

Lintner, B. 2015. *Great Game East: India, China, and the Struggle for Asia's Most Volatile Frontier.* New Haven, CT: Yale University Press.

Lottholz, P., and N. Lemay-Hébert. 2016. "Re-reading Weber, Reconceptualizing State-Building: From Neo-Weberian to Post-Weberian Approaches to State, Legitimacy and State-Building." *Cambridge Review of International Affairs* 29 (4): 1467–1485.

Lund, C. 2006. "Twilight Institutions: Public Authority and Local Politics in Africa." *Development and Change* 37:685–705.

Lund, C. 2011. "Fragmented Sovereignty: Land Reform and Dispossession in Laos." *Journal of Peasant Studies* 38 (4): 885–905.

Lund, C. 2016. "Rule and Rupture: State Formation through the Production of Property and Citizenship." *Development and Change* 47 (6): 1199–1228.

MacLean, K. 2008. "Sovereignty after the Entrepreneurial Turn: Mosaics of Control, Commodified Spaces, and Regulated Violence in Contemporary Burma." In *Taking Southeast Asia to Market: Commodities, Nature, and People in a Neoliberal Age,* edited by N. Peluso and J. Nevins, 140–158. Ithaca, NY: Cornell University Press.

MacLean, K. 2016. "Humanitarian Mine Action in Myanmar and the Reterritorialization of Risk." *Focaal—Journal of Global and Historical Anthropology* 74:83–96.

MacLean, K. 2022. *Crimes in Archival Form: Human Rights, Fact Production, and Myanmar.* Berkeley: University of California Press.

Malthaner, S., and S. Malešević. 2022. "Violence, Legitimacy, and Control: The Dynamics of Rebel Rule." *Partecipazione e Conflitto* 15 (1): 1–16.

Mampilly, Z. 2011. *Rebel Rulers: Insurgent Governance and Civilian Life during War.* Ithaca, NY: Cornell University Press.

Mark, S. 2016. "Are the Odds of Justice 'Stacked' against Them? Challenges and Opportunities for Securing Land Claims by Smallholder Farmers in Myanmar." *Critical Asian Studies* 48 (3): 443–460.

Marsh, N. 2007. "Conflict Specific Capital: The Role of Weapons Acquisition in Civil War." *International Studies Perspectives* 8 (1): 54–72.

Marsh, N. 2012. "The Tools of Insurgency: A Review of the Role of Small Arms and Light Weapons in Warfare." In *Small Arms, Crime and Conflict: Global Governance and the Threat of Armed Violence,* edited by O. Greene and N. Marsh, 13–28.

Marsh, N. 2018. "The Availability Puzzle: Considering the Relationship between Arms and Violence Taking Place within States." *History of Global Arms Transfers* 6:3–21.

Marsh, N. 2020. *Because We Have the Maxim Gun: The Relationship between Arms Acquisition by Non-State Groups and Violence.* PhD diss. Oslo: University of Oslo.

Massey, D. 1992. "Politics and Space/Time." *New Left Review* 196:65–84.

Maung Maung. 1980. *From Sangha to Laity: Nationalist Movements of Burma 1920–1940.* Delhi: Manohar.

Maxim, H. 1915. *My Life.* London: Methuen & Co.

Mbembe, A. 2019. *Necropolitics.* Durham: Duke University Press.

McCabe, S. 2019. "Explosive Evidence." *Southeast Asia Globe.* December 9, 2019. https://southeastasiaglobe.com/why-are-myanmars-landmine-casualties-growing/.

McCarthy, G., and N. Farrelly. 2020. "Peri-conflict Peace: Brokerage, Development and Illiberal Ceasefires in Myanmar's Borderlands. *Conflict, Security & Development* 20 (1): 141–163.

McCarthy, G. 2023. *Outsourcing the Polity: Non-State Welfare, Inequality, and Resistance in Myanmar.* Ithaca, NY: Cornell University Press.

McCracken, D. 2001. "Thailand: The Land of Smiles (Until You Take Your First Step)." *Journal of Mine Action* 5 (1): 60–63.

McEvoy, C., and G. Hideg. 2017. *Global Violent Deaths*. Small Arms Survey. Geneva: Small Arms Survey.

Mac Ginty, R., and O. P. Richmond. 2016. "The Fallacy of Constructing Hybrid Political Orders: A Reappraisal of the Hybrid Turn in Peacebuilding." *International Peacekeeping* 23 (2): 219–239.

Meehan, P. 2011. "Drugs, Insurgency and State-Building in Burma: Why the Drugs Trade Is Central to Burma's Changing Political Order." *Journal of Southeast Asian Studies* 42 (3): 376–404.

Meehan, P. 2016a. "The Continuation of War by Other Means: An Anatomy of the Palaung Ceasefire in Northern Shan State." In *War and Peace in the Borderlands of Myanmar: The Kachin Ceasefire, 1994–2011*, edited by M. Sadan, 361–387. Copenhagen: NIAS Press.

Meehan, P. 2016b. *The Political Economy of the Opium/Heroin Trade in Shan State, Myanmar, 1988–2012*. PhD diss. London: SOAS University of London.

Meehan, P., and Dan Seng Lawn 2022. "Brokered Rule: Militias, Drugs, and Borderland Governance in the Myanmar-China Borderlands." *Journal of Contemporary Asia* 53 (4): 561–583.

Min Zaw Oo and Win Min. 2007. *Assessing Burma's Ceasefire Accords*. Washington, DC: East-West Center and ISEAS Publishing.

Mitchell, A. 2013. "Escaping the 'Field Trap': Exploitation and the Global Politics of Educational Fieldwork in 'Conflict Zones.'" *Third World Quarterly* 34 (7): 1247–1264.

Moser-Puangsuwan, Y. 2000. "Seeds of Destruction." *Burma Debate* 7 (4): 4–10.

Most, B. A., and H. Starr. 1989. *Inquiry, Logic and International Politics*. Columbia: University of South Carolina Press.

Myanmar Witness. 2023. "Eyes on the Skies: The Dangerous and Sustained Impact of Airstrikes on Daily Life in Myanmar." *Myanmar Witness*, January 31, 2023. https://www.myanmarwitness.org/reports/eyes-on-the-skies.

Myat Thura. 2020. "Ministry to Replace Colonial Era Arms Laws." *Myanmar Times*, February 6. https://www.mmtimes.com/news/ministry-replace-colonial-era-arms-laws.html.

Myoe, M. 2009. *Building the Tatmadaw: Myanmar Armed Forces Since 1948*. Singapore: ISEAS—Yusof Ishak Institute.

Nehginpao Kipgen. 2016. *Myanmar: A Political History*. Oxford: Oxford University Press.

O'Morchoe, F. 2023. "Teak & Lead: Making Borders, Resources, and Territory in Colonial Burma." *Geopolitics* 28 (1): 23–46.

Ong, A. 2018. "Producing Intransigence: (Mis) understanding the United Wa State Army in Myanmar." *Contemporary Southeast Asia* 40 (3): 449–474.

Ong, A. 2023. *Stalemate: Autonomy and Insurgency on the China-Myanmar Border*. Ithaca, NY: Cornell University Press.

Ong, A., and S. J. Collier, eds. 2005. *Global Assemblages: Technology, Politics, and Ethics as Anthropological Problems*. Oxford, UK: Blackwell.

Osada, Noriyuki. 2011. "An Embryonic Border: Racial Discourses and Compulsory Vaccination for Indian Immigrants at Ports in Colonial Burma, 1870–1937." *Moussons: Recherche en sciences humaines sur l'Asie du Sud-Est* 17:145–164.

Ozguc, U., and A. Burridge. 2023. "More-Than-Human Borders: A New Research Agenda for Posthuman Conversations in Border Studies." *Geopolitics* 28 (2): 1–19.

Ozguc, U., and A. Little. 2023. "The Politics of Temporality and the Ethos of Open Societies: Transfrontier Conservation Areas as Spatio-temporal Chokepoints." *Geopolitics* 28 (2): 570–592.

Painter, J. 2010. "Rethinking Territory." *Antipode* 42 (5): 1090–1118.

Peluso, N. L. 2008. "A Political Ecology of Violence and Territory in West Kalimantan." *Asia Pacific Viewpoint* 49 (1): 48–67.

de la Perriere, B. 2009. "An Overview of the Field of Religion in Burmese Studies." *Asian Ethnology* 68 (2): 185–210.

Phongpaichit, P., S. Piriyarangsan, and N. Treerat. 1998. *Guns, Girls, Gambling, Ganja: Thailand's Illegal Economy and Public Policy*. Chiang Mai, Thailand: Silkworm Books.

Picard, M., P. Holtom, and F. Mangan. 2019. *Trade Update 2019: Transfers, Transparency, and South-east Asia Spotlight*. Small Arms Survey Report. Geneva: Small Arms Survey.

Proto, M. 2023. "Applying Friedrich Ratzel's Political and Biogeography to the Debate on Natural Borders in the Italian Context (1880–1920)." *Geographica Helvetica* 78:41–52.

Prasse-Freeman, E. 2020. "Resistance/Refusal: Politics of Maneuver under Diffuse Regimes of Governmentality. *Anthropological Theory* 22 (1): 102–127.

Prasse-Freeman, E. 2021. "Necroeconomics: Dispossession, Extraction, and Indispensable/Expendable Laborers in Contemporary Myanmar." *Journal of Peasant Studies*, in press.

Prasse-Freeman, E. 2023a. *Rights Refused: Grassroots Activism and State Violence in Myanmar*. Redwood City, CA: Stanford University Press.

Prasse-Freeman, E. 2023b. "Reassessing Reification: Ethnicity amidst 'Failed' Governmentality in Burma and India." *Comparative Studies in Society and History* 65 (33): 670–701.

Prasse-Freeman, E. 2023c. "Refusing Rohingya: Reformulating Ethnicity amidst Blunt Biopolitics." *Current Anthropology* 64 (4): 432–453.

Prasse-Freeman, E., and A. Ong. 2021. "Expulsion/Incorporation: Valences of Mass Violence in Myanmar." In *Political Violence in Southeast Asia Since 1945*, edited by M. E. Zucker and B. Kiernan, 41–55. London: Routledge.

"PSLA Led by U Aik Mone Unconditionally Exchanges Arms for Peace." 2005. *The New Light of Myanmar* 13, no. 14 (April 30). https://www.burmalibrary.org/sites/burmalibrary.org/files/obl/docs3/NLM2005-04-30.pdf.

PWO (Palaung Women Organization). 2006. *Poisoned Flowers: The Impacts of Spiraling Drug Addiction on Palaung Women in Burma*. Mae Sot, Thailand: PWO.

PWO (Palaung Women Organization). 2010. *Poisoned Hills: Opium Cultivation Surges under Government Control in Burma*. Mae Sot, Thailand: PWO.

PWO (Palaung Women Organization). 2011. *Still Poisoned: Opium Cultivation Soars in Palaung Areas under Burma's New Regime*. Mae Sot, Thailand: PWO.

PYNG (Palaung Youth Network Group). 2007. *Under the Boot: The Burma Army Clears the Way for Chinese Dams on the Shweli River*. Mae Sot, Thailand: PYNG.

Rasmussen, M. B., and C. Lund. 2018. "Reconfiguring Frontier Spaces: The Territorialization of Resource Control." *World Development* 101:388–399.

Roberts, R. 2008. "Cambodia: Surplus Destruction after War and Genocide." *Contemporary Security Policy* 29 (1): 103–128.

Robinne, F., and M. Sadan. 2007. "Postscript: Reconsidering the Dynamics of Ethnicity through Foucault's Concept of 'Spaces of Dispersion.'" In *Social Dynamics in the Highlands of Southeast Asia*, edited by F. Robinne and M. Sadan, 299–308. Herndon, VA: Brill.

Rose, N. 1999. *Powers of Freedom: Reframing Political Thought*. Cambridge: Cambridge University Press.

Rose, N., and P. Miller. 1992. "Political Power beyond the State: Problematics of Government." *British Journal of Sociology* 43 (2): 173–205.

M. Quadrini. 2019. "'They Want to Grow Their Armies': Shan Armed Groups Obstruct Family Planning Efforts." Frontier, May 22. https://www.frontiermyanmar.net/en/they-want-to-grow-their-armies-shan-armed-groups-obstruct-family-planning-efforts/.

Sadan, M. 2013a. *Being and Becoming Kachin: Histories beyond the State in the Borderworlds of Burma.* Oxford: British Academy and Oxford University Press.

Sadan, M. 2013b. "Ethnic Armies and Ethnic Conflict in Burma: Reconsidering the History of Colonial Militarization in the Kachin Region of Burma during the Second World War." *South East Asia Research* 21 (4): 601–626.

Sai Aung Tun. 2009. *The History of Shan State: From Its Origins to 1962.* Chiang Mai, Thailand: Silkworm Books.

Sai Latt. 2016. *Depoliticization, Securitization and Violent Accumulation in the Integration of the Greater Mekong Sub-region.* PhD diss. Burnaby, BC: Simon Fraser University.

Sakhong, L. 2014. "The Dynamics of Sixty Years of Ethnic Armed Conflicts in Burma." In *Ending Ethnic Armed Conflict in Burma: A Complicated Peace Process; A Collection of BCES Analysis and Briefing Papers,* edited by L. Sakhong and P. Keenan, 1–29. Chiang Mai, Thailand: Wanida Press.

Sakhong, L. 2014. "The Kachin's Dilemma: Become a Border Guard Force or Return to Warfare." In *Ending Ethnic Armed Conflict in Burma: A Complicated Peace Process; A Collection of BCES Analysis and Briefing Papers,* edited by L. Sakhong and P. Keenan, 127–142. Chiang Mai, Thailand: Wanida Press.

Sarma, J. 2020. "Seeing like a Border—Resource Frontiers, Voices and Visions on Myanmar's Borderlands with India and China." PhD diss. Singapore: National University of Singapore.

Sarma, J., H. O. Faxon, and K. B. Roberts. 2022. "Remaking and Living with Resource Frontiers: Insights from Myanmar and Beyond." *Geopolitics* 28 (1): 1–22.

Schlichte, K. 2009. *In the Shadow of Violence: The Politics of Armed Groups.* Frankfurt: Campus Verlag.

Scott, J. C. 2009. *The Art of Not Being Governed: An Anarchist History of Upland Southeast Asia.* New Haven, CT: Yale University Press.

Selth, A. 2000a. "Landmines in Burma: The Military Dimension." *Burma Debate* 7 (4): 10–20.

Selth, A. 2000b. "Burma's Order of Battle: An Interim Assessment." Strategic & Defense Studies Center (ANU). Working paper no. 351.

Selth, A. 2001. "Landmines in Burma: Forgotten Weapons in a Forgotten War." *Small Wars and Insurgency* 12 (2): 19–50.

Selth, A. 2019. Myanmar's Intelligence Apparatus and the Fall of General Khin Nyunt. *Intelligence and National Security* 34 (5): 619–636.

Shah, N. 2017a. "Death in the Details: Finding Dead Bodies at the Canadian War Museum." *Organization* 24 (4): 549–569.

Shah, N. 2017b. "Gunning for War: Infantry Rifles and the Calibration of Lethal Force." *Critical Studies on Security* 5 (1): 81–104.

Shan Human Rights Foundation. 2016. "Save the Namtu River: Impacts of the Upper Yeywa and Other Planned Dams on the Namtu in Shan State." Shan State Farmers Network, Shan Sapawa Environmental Organization.

da Silva, F. D. 2007. *Toward a Global Idea of Race.* Minneapolis: Minnesota University Press.

Sislin, J., and F. S. Pearson. 2001. *Arms and Ethnic Conflict.* Lanham, MD: Rowman & Littlefield.

Sislin, J., and F. S. Pearson. 2006. "Arms and Escalation in Ethnic Conflicts: The Case of Sri Lanka." *International Studies Perspectives* 7 (2): 137–158.

Small Arms Survey, ed. 2013. "Armed Groups' Holdings of Guided Light Weapons." *Research Notes*, no. 31. https://www.smallarmssurvey.org/resource/armed-groups-holdings-guided-light-weapons-research-note-31.

Smith, N. 1992. Geography, Difference and the Politics of Scale. In *Postmodernism and the Social Sciences*. Doherty, J., Graham, E., and M. Malek eds. London: Palgrave Macmillan. Ch.3, 57–79.

Smith, M. 1999. *Burma: Insurgency and the Politics of Ethnicity*. London: Zed Books.

Smith, M. 2007. *State of Strife: The Dynamics of Ethnic Conflict in Burma*. Policy Studies No. 36. Washington, DC: East-West Center.

Smith, M. 2016. "Reflections on the Kachin Ceasefire: A Cycle of Hope and Disappointment." In *War and Peace in the Borderlands of Myanmar: The Kachin Ceasefire, 1994–2011*, edited by M. Sadan, 57–91. Copenhagen: NIAS Press.

Smith, M. 2019. *Arakan (Rakhine State): A Land in Conflict on Myanmar's Western Frontier*. Amsterdam: Transnational Institute.

Steinmüller, H. 2022. "Sovereignty as Care: Acquaintances, Mutuality, and Scale in the Wa State of Myanmar." *Comparative Studies in Society and History* 64 (4): 910–933.

Strazzari, F., and S. Tholens. 2010. "Another Nakba: Weapons Availability and the Transformation of the Palestinian National Struggle, 1987–2007." *International Studies Perspectives* 11:112–30.

Strazzari, F., and S. Tholens. 2014. "'Tesco for Terrorists' Reconsidered: Arms and Conflict Dynamics in Libya and in the Sahara-Sahel Region." *European Journal on Criminal Policy and Research* 20:343–60.

Suhardiman, D., J. Bright, and C. Palmano. 2019. "The Politics of Legal Pluralism in the Shaping of Spatial Power in Myanmar's Land Governance." *The Journal of Peasant Studies* 48 (2): 411–435.

Suriyavorapunt, A. 2018. "The Development of Appropriate Gun Control Measures for Thailand: District Chief and Police Perspectives." *International Journal of Crime, Law and Social Issues* 5 (1): 12–26.

Swyngedouw, E. 1997. "Neither Global nor Local: 'Glocalization' and the Politics of Scale." In *Spaces of Globalization: Reasserting the Power of the Local*, edited by K. R. Cox, 137–166. New York: The Guilford Press.

Ta'ang (Palaung) Working Group. 2011. *Monopoly Tea Farms*. Ta'ang (Palaung) Working Group. https://www.burmalibrary.org/sites/burmalibrary.org/files/obl/docs13/Monopoly_Tea_Farms%28en%29-red.pdf.

TSYO (Ta'ang Student and Youth Organization). 2010. *Lightless Life: The Negligence of the Military Regime, Lack of Opportunity in Education and the Futureless Lives of Ta'ang People*. Mae Sot, Thailand: TSYO.

TSYO (Ta'ang Student and Youth Organization). 2011. *Grabbing Land: Destructive Development in Ta'ang Region*. Mae Sot, Thailand: TSYO.

TSYO (Ta'ang Student and Youth Organization). 2012. *Pipeline Nightmare: Shwe Gas Fuels Civil War and Human Rights Abuses in Ta'ang Communities in Northern Burma*. Mae Sot, Thailand: TSYO.

TWO (Ta'ang Women Organization). 2016. *Trained to Torture: Systematic War Crimes by the Burma Army in Ta'ang Areas of Northern Shan State (March 2011–March 2016)*. https://www.burmalink.org/wp-content/uploads/2016/10/Trained-to-Torture-English_for-Web.pdf

Tagliacozzo, E. 2004. "Ambiguous Commodities, Unstable Frontiers: The Case of Burma, Siam, and Imperial Britain, 1800–1900." *Comparative Studies in Society and History* 46 (2): 354–377.

Tagliacozzo, E. 2005. *Secret Trades, Porous Borders: Smuggling and States along a Southeast Asian Frontier, 1865–1915*. New Haven, CT: Yale University Press.

Taylor, R. H. 1973. *Foreign and Domestic Consequences of the KMT Intervention in Burma. Southeast Asia Program.* Paper No. 93. Department of Asian Studies, Cornell University.

Thant Myint U. 2006. *The River of Lost Footsteps: Histories of Burma.* New York: Farrar, Straus and Giroux.

Thant Myint U. 2019. *The Hidden History of Burma: Race, Capitalism, and the Crisis of Democracy in the 21st Century.* New York: W. W. Norton.

Thawnghmungh, A. 2019. *Everyday Economic Survival in Myanmar.* Madison: University of Wisconsin Press.

Thawnghmungh, A. 2022. "National Races" In *Myanmar: Oxford Research Encyclopedia of Asian History.* Oxford: Oxford University Press.

Than, T. 2015. "Nationalism, Religion, and Violence: Old and New Wunthanu Movements in Myanmar." *Review of Faith and International Affairs* 13 (4): 12–24.

Than, T. 2016. Mongla and the Borderland Politics of Myanmar. *Asian Anthropology* 15 (2): 152–168.

Than, T., and H. May. 2022. "Multiple Identities of Young Sittwe Muslims and Becoming Rohingya." In *Flow and Frictions in Trans-Himalayan Spaces: Histories of Networking and Border Crossing,* edited by G. Cederlöf and W. Van Schendel. Amsterdam: Amsterdam University Press.

Tholens, S. 2012. *International Norms and Local Agents in Peacebuilding: Small Arms Control in Post-War Kosovo and Cambodia.* PhD diss. Florence: European University Institute.

Tilly, C. 1990. *Coercion, Capital and European States, AD 990–1992.* Cambridge: Blackwell.

Tilly, C. 2003. *The Politics of Collective Violence.* Cambridge: Cambridge University Press.

Thongchai Winichakul. 1994. *Siam Mapped: A History of the Geo-body of a Nation.* Chiang Mai, Thailand: Silkworm Books.

Tosa, K. 2009. "The Cult of Thamanya Sayadaw: The Social Dynamism of a Formulating Pilgrimage Site." *Asian Ethnology* 68 (2): 239–264.

Tønnesson, Ne Lynn Aung, and M. Nilsen. 2019. *Will Myanmar's Northern Alliance Join the Peace Process?* Peace Research Institute Oslo (PRIO) Policy Brief 2.

TNI (Transnational Institute). 2014. *Ethnicity without Meaning, Data without Context. The 2014 Census, Identity and Citizenship in Burma/Myanmar.* Burma Policy Briefing No.13, February 2014. Amsterdam: Transnational Institute.

TNI (Transnational Institute). 2015. *Military Confrontation or Political Dialogue: Consequences of the Kokang Crisis for Peace and Democracy in Myanmar.* Myanmar Policy Briefing No 15, July 2015. Amsterdam: Transnational Institute. https://www.tni.org/files/publication-downloads/military_confrontation_or_political_dialogue_w.pdf.

TNI (Transnational Institute). 2018. *From War to Peace in Kayah (Karenni) State: A Land at the Crossroads in Myanmar.* Amsterdam: Transnational Institute. https://www.tni.org/files/publication-downloads/tni-2018_karenni_eng_web_def.pdf.

Tsing, A. L. 2003. "Natural Resources and Capitalist Frontiers." *Economic & Political Weekly* 38 (48): 5100–5106.

Tsing, A. L. 2005. *Friction: An Ethnography of Global Connection.* Princeton, NJ: Princeton University Press.

Tsing, A. L. 2015. "The Mushroom at the End of the World: On the Possibility of Life in Capitalist Ruins." Princeton, NJ: Princeton University Press.

Tsing, A. L. 2019. "When the Things We Study Respond to Each Other: Tools for Unpacking the Material." In *Anthropos and the Material*, edited by P. Harvey, C. Krohn-Hansen, and K. G. Nustad, 221–245. Durham: Duke University Press.

Turner, A. 2014. *Saving Buddhism: The Impermanence of Religion in Colonial Burma*. Honolulu: University of Hawai'i Press.

UNODC (United Nations Office on Drugs and Crime). 2003. *Myanmar Opium Survey 2003*. UNODC. https://www.unodc.org/pdf/publications/myanmar_opium_survey_2003.pdf.

Vandergeest, P., and N. L. Peluso. 1995. "Territorialization and State Power in Thailand." *Theory and Society* 24:385–426.

Vandergeest, P., and N. L. Peluso. 2011. "Political Ecologies of War and Forests: Counterinsurgencies and the Making of National Natures." *Annals of the Association of American Geographers* 101 (3): 587–608.

Van Schendel, W. 2002. "Geographies of Knowing, Geographies of Ignorance: Jumping Scale in Southeast Asia." *Environment and Planning D: Society and Space* 20:647–668.

Van Schendel, W. 2005. "Spaces of Engagement: How Borderlands, Illegal Flows, and Territorial States Interlock." In *Illicit Flows and Criminal Things: States, Borders, and the Other Side of Globalization*, edited by I. Abraham and W. van Schendel, 38–68. Bloomington: Indiana University Press.

Van Schendel, W. 2006. "Guns and Gas in Southeast Asia: Transnational Flows in the Burma-Bangladesh Borderland." *Kyoto Review of Southeast Asia* 7:1–19.

Van Schendel, W. 2020. "Fragmented Sovereignty and Unregulated Flows: The Bangladesh-China-India-Myanmar Corridor." In *Shadow Exchanges along the New Silk Roads,* edited by E. P. W. Hung and Tak-Wing Ngo, 37–73. Amsterdam: Amsterdam University Press.

Vining, M. 2019. "State SALW Production and Transfers in Myanmar." Unpublished background paper for the Small Arms Survey. March.

Vining, M. 2020. "The Yat-Thai Rifles of Shan State, Myanmar. *The Hoplite*. February 19. http://armamentresearch.com/the-yat-thai-rifles-of-shan-state-myanmar/.

Walton, M. 2008. "Ethnicity, Conflict and History in Burma: The Myths of Panglong." *Asian Survey* 48 (6): 889–910.

Walton, M. 2013. "The 'Wages of Burman-ness': Ethnicity and Burman Privilege in Contemporary Myanmar. *Journal of Contemporary Asia* 43 (1): 1–27.

Walton, M. 2015. "The Disciplining Discourse of Unity in Burmese Politics." *Journal of Burma Studies* 19 (1): 1–26.

Waterman, A., and J. Worrall. 2020. "Spinning Multiple Plates under Fire: The Importance of Ordering Processes in Civil Wars." *Civil Wars* 22 (4): 567–590.

Watkins, A., J. Ismay, and T. Gibbons-Neff. 2018. "Once Banned, Now Loved and Loathed: How the AR-15 Became 'America's Rifle.'" *New York Times*, March 3. https://www.nytimes.com/2018/03/03/us/politics/ar-15-americas-rifle.html.

Watts, M. J. 2003. "Development and Governmentality." *Singapore Journal of Tropical Geography* 24 (1): 6–34.

Watts, M. J. 2004. "Antinomies of Community: Some Thoughts on Geography, Resources and Empire." *Transactions of the Institute of British Geographers* 29 (2): 195–216.

Weizman, E. 2007. *Hollow Land: Israel's Architecture of Occupation*. London: Verso.

Weng, L. 2008. "Armed Insurgents in Burma Face Shortage of Ammunition." *The Irrawaddy*, December 22. http://www.irrawaddy.org/highlight.php?art_id=14829.

Weng, L. 2015. "In Shan State Rebel Army Seeks Divine Intervention from Guardian Spirits." *The Irrawaddy*, December 19, 2015. https://www.irrawaddy.com/news/

burma/in-shan-state-rebel-army-seeks-divine-intervention-from-guardian-spirits.html.

Weng, L. 2016. "Ethnic Ta'ang Rights Groups Accuse Shan Armed Group of Abuse." *The Irrawaddy*, September 29, 2016. https://www.irrawaddy.com/news/burma/ethnic-taang-rights-groups-accuse-shan-armed-group-of-abuse.html.

Woods, K. 2011. "Ceasefire Capitalism: Military–Private Partnerships, Resource Concessions and Military–State Building in the Burma–China Borderlands." *Journal of Peasant Studies* 38 (4): 747–770.

Woods, K. 2016. "The Commercialization of Counterinsurgency: Battlefield Enemies, Business Bedfellows in Kachin State, Burma." In *War and Peace in the Borderlands of Myanmar: The Kachin Ceasefire, 1994–2011*, edited by M. Sadan, 114–145. Copenhagen: NIAS Press.

Woods, K. 2019a. "Green Territoriality: Conservation as State Territorialization in a Resource Frontier." *Human Ecology* 47 (2): 217–232.

Woods, K. 2019b. "Rubber out of the Ashes: Locating Chinese Agribusiness Investments in Armed Sovereignties' in the Myanmar-China Borderlands." *Territory, Politics, Governance* 7 (1): 79–95.

Woods, K., and J. Naimark. 2020. "Conservation as Counterinsurgency: A Case of Ceasefire in a Rebel Forest in Southeast Myanmar." *Political Geography* 83:1–11.

Xiaobo, S. 2018. "Fragmented Sovereignty and the Geopolitics of Illicit Drugs in Northern Burma." *Political Geography* 63:20–30.

Yawnghwe, C. T. 1993. "The Political Economy of the Opium Trade: Implications for Shan State." *Journal of Contemporary Asia* 23 (3): 306–326.

Yawnghwe, C. T. 2010. *The Shan of Burma: Memoirs of an Exile*. Singapore: Institute of Southeast Asian Studies.

Zaw, Y. K. 2022. *An Ethnographic Study of Medicines, Care, and Antimicrobial Resistance amidst Disorder and Decline in Yangon, Myanmar*. PhD diss. London: London School of Hygiene & Tropical Medicine.

Index

105th Mile Trade Zone 71, 208
8888 uprisings 61, 76, 82, 155, 257n57, 257n64

Arakan. *See* Rakhine
assemblage 15, 18–20

Bay of Bengal 11, 60, 264n105
Bawdwin mines 80, 91, 260n32
becoming and being a weapon 6, 15, 234
Beijing 65
Benin 21, 253n11
biopolitics 15, 21–23, 198, 251–2n8
  blunt 22–24, 30, 145, 225–229
Boal, Augusto 222
borderland 19–21, 29–30
borderworld 11, 36, 83, 86, 109, 152–159,
  162–176, 193
Burma Road 42, 71

ceasefire capitalism 5, 79, 162, 182
Chiang Mai 12, 95, 98, 149, 150, 164, 192,
  263n93
Chinshwehaw 269n15
coup d'etat 16–7, 49–52, 144–50, 173–176,
  222–28, 234
compound communities 55, 103, 123, 225
craft-manufactured weapons 14, 131–34,
  152–54, 156–160, 173, 177, 187–189

de Castro, Viveiros 111
Demoso 12, 221, 253–4n20
Deng Xiaoping 60
Directorate of Defence Industries (Ka-Pa-Sa)
  50, 154, 255n21, 257n64, 269n9
dualism. *See* instrumentalism and
  substantivism

enterprise nationalization law 50
ethnocentr(ism) 4, 66, 124
exchanging arms for peace 11, 86–88

FN-6. *See* MANPADS
four cuts 28, 35–6, 56–63, 72–4, 120–26, 144,
  184–194, 216

Free Burma Rangers 159, 190, 271n27, 272n43
French foreign legion 45
frontier 18–21, 29–30
frontierization and territorialization 5–10,
  16–21, 70–9, 106–17, 123–32, 144–57,
  169–79, 183–195, 225

genocide 17, 93, 106, 112, 113–114, 225
geopolitic(s) 23, 112, 145, 226
Ghana 21, 253n11

Heckler & Koch 49, 168, n21p255
Homein 1, 4–6, 24–26, 150, 258n78, 260n36,
  261n54
Ho Mong 59, 156
Hpakant 24, 102
Hsipaw 12, 54, 72, 102, 116–7, 120, 126, 137,
  173, 208
Human minesweeping 28, 124
Hsenwi 71, 96, 157, 253n8, 254n16, 258n78

instrumentalism and substantivism 7–8, 161,
  230–31, 252n11
Irrawaddy (river) 38, 50, 260n34

just war 16–17

Kachinland 96
Kachin substate 53, 70, n16p254
Kalaw 70, 118
Kawngkha 208, n8p259, n71p262, 270n18
Keng Tung 44, 70
Khin Nyunt 73, 86, 261n48
Kokang 59, 71, 73, 99, n60p257, 265n118
Kyaukme 24–36, 68–70, 79–87, 109–23, 137–41,
  173–78, 195–6, 206–10, 213–16
Kyauk Phyu 71
Kutkai 53, 60, 70–1, 74, 79–80, 96, 117, 120,
  123, 208

Laiza 100, 156, 165, n64p258, 264n100, 269n15
Lao Gai 99
limited access orders 53, 159, 182
Lo Hsing Han 52, 57